Biomaterial Fabrication Techniques

Edited by

Adnan Haider
Department of Biological Sciences
National University of Medical Sciences
Rawalpindi
Pakistan

&

Sajjad Haider
Department of Chemical Engineering
King Saud University
Riyadh
Saudi Arabia

Biomaterial Fabrication Techniques

Editors: Adnan Haider and Sajjad Haider

ISBN (Online): 978-981-5050-47-9

ISBN (Print): 978-981-5050-48-6

ISBN (Paperback): 978-981-5050-49-3

First published in 2022.

need for a court order if at any point you breach any terms of this License Agreement. In no event will any delay or failure by Bentham Science Publishers in enforcing your compliance with this License Agreement constitute a waiver of any of its rights.

3. You acknowledge that you have read this License Agreement, and agree to be bound by its terms and conditions. To the extent that any other terms and conditions presented on any website of Bentham Science Publishers conflict with, or are inconsistent with, the terms and conditions set out in this License Agreement, you acknowledge that the terms and conditions set out in this License Agreement shall prevail.

Bentham Science Publishers Pte. Ltd.
80 Robinson Road #02-00
Singapore 068898
Singapore
Email: subscriptions@benthamscience.net

BENTHAM SCIENCE

CONTENTS

FOREWORD

This book provides an up-to-date, comprehensive, and authoritative overview of advancements in scaffold manufacture and uses in tissue engineering, combining the foundations for a wide understanding of scaffolds for tissue growth and development. The chapters cover a wide range of issues, including innovative materials and methodologies for scaffold preparation, difficulties, and future prospects. The chapters include topics such as novel materials and techniques for scaffold preparation, challenges, future prospects, and much more. The authors have carefully analyzed and summarized recent research findings in the aforementioned areas, providing an in-depth understanding of scaffold that maintains a balance among a variety of topics related to tissue engineering, including biology, chemistry, material science, and engineering, among others, while prioritizing study topics that are likely to be useful in the future.

Professor Inn-Kyu Kang
Department of Polymer Science and Engineering,
Kyungpook National University,
Daegu, South Korea

PREFACE

This book is a collection of research and review articles from various parts of the world, highlighting the pivotal importance of biomaterials and their potential biomedical application. The articles link new findings and critically review the fundamental concepts and principles that are making the base of innovation. The book comprises ten chapters; the first two chapters deal with vital information about biomaterials and the strategies used for their fabrication. The rest of the chapters highlight the most widely used technique, their principle and their application in detail. The book contains up-to-date knowledge of biomaterials, their fabrication technique and their potential application, which is beneficial both for the experience as well as new researchers.

Adnan Haider
Department of Biological Sciences
National University of Medical Sciences
Rawalpindi
Pakistan

Sajjad Haider
Department of Chemical Engineering
King Saud University
Riyadh
Saudi Arabia

List of Contributors

Abdul Muhaymin   Preston Institute of Nanoscience and Technology, Preston University Kohat, Islamabad Campus, Islamabad, Pakistan

Adeeb Shehzad   Department of Biomedical Engineering & Sciences, School of Mechanical and Manufacturing Engineering, National University of Science & Technology, Islamabad, Pakistan

Aliyah Putranto   Faculty of Pharmacy, Hasanuddin University, Jl. Perintis Kemerdekaan Km 10, Makassar 90245, Republic of Indonesia

Ammara Safdar   Preston Institute of Nanoscience and Technology, Preston University Kohat, Islamabad Campus, Islamabad, Pakistan

Amjad Khan   Department of Pharmacy, Kohat University of Science and Technology, Kohat, Pakistan

Anwarul Hassan   Department of Mechanical and Industrial Engineering, College of Engineering, Qatar
Biomedical Research Center, Qatar University, Doha, Qatar

Andi Arjuna   Faculty of Pharmacy, Hasanuddin University, Jl. Perintis Kemerdekaan Km 10, Makassar 90245, Republic of Indonesia

Arun Kumar Jaiswal   PG Program in Bioinformatics, Institute of Biological Sciences, Federal University of Minas Gerais, Belo Horizonte, MG, Brazil
Department of Immunology, Microbiology and Parasitology, Institute of Biological Sciences and Natural Sciences, Federal University of Triângulo Mineiro (UFTM), Uberaba, MG, Brazil

Aref Ahmad Wazwaz   Department of Chemical Engineering, College of Engineering, Dhofar University, Salalah 211, Sultanate of Oman

Asmat Ullah Khan   Department of Zoology, Shaheed Benazir Bhutto University, Dir Upper, KPK, Pakistan

Atiya Fatima   Department of Chemical Engineering , College of Engineering, Dhofar University, Salalah 211, Sultanate of Oman

Fakhar-ud-Din   Department of Pharmacy, Quaid-i- Azam University, Islamabad, Pakistan

Fazli Khuda   Department of Pharmacy, University of Peshawar, Khyber Pakhtoonkhwa, Pakistan

Fazli Subhan   Department of Biological Sciences, National University of Medical Sciences, Rawalpindi, Pakistan

Haseeb Ahsan   School of Life Sciences, College of Natural Sciences, Kyungpook National University, 41566, Korea

Jong Kyung Sonn   School of Life Sciences, College of Natural Sciences, Kyungpook National University, 41566, Korea

Kainat Tufail   Department of Pharmacy, University of Peshawar, Khyber Pakhtoonkhwa, Pakistan

Khalil K. Hussain   Medical Research Council Centre for Medical Mycology, University of Exeter, Geoffrey Pope Building, Stocker Road, EX4 4QD, Exeter, UK

Mazhar-Ul-Islam — Department of Chemical Engineering, College of Engineering, Dhofar University, Salalah 211, Sultanate of Oman

Md. Wasi Ahmed — Department of Chemical Engineering, College of Engineering, Dhofar University, Salalah 211, Sultanate of Oman

Meshal Gul — Department of Pharmacy, University of Peshawar, Khyber Pakhtoonkhwa, Pakistan

Mohammad Sherjeel Javed Khan — Department of Chemistry, King Abdulaziz University, Jeddah 21589, Saudi Arabia

Mohsin Ali Raza — Nanoscience and Nanotechnology Department (NS & TD)People, National Centre for Physics, Islamabad, Pakistan, and School of Biomedical Engineering, Medix research Institute, Shangai Jiao Tong University (SJTU), 1954 Huashan road, Shangai 200030, Republic of China

Muhammad Naeem — Department of Biological Sciences, National University of Medical Sciences, Rawalpindi, Punjab, Pakistan

Muhammad Umar Aslam Khan — BioInspired Device and Tissue Engineering Research Group, School of Biomedical Engineering and Health Sciences, Faculty of Engineering, Universiti Teknologi Malaysia, 81300 Skudai, Johor, Malaysia

Muhammad Faheem — Department of Biological Sciences, National University of Medical Sciences, Rawalpindi, Pakistan

Muhammad Wajid Ullah — Biofuels Institute, School of the Environment and Safety Engineering, Jiangsu University, Zhenjiang 212013, China

Muhammad Bilal Ahmed — School of Life Sciences, College of Natural Sciences, Kyungpook National University, 41566, Korea

Murtada A. Oshi — Department of Pharmaceutics, Faculty of Pharmacy, Omdurman Islamic University, Omdurman, Sudan

Naeem Khan — Department of Chemistry, Kohat University of Science and Technology, Kohat, Pakistan

Rabail Zehra Raza — Department of Biological Sciences, National University of Medical Sciences, Rawalpindi, Pakistan

Raees Khan — Department of Biological Sciences, National University of Medical Sciences, Rawalpindi, Pakistan

Rashid Amin — Department of Biology, Hafar Al-Batin 39524, Saudi Arabia, College of Sciences, University of Hafr Al Batin, Hafar Al-Batin 39524, Saudi Arabia

Rawaiz Khan — BioInspired Device and Tissue Engineering Research Group, School of Biomedical Engineering and Health Sciences, Faculty of Engineering, Universiti Teknologi Malaysia, 81300 Skudai, Johor, Malaysia
Department of Polymer Engineering, School of Chemical and Energy, Faculty of Engineering, Universiti Teknologi Malaysia, 81310 UTM Johor Bahru, Johor, Malaysia

Ronan R. McCarthy — Division of Biosciences, Department of Life Sciences, College of Health and Life Sciences, Brunel University London, Uxbridge UB8 3PH, UK

Saiful Izwan Abd. Razak BioInspired Device and Tissue Engineering Research Group, School of Biomedical Engineering and Health Sciences, Faculty of Engineering, Universiti Teknologi Malaysia, 81300 Skudai, Johor, Malaysia
Centre for Advanced Composite Materials, Universiti Teknologi Malaysia, 81300 Skudai, Johor, Malaysia

Salman-Ul-Islam School of Life Sciences, College of Natural Sciences, Kyungpook National University, 41566, Korea

Sajjad Haider Department of Chemical Engineering, College of Engineering, King Saud University, P.O. BOX 800, Riyadh 11421, KSA, Saudi Arabia

Saqlain A Shah Department of Physics, Forman Christian College (University), Lahore, Pakistan

Sandeep Tiwari PG Program in Bioinformatics, Institute of Biological Sciences, Federal University of Minas Gerais, Belo Horizonte, MG, Brazil

Sehrish Manan Biofuels Institute, School of the Environment and Safety Engineering, Jiangsu University, Zhenjiang 212013, China

Sumarheni Faculty of Pharmacy, Hasanuddin University, Jl. Perintis Kemerdekaan Km 10, Makassar 90245, Republic of Indonesia

Shaukat Khan Materials Science Institute, the PCFM and GDHPRC Laboratory, School of Chemistry, Sun Yat-sen University, Guangzhou 510275, PR China

Siomar de Castro Soares Department of Immunology, Microbiology and Parasitology, Institute of Biological Sciences and Natural Sciences, Federal University of Triângulo Mineiro (UFTM), Uberaba, MG, Brazil

Sultan Ullah Department of Molecular Medicine, The Scripps Research Institute, Florida, USA

Syed Babar Jamal Department of Biological Sciences, National University of Medical Sciences, Rawalpindi, Pakistan

Vasco Azevedo PG Program in Bioinformatics, Institute of Biological Sciences, Federal University of Minas Gerais, Belo Horizonte, MG, Brazil

Young Sup Lee School of Life Sciences, College of Natural Sciences, Kyungpook National University, 41566, Korea

Zahoor Ullah Department of Chemistry, Balochistan University of Information Technology, Engineering and Management Sciences (BUITEMS), Takatu campus, Quetta 87100, Pakistan

CHAPTER 1

Introduction to Biomaterials and Scaffolds for Tissue Engineering

Khalil K. Hussain[1,*] and **Muhammad Naeem**[2]

[1] *Medical Research Council Centre for Medical Mycology, University of Exeter, Geoffrey Pope Building, Stocker Road, EX4 4QD, Exeter, UK*

[2] *Department of Biological Sciences, National University of Medical Sciences, Rawalpindi, Punjab, Pakistan*

Abstract: Biomaterials are essential elements in various fields, especially medicine. They can help restore biological functions and speed up the healing process after injury or disease. Natural or synthetic biomaterials are used in clinical applications to provide support, replace damaged tissue, or restore biological function. The study of such types of biomaterials is an active area of research, particularly in the field of tissue engineering (TE). In general, the term TE describes the regeneration, growth, and repair of damaged tissue due to disease or injury. TE is a modern science that combines biology, biochemistry, clinical medicine and biomaterials, which led to the research and development of various applications. For example, in the field of regenerative medicine, biomaterials can serve as a support (scaffold) to promote cell growth and differentiation, which ultimately facilitates the healing process of tissues. This chapter describes the various properties of biomaterials, a detailed discussion of scaffolds in terms of design, properties and production techniques, and future directions for TE.

Keywords: Biomaterials, Scaffold, Tissue engineering.

INTRODUCTION

The U.S. National Institute of Health defines biomaterials as "any substance or combination of substances, other than drugs, of synthetic or natural origin, that can be used for any period of time, partially or completely augments or replaces a tissue, organ, or function of the body to maintain or improve the quality of life of the individual" [1]. Interestingly, the use of biomaterials dates back to ancient times, when the Romans and Egyptians used plant fibres to suture skin wounds and made prosthetic limbs from wood [2]. Since then, the use of biomaterials has gone through different phases. In the industrial era, biomaterials have changed dramatically, leading to the synthesis of novel biomaterials for various applic-

* **Corresponding author Khalil K. Hussain:** Medical Research Council Centre for Medical Mycology, University of Exeter, Geoffrey Pope Building, Stocker Road, EX4 4QD, Exeter, UK; E-mail: k.Hussain@exeter.ac.uk

ations, especially in regenerative medicine and tissue engineering strategies. In general, biomaterials can be divided into three groups: Ceramics, synthetic polymers and natural polymers. However, each group has advantages and disadvantages [3]. In humans, the extracellular matrix (ECM) is considered a natural template biomaterial that provides support, spatial organisation, and maintenance of a biologically active microenvironment. The matrix is composed of different proteins that serve different functions, *e.g.*, structural support proteins such as collagen and elastin, adhesion proteins such as fibronectin and laminin, and swellable proteins that contain polysaccharides such as glycosaminoglycans (GAGs) and proteoglycans [4]. The restructuring and remodelling of the ECM support tissue regeneration, cell survival, proliferation, and other functions [5]. Based on the functions of ECM, researchers are working to synthesise biomaterials that can mimic the role of ECM, which is currently not possible. Therefore, the most typical approach in the field of biomaterials is to understand the ECM mechanisms at the cellular level [6]. The approach has led to the emergence of a new field called tissue engineering (TE), which enables the formation of functional tissues. However, the equation is not simple, as the host response to biomaterials is complex and can trigger a proinflammatory response [7, 8]. TE is a multifaceted field that connects many disciplines, as shown in Fig. (**1**). Interestingly, in recent studies, macrophages play a crucial positive role in remodelling by secreting cytokines and/or scaffold degradation products [9 - 12].

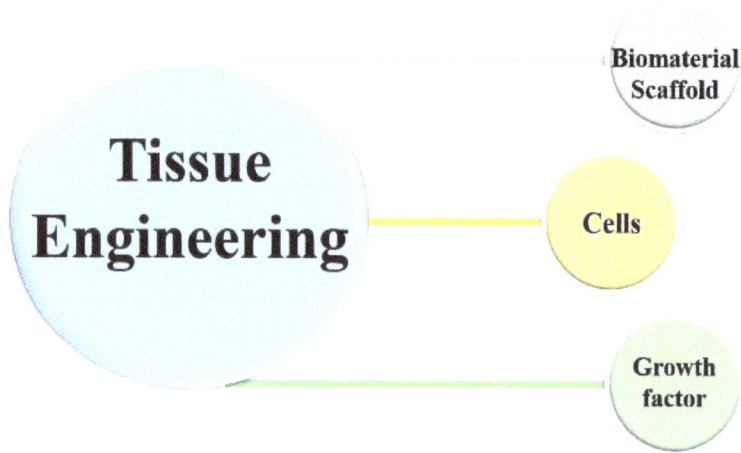

Fig. (1). Basic components in TE: Biomaterial scaffold serving as a template for tissue formation. Cells for regeneration, and signal either chemically from growth factors or physically from bioreceptor.

BIOMATERIALS FOR SCAFFOLD FABRICATION

As mentioned earlier, biomaterials play an important role in tissue replacement and regeneration. So far, various types of materials have been synthesised and used as scaffolds in TE. In the following section, these biomaterials are described in detail.

Ceramics

Ceramic-based biomaterials are inorganic compounds of natural or synthetic origin that can be doped or un-doped with metals. Ceramics are an ideal choice as biomaterials because they have excellent properties, such as biocompatibility and osteoinductivity. This type of material has a similar chemical composition to natural human bone and hardly triggers any immune response. They also help in cell migration and facilitate osteogenic differentiation. Therefore, these types of biomaterials are popular to rebuild injured body parts, especially in bone regeneration. However, ceramics have some disadvantages that limit their use in scaffold fabrication, such as fragility and slow degradation [13 - 15]. There are three types of ceramic biomaterials: (I) inert to the biological environment; (II) resorbable: subject to *in vivo* degradation by phagocytosis; and (III) bioactive by chemically bonding with the cell surface [16]. Commonly used ceramic biomaterials include (a) calcium phosphate (CaP) biomaterials such as hydroxyapatite (HA), beta-tricalcium phosphate (BTP), a mixture of HA and BTP, (b) bioactive glass, (c) alumina, and (d) zirconia.

Natural HA is derived from a certain type of bovine ribbon phosphate and contains minute amounts of magnesium, sodium, carbon trioxide and fluorine. Synthetic HA, on the other hand, is prepared by various methods, including chemical deposition, biomimetic deposition and wet chemical precipitation [17]. Several reports have been published on synthetic HA. For example, Ray and colleagues reported synthetic HA with biocompatible and biomimetic properties. The prepared material was used for bone tissue engineering and iliac wings of dogs [18]. Similarly, Calabrese prepared a bilayer type 1 collagen HA /Mg scaffold and used it for osteochondral regeneration *in vitro* and *in vivo* [19 - 21]. Bioglass is composed of different elements with different weight percentages in the following order: SiO_2, CaO, Na_2O, and P_2O_5 with weight percentages of 45, 24.5, 24.5, and 6.0, respectively. It was first described by Hench and named 45S5 Bioglass, which has been used in biomedical applications [22]. Since then, various methods for the synthesis of bioglass have been reported, such as polymer foam replication, thermal bonding, and sol-gel. Bioglass and HA have similar properties, such as higher Ca to P content, making them ideal for bone grafts. The role of Bioglass in bone regeneration is outlined in Fig. (**2**). Moreover, bioglass has excellent osteoinductivity, controlled degradation rate and high bioactivity. However, they suffer from low strength and toughness [23, 24].

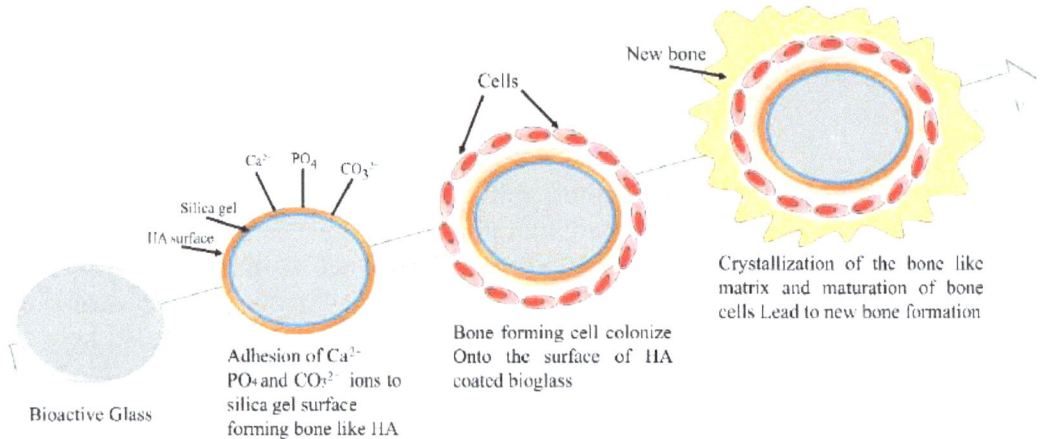

Fig. (2). Schematic representation of the integration of bioglass with bone. The two stages show the interaction of the bioglass with the physiological environment and later the two stages of the process of new bone formation.

Alumina (Al_2O_3) is a polycrystalline ceramic material with low porosity and small grain size. Zirconia, on the other hand, is a polymorphic structure with low thermal conductivity and a higher coefficient of thermal expansion. They are also an excellent choice for bone grafts and prostheses due to their higher biocompatibility and fracture load. So far, different ceramics have been used for different applications. The clinical application of such materials is limited due to their brittleness, mechanical flexibility and low mechanical strength, as well as the difficulty in controlling the rate of degradation.

Polymers

Polymers are a popular material for scaffold fabrication because they are biocompatible and bioactive. Therefore, different types of polymers have been introduced, either natural or synthetic. Among the naturally derived polymers, collagen and its derivatives are considered as an ideal choice for the regeneration or reconstruction of osteochondral lesions and ligaments [25, 26]. The bioactivity of such materials facilitates cellular adhesion. However, the resistance to mechanical stress is low, so they are coupled with other support materials. A number of reports have been published using collagen for TE. Aravamudhan reported the synthesis of micro-nanostructured scaffolds based on cellulose and collagen. The results showed good adhesion of human osteoblasts to their surface and progressive calcium deposition [27]. Collagen I/ III hydrogel scaffold was prepared by Schneider *et al.* The group isolated human mesenchymal stem cells from bone marrow and used them for seeding. After stimulation, the cells showed comparable osteogenic gene expression and migration [28]. Another application of the collagen scaffold was reported by Lu H., who used it to deliver osteogenic differentiation factors to promote osteogenesis [29]. Chitin, chitosan and alginate are polysaccharides used in tissue regeneration and can be prepared by various methods. For example, chitosan scaffolds are prepared by freeze-drying, which produces porous material. Chitosan scaffolds are positively charged and can therefore interact with swellable proteins in tissues and provide good cell adhesion. Several applications have been reported so far, including the successful cultivation of human bone marrow MSC cells on porous chitosan scaffolds by Costa-Pinto *et al.* The results showed an increase in cell viability [30]. In addition, chitosan was also used as an injectable biomaterial for scaffold fabrication. The scaffold consisted of tricalcium phosphate, chitosan, and platelet-rich plasma and was used *in vivo* to test its ability to repair bone fractures in animals [31]. In short, naturally derived polymers have numerous advantages. However, the fabrication of scaffolds from biological materials is not very homogeneous and reproducible.

Synthetic polymers generally consist of a sequence of monomeric components. Structurally, they can be linear, branched or cross-linked and can be produced in various forms such as fibers, films, rods and viscous liquids. To date, several types of synthetic polymers such as polystyrene, polylactic acid (PLA), polyglycolic acid (PGA), polylactic acid-co-glycolic acid (PLGA), and polycaprolactone (PCL) have been reported. Biodegradable is the most important property to be considered in these biomaterials. Fortunately, these types of synthetic polymers degrade on the surface and produce non-toxic compounds. Therefore, these polymers are used in various applications, such as sutures, orthopedic screws and prostheses, and scaffold fabrication for TE. The combination of PLA and PEG for the fabrication of PLA-PEG-PLA scaffolds was reported by Eğri *et al.* The obtained material was able to release vascular endothelial growth factors in bone tissue lesions [32]. However, PLLA- and PGA-containing materials generate CO_2 during hydrolysis, which decreases the pH and may lead to cell and tissue necrosis [33].

Metals

Metals are a popular choice for TE because of their excellent mechanical properties. However, they suffer from poor cell adhesion and the release of toxic ions during implantation, which limits their use [34]. Various metals, such as iron, cobalt and titanium, have been used for scaffold fabrication. In iron, carbon and chromium-based materials, carbon provides good mechanical properties, but iron causes corrosion in a biological environment. Cobalt alloys, on the other hand, are divided into two categories: Co/Cr/Mo alloys can be synthesised by casting or melting, while Co/Ni/Cr/Mo alloys are produced by forging. Higher concentrations of Cr and Mo increase the granule size and improve the mechanical properties. Titanium-based frameworks can be divided into three types: Alpha, Beta, and combinations of Alpha and Beta. The first type contains aluminium and gallium and has excellent mechanical properties such as strength, hardness and sliding resistance. Beta stabiliser alloys contain vanadium and niobium and have better ductility. The third category contains a mixture of stabilisers and is ductile, such as Ti 6Al 4 V, which is used in biomedical applications. Ti and TiO_2 scaffolds were tested by Wohlfahrt for osteoinductivity and osteointegration in rabbits, with new bone observed after 4 weeks of implantation [35]. In another study, a scaffold was fabricated from Cr-Co-Mo membranes and implanted into the tibias of rabbits. The results showed that a significant number of cells were observed on the scaffold [36].

Composites

This type of framework is made by combining one, two, or even three materials, such as combining polymers with ceramics or metals to make composite scaffolds. The composites have higher biocompatibility, biodegradability and considerable mechanical strength so that they can be used for soft and hard tissue regeneration. A number of studies have been reported with promising results in which scaffolds were fabricated from the combination of polymers and ceramics. Recently, scaffolds made of coated polymer, metal or ceramic have been reported [37, 38].

Hydrogels

Hydrogels are hydrophilic polymers with functional groups, such as carboxyl, amide, amino, and hydroxyl. In general, hydrogels are synthesised either by chemical or physical interactions. Interestingly, they are excellent water absorbers and swell without dissolving. Such materials were first described by Wichterle and Lim, who synthesised a poly(2-hydroxyethyl methacrylate) and used it in contact lenses [39]. Hydrogels are broadly classified into three main groups based on their origin: natural: derived from polypeptides and polysaccharides, synthetic: produced by polymerization, and semi-synthetic [40]. Usually, they have a soft and rubbery structure similar to ECM. In addition, hydrogels have been explored for TE scaffold fabrication and used as injectable hydrogels with cells. The mechanism for cartilage- and bone TE applications is sketched in Fig. (3).

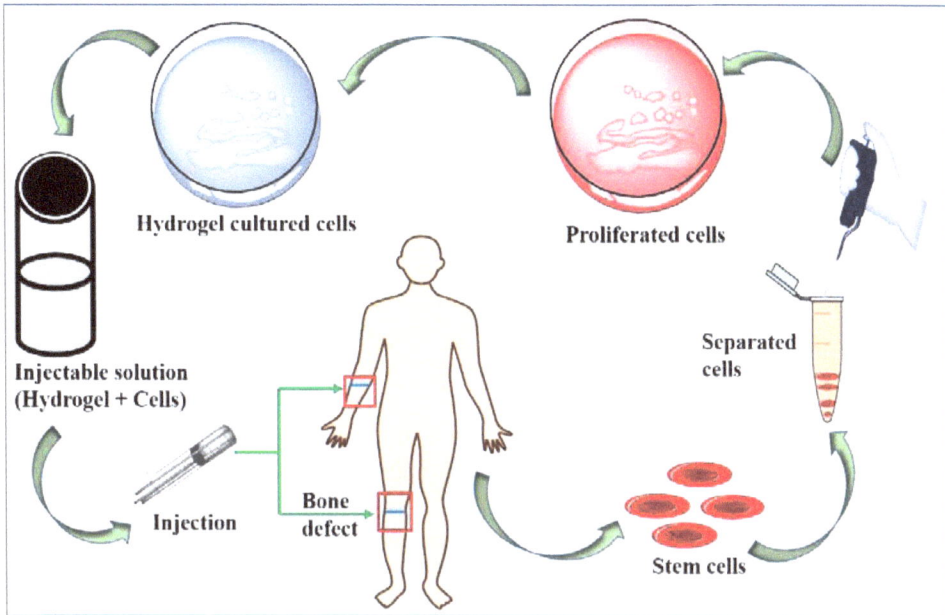

Fig. (3). Schematic representation for injectable hydrogels applied in cartilage- and bone tissue engineering.

Table 1. Summary of Biomaterials for scaffold production.

Biomaterials Characteristics	Ceramics	Natural Polymer	Synthetic Polymer	Metal	Composite	Hydrogel
Advantages	•Hard Surface •Excellent biocompatibility, mechanical strength and osteoinductivity	•Good biocompatibility and bioactivity	•Porous scaffold can be produced	•High mechanical properties and ductility	•High biocompatibility and mechanical properties	•Biocompatible •Controlled biodegradation •Tuneable properties
Disadvantages	•Slow degradation •Processing difficulty	•Rapid degradation •Low mechanical properties	•Low biocompatibility and mechanical strength •release of toxic ions	•Poor cell adhesion •Possibility of corrosion	•Complicated procedure	•Lengthy and complex procedure
Applications	•Bone and cartilage •Prosthesis	•Bond and cartilage •Tendons	•Sutures •Catheters •Prosthesis	•Orthopaedic •Prosthesis	•Hard and soft tissues	•Hard and soft tissues

The most important factors in the synthesis of hydrogels are the determination and control of crosslink density, pore size, and structural interconnectivity for cellular functions. This type of material can be easily modified on its surface by various biological molecules such as growth factors. Recently, researchers have been developing smart hydrogels that interact and change their structure depending on the microenvironment, pH or temperature. For example, Pasqui *et al.* reported a natural cellulose-hydroxyapatite hybrid hydrogel used for bone TE. The hydrogel was mixed with microcrystals of HA and successfully applied *in vitro* [41]. Synthetic hydrogels offer the possibility of modifying mechanical and biological properties, such as the rate of degradation. For example, biodegradable oligo [poly(ethylene glycol) fumarate] was synthesised by Kinard to deliver demineralized bone matrix (DBM) into a bone defect in rats. The results showed that the higher the DBM concentration, the faster the degradation rate [42]. Summary of biomaterials for scaffold production in Table **1**.

Methodologies for Scaffold Production

Scaffold fabrication has become an active area of research over the last two decades. To date, several techniques have been described for the development of scaffolds. These techniques are mainly focused on the fabrication of porous scaffolds for cell seeding [43 - 45]. In general, these fabrication techniques are classified into three categories: (I) porogen-based techniques, (II) woven or nonwoven fiber fabrication techniques, and (II) rapid prototyping [46]. The first two techniques are largely used for scaffold fabrication. In the first category, the biomaterial is combined with porogens (CO_2, H_2O or paraffin), and the combined mixture is then processed by casting or extrusion. Finally, the porogen is removed from the biomaterial by sublimation, evaporation or leaching to obtain a porous material. Porogens based techniques include gas foaming, freeze drying and solvent casting [47 - 49]. Nonwoven fibers are produced by spinning. Fibers produced by spinning are called nonwoven fibers. Spinning processes include electrospinning, wet spinning, microfludic spinning, melt spinning and bio-spinning. Owven fibers are produced by weaving, a textile technique in which two different sets of warp or weft threads are interwoven at right angles to form a fabric with controlled strength, porosity, morphology, and geometry, and by knotting, in which a fabric is formed by interlacing yarns or threads in a series of interconnected loops [50]. It is worth noting that the scaffold produced by these techniques has numerous advantages. For example, any type of biomaterial can be used for fabrication with precise architectural designs. Moreover, the scaffold structure can be combined with different materials that provide better mechanical support. However, some of these approaches are time-consuming and suffer from limited cell penetration into the scaffolds. One of the techniques to overcome these drawbacks is decellularized ECM from allogeneic or xenogeneic tissues for cell colonization. This method has been applied to various tissues in TE [51, 52]. This approach uses different strategies to decellularize ECM, such as physical, chemical and enzymatic methods [53, 54]. Decellularized ECM is useful in many other applications, such as small intestine submucosa, skin, and other body tissues

[51, 55 - 57]. These natural scaffolds are biocompatible and can provide promising results when combined with growth factors [58]. However, incomplete removal of cellular components during implantation may trigger immune reactions [59]. In general, some basic requirements should be considered before the fabrication of scaffolds, such as (I) biocompatibility, to facilitate cell adhesion, proliferation and migration, (II) bioactivity, the ability of the biomaterial to interact with the microenvironment, and (III) biodegradability, which allows cells to produce their own extracellular matrix [60 - 62].

Scaffold Design and Properties Relationship

Structural and mechanical properties are of great importance in the development of scaffolds for TE. These properties ensure that the ECM has good strength at the anatomical site so that it can perform its functions. The precise structure of the scaffold facilitates cell survival, adhesion, proliferation, differentiation, vascularization, and specific gene expression [63, 64]. In addition, the engineered structure should support a physiological load with appropriate porosity to allow cellular functions and avoid cell colonisation. However, fabricating a scaffold with a porous structure and mechanically strong properties is a challenging task. For example, an appropriate pore size allows cell penetration and migration into the scaffold and cell attachment. The size of the macropores of the scaffold in TE depends on the host tissues and is usually > 50 nm. For example, the ideal pore size for hepatocyte and fibroblast growth is 20 microns, and for soft tissue, healing is 20-150 microns. A higher pore size between 200 and 400 microns is used for bone TE.

Adequate mechanical properties are equally important in the fabrication of scaffolds as they facilitate the remodelling process and serve as support [65]. These properties depend on the type of bond or forces that hold the atoms together in the scaffold architecture. The rigidity of the scaffold is also an important parameter measured by Young's modulus. It is worth noting that the cellular response to scaffold stiffness is controlled by the activation of ion channels or protein unfolding mechanisms. Therefore, stiffness affects cell proliferation and differentiation. For example, it has been reported that the higher the stiffness of the free-floating collagen matrix, the greater the proliferation of dermal fibroblasts [64]. The general properties and *in vivo*, *in vitro* application characteristics are shown in Fig. (**4**).

Current Scaffold Fabrication Technologies

Tissues are a 3-dimensional (3D) like entity with variable sizes. This 3D structure allows them to perform physiological functions such as organizing biological processes, facilitating mechanical properties, and connecting with other body organs. The main goal of scaffolds is to achieve this development in 3D cells by supporting and ensuring mechanical properties in the process of tissue regeneration [66]. Current scaffold fabrication methods should mainly address two main aspects: 1) At the microscale level, the environment should facilitate cellular functions and survival. 2) At the macroscale, the tissue design should allow the organization of multicellular processes, facilitate nutrient transport, and have better mechanical properties. A suitable technique is the fabrication of scaffolds using 3D technology, which can be either conventional or rapid prototyping (RP) [67]. Conventionally fabricated scaffolds use porous polymer structures to support cell adhesion, but suffer from heterogeneous structures and difficulties in maintaining 3D structure at micro and macro levels [68]. In short, combining clinical knowledge and 3D fabrication techniques to produce a tailored scaffold that accelerates healing and improves scaffold properties [69].

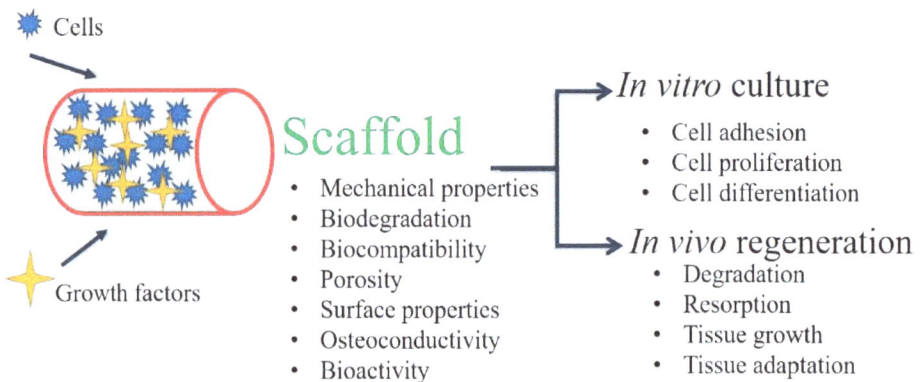

Fig. (4). Illustration of Tissue engineering process using biomaterial scaffold.

Conventional approaches to scaffold fabrication have been applied in various fields, such as drug delivery and 3D cell culture at TE [70]. The scaffold is fabricated using traditional methods such as casting/particle leaching, which control the shape and pore size. Generally, in this method, the solvent is mixed with salt particles that have a certain particle size to dissolve the polymer solution. Then the solvent is evaporated, and the remaining matrix is dissolved in water. In this way, the salt is leached out, and a porosity of 50% and 90% is achieved. This technique is best suited for thin-walled 3D specimens [71, 72]. The method is simple, cheap and, most importantly, adjustable pore size can be obtained [73]. There are numerous reports on the application of this method. For example, a porous PLA/MD- HAP /PEO nanocomposite was prepared and used in bone engineering. PLA was mixed with different concentrations of NH_4HCO_3 and

Progenia. The combination of natural polymers has also been reported [74, 75]. The addition of bioactive material into the scaffold structure has also been extensively described [76, 77]. Among its numerous advantages, this technique is time-consuming and uses toxic solvents [76]. In freeze-drying or lyophilization, the synthetic polymer is dissolved and cooled below the freezing point, causing the solvent to evaporate by sublimation and form pores [78, 79]. This technique does not require high temperatures and allows controllable pore size by simply changing the freezing method [80, 81]. Various types of scaffold fabrication materials have been produced using this technique, such as chitosan nanoparticles [78, 82]. This technique is popular partly because it avoids the use of toxic solvents. However, it requires high energy consumption, is a lengthy process, and causes irregular pore size [83]. To overcome these drawbacks, G'eraldine *et al.* suggested lowering the temperature to -70°C and performing an additional annealing step [84]. Thermally-induced phase separation (TIPS) is a conventional method that operates at lower temperatures to force phase separation. The method has been used in the preparation of thermoplastic crystalline polymer scaffolds [85, 86]. Incorporation of proteins into the scaffold prepared using TIPS has been used, for example, for drug delivery and the preparation of microspheres for TE [87, 88]. The gas foam method uses inert gasses, such as CO_2 or N_2, to pressurize the polymer with water or fluoroform until it is saturated and a sponge-like structure is formed. The pore size ranges from 30 to 700 microns with porosity up to 85%. However, the method suffers from closed pores. Harris reported an improved method, and the results showed good cell adhesion [89]. In the electrospinning method, scaffolds of nanofibrous material are fabricated using electricity [90]. The technique is complicated and high voltage is applied to obtain the desired scaffold. However, this technique does not succeed in obtaining a homogeneous distribution of pores. Examples of this method include the preparation of spider silk proteins and collagen [91]. Similarly Sarhan *et al.* [92] successfully incorporated antimicrobial molecules into the construction of nanofibers and used them in wound dressings. Surprisingly, the fabricated scaffold was active against a range of bacteria. In addition, chitosan-based nanofibers have been produced by electrospinning [93, 94]. Although electrospinning is a popular method for fabricating scaffolds, it lacks homogeneous pore distribution and pore size.

On the other hand, rapid prototyping technologies are often used in scaffold development. These methods are also known as solid free-form fabrication (SFF). These techniques are combined with computer-aided design (CAD) to obtain precise spatial control over the polymer structure during scaffold fabrication [95, 96]. In addition, these techniques allow the fabrication of customizable and patient-specific scaffolds. These methods mainly include stereolithography, fused deposition modelling (FDM), selective laser sintering (SLS), and 3D printing

(3DP) [71, 97]. The stereolithography method creates solid 3D objects by successively printing a smooth, thin layer of ultraviolet material. The components work in flux to create a 3D scaffold structure without wasting starting materials. For example, a scaffold based on poly(D, L- lactate) or poly(D,L-lactide-co-e-caprolactone) resin was fabricated using the stereolithographic method [98]. Other researchers used gelatin methacrylate to fabricate precisely designed scaffolds [98, 99], and significant results have been reported proving that this method is suitable for TE. However, the method has its limitations as it requires a large number of monomers and polymerization treatment [100]. This method has great potential for many other applications such as biosensors, drug development and energy generation [100].

Hutmacher *et al*. reported a nonwoven scaffold [101]. The scaffold was made up of polyester to facilitate the growth of cells. Similarly, the fused deposition modeling(FDM) method was used to fabricate bioresorbable poly(e-caprolactone) to create porosity in the scaffold [101, 102]. The results for the scaffold fabricated were biocompatible, biodegradable, and better conductive in bone repair experiments and TE [102]. The SLS method is based on a laser that delivers energy to sinter powdered material in thin layers. Different types of materials such as polymers, metals or ceramics can be produced [103, 104]. PLLA-based biomaterial for TE scaffolds has also been reported [105]. 3DP uses a computer to create a complex 3D structure. Different biomaterials can be printed with a specific design for TE [106]. Previously, various scaffolds based on 3DP have been reported, such as poly(dopamine) and poly(lactic acid) scaffolds for bone TE [107]. A better 3DP scaffold was prepared using cold atmospheric plasma [108]. The proposed method showed better nanoscale roughness and chemical composition. Another study reported high-quality ceramic scaffolds fabricated by controlling the extrusion pressure [109]. The results showed that the frameworks were uniform. In addition, a low-cost scaffold based on bioactive (polyphosphazene, polyetheretherketone/hydroxyapatite (PEEK / HA) composites was prepared for bone tissue engineering [110]. Recently, an advanced method for 3DP, called bioprinting, has been developed. This technique is based on the assembly of relevant biological materials, such as cells or tissues [111]. This method can print biomaterials directly into a 3D structural scaffold without solvents. Bioprinting can be either acellular (without cells) or with cells, with the acellular method being superior due to its flexibility and lower fabrication requirements. To date, several printing methods have been developed, including autonomous self-assembly, biomimicry, mini-tissue building blocks, inkjet printing, microextrusion, and laser-assisted printing, with the latter three methods most commonly used for biomaterial printing [112, 113]. Inkjet bioprinting is a non-contact method that uses material jet techniques, such as thermal ink, acoustic waves, and electrohydrodynamic techniques [114]. The method has the advantage of being fast, inexpensive, and has cell viability of 70-90% [114]. However, it suffers from limited material selectivity and clogging [115]. Microextrusion is a temperature-controlled material handling and delivery system and a stage that moves in 3D. The technique consists of a fibre optic light source, a piezoelectric humidifier, and a video camera. The method produces continuous beads of material instead of liquid droplets [116]. The laser-assisted bioprinting method is nozzle-free and compatible with biomaterials of a certain viscosity (1-300 mPa/s) [117]. To date, considerable results have been achieved in the field of *in vivo* bioprinting [118, 119]. A summary of current scaffold fabrication methods can be found in Table **2**.

Table 2. Comparison between conventional and rapid prototyping fabrication methods.

Fabrication Techniques	Production Method	Advantages	Disadvantages
Conventional Fabrication Techniques	Freeze-drying	•Multipurpose technique •Controllable pore size	•Cytotoxic solvent •Irregular pore size •Lengthy and expensive
	Solvent casting	•Excellent porosity •Inexpensive	• Cytotoxic solvent •Time consuming
	Gas foaming	•Good porosity	•Closed pore
	Electrospinning	•Nanofibrous scaffolds •High tensile strength	•Complex procedure •Toxic solvent •Difficult to obtain 3D structure
	Thermal-induced phase separation	•Integration of bioactive molecules •Excellent porosity •Thermoplastic scaffold	•Only used for thermoplastic
Rapid Prototyping Techniques	Stereolithogrphy (SLA)	•Economical •Uniformity in pores	•Limitations in the process of photopolymerization
	Selective laser sintering (SLS)	•Controlled scaffold production •Use of ultrahigh-molecular-weight	•Additional cleaning steps •Require higher temperature
	Solvent-based extrusion free forming (SEF)	•Production of composite materials •Tuneable with precise structure	•Temperature extrusion
	Bioprinting	•Accuracy with excellent cell viability	•Depends on existence of cells
	Fused deposition modelling (FDM)	•Produced controlled design at lower temperature	•Limitation with biodegradable polymers

CONCLUSION AND FUTURE DIRECTIONS

In recent decades, TE has become an active area of research that has led to the discovery of novel biomaterials for scaffold fabrication. Considerable success has been achieved in both *in vitro* and *in vivo* applications. Recently, research in the field of TE has greatly expanded. In particular, promising results have been achieved in the regeneration of various organs of the body, with the greatest success in skin, bladder, respiratory tract and bone. Some products have already been launched in various countries. One of the biggest challenges in the field of TE and scaffold fabrication is the lack of vascularity, so more attention needs to be paid to this research area. Introducing artificial microvasculature into the scaffolds before implantation could help to overcome the above-mentioned drawbacks, and this is only possible through multidisciplinary research approaches. Undoubtedly, TE offers many opportunities to improve human health, which is only possible if different fields are combined.

CONSENT FOR PUBLICATION

Not applicable.

CONFLICT OF INTEREST

The author declares no conflict of interest, financial or otherwise.

ACKNOWLEDGEMENT

Declared none.

REFERENCES

[1]   M.F. Bandeira, "Biopharmaceutical products and biomaterials of the Amazon region used in dentistry", *Biomaterial-supported Tissue Reconstruction or Regeneration,* pp. 113-126, 2018.

[2]   M. Barbeck, O. Jung, R. Smeets, and T. Koržinskas, *Biomaterial-Supported Tissue Reconstruction or Regeneration.* BoD–Books on Demand, 2019.
[http://dx.doi.org/10.5772/intechopen.77756]

[3]   M. Alizadeh-Osgouei, Y. Li, and C. Wen, "A comprehensive review of biodegradable synthetic polymer-ceramic composites and their manufacture for biomedical applications", *Bioact. Mater.,* vol. 4, no. 1, pp. 22-36, 2018.
[PMID: 30533554]

[4]   G.S. Hussey, J.L. Dziki, and S.F. Badylak, "Extracellular matrix-based materials for regenerative medicine", *Nat. Rev. Mater.,* vol. 3, no. 7, pp. 159-173, 2018.
[http://dx.doi.org/10.1038/s41578-018-0023-x]

[5]   C. Bonnans, J. Chou, and Z. Werb, "Remodelling the extracellular matrix in development and disease", *Nat. Rev. Mol. Cell Biol.,* vol. 15, no. 12, pp. 786-801, 2014.
[http://dx.doi.org/10.1038/nrm3904] [PMID: 25415508]

[6]   F.J. O'Brien, "Biomaterials & scaffolds for tissue engineering", *Mater. Today,* vol. 14, no. 3, pp. 88-

95, 2011.
[http://dx.doi.org/10.1016/S1369-7021(11)70058-X]

[7]   S.A. Eming, P. Martin, and M. Tomic-Canic, "Wound repair and regeneration: mechanisms, signaling, and translation", *Science translational medicine*, vol. 6, no. 265, pp. 265sr6-265sr6, 2014.

[8]   R.M. Boehler, J.G. Graham, and L.D. Shea, "Tissue engineering tools for modulation of the immune response", *Biotechniques*, vol. 51, no. 4, pp. 239-254, 2011.
[http://dx.doi.org/10.2144/000113754] [PMID: 21988690]

[9]   A. Nieponice, T.W. Gilbert, S.A. Johnson, N.J. Turner, and S.F. Badylak, "Bone marrow–derived cells participate in the long-term remodeling in a mouse model of esophageal reconstruction", *J. Surg. Res.*, vol. 182, no. 1, pp. e1-e7, 2013.
[http://dx.doi.org/10.1016/j.jss.2012.09.029] [PMID: 23069684]

[10]   S. Tottey, M. Corselli, E.M. Jeffries, R. Londono, B. Peault, and S.F. Badylak, "Extracellular matrix degradation products and low-oxygen conditions enhance the regenerative potential of perivascular stem cells", *Tissue Eng. Part A*, vol. 17, no. 1-2, pp. 37-44, 2011.
[http://dx.doi.org/10.1089/ten.tea.2010.0188] [PMID: 20653348]

[11]   A.V. Boruch, A. Nieponice, I.R. Qureshi, T.W. Gilbert, and S.F. Badylak, "Constructive remodeling of biologic scaffolds is dependent on early exposure to physiologic bladder filling in a canine partial cystectomy model", *J. Surg. Res.*, vol. 161, no. 2, pp. 217-225, 2010.
[http://dx.doi.org/10.1016/j.jss.2009.02.014] [PMID: 19577253]

[12]   F. Ambrosio, S.L. Wolf, A. Delitto, G.K. Fitzgerald, S.F. Badylak, M.L. Boninger, and A.J. Russell, "The emerging relationship between regenerative medicine and physical therapeutics", *Phys. Ther.*, vol. 90, no. 12, pp. 1807-1814, 2010.
[http://dx.doi.org/10.2522/ptj.20100030] [PMID: 21030663]

[13]   L.G. Griffith, "Emerging design principles in biomaterials and scaffolds for tissue engineering", *Ann. N. Y. Acad. Sci.*, vol. 961, no. 1, pp. 83-95, 2002.
[http://dx.doi.org/10.1111/j.1749-6632.2002.tb03056.x] [PMID: 12081872]

[14]   P. Lichte, H.C. Pape, T. Pufe, P. Kobbe, and H. Fischer, "Scaffolds for bone healing: Concepts, materials and evidence", *Injury*, vol. 42, no. 6, pp. 569-573, 2011.
[http://dx.doi.org/10.1016/j.injury.2011.03.033] [PMID: 21489531]

[15]   A.K. Saxena, "Tissue engineering and regenerative medicine research perspectives for pediatric surgery", *Pediatr. Surg. Int.*, vol. 26, no. 6, pp. 557-573, 2010.
[http://dx.doi.org/10.1007/s00383-010-2591-8] [PMID: 20333389]

[16]   S.M. Best, A.E. Porter, E.S. Thian, and J. Huang, "Bioceramics: Past, present and for the future", *J. Eur. Ceram. Soc.*, vol. 28, no. 7, pp. 1319-1327, 2008.
[http://dx.doi.org/10.1016/j.jeurceramsoc.2007.12.001]

[17]   R.Z. LeGeros, "Properties of osteoconductive biomaterials: calcium phosphates", *Clin. Orthop. Relat. Res.*, vol. 395, no. 395, pp. 81-98, 2002.
[http://dx.doi.org/10.1097/00003086-200202000-00009] [PMID: 11937868]

[18]   "A preliminary report on studies of basic calcium phosphate in bone replacement", In: *in Surgical forum*, 1951, pp. 429- 434.

[19]   G. Calabrese, R. Giuffrida, C. Fabbi, E. Figallo, D. Lo Furno, R. Gulino, C. Colarossi, F. Fullone, R. Giuffrida, R. Parenti, L. Memeo, and S. Forte, "Collagen-hydroxyapatite scaffolds induce human adipose derived stem cells osteogenic differentiation *in vitro*", *PLoS One*, vol. 11, no. 3, p. e0151181, 2016.
[http://dx.doi.org/10.1371/journal.pone.0151181] [PMID: 26982592]

[20]   G. Calabrese, R. Giuffrida, S. Forte, L. Salvatorelli, C. Fabbi, E. Figallo, M. Gulisano, R. Parenti, G. Magro, C. Colarossi, L. Memeo, and R. Gulino, "Bone augmentation after ectopic implantation of a cell-free collagen-hydroxyapatite scaffold in the mouse", *Sci. Rep.*, vol. 6, no. 1, p. 36399, 2016.

[http://dx.doi.org/10.1038/srep36399] [PMID: 27821853]

[21] G. Calabrese, R. Giuffrida, S. Forte, C. Fabbi, E. Figallo, L. Salvatorelli, L. Memeo, R. Parenti, M. Gulisano, and R. Gulino, "Human adipose-derived mesenchymal stem cells seeded into a collagen-hydroxyapatite scaffold promote bone augmentation after implantation in the mouse", *Sci. Rep.,* vol. 7, no. 1, p. 7110, 2017.
[http://dx.doi.org/10.1038/s41598-017-07672-0] [PMID: 28769083]

[22] L.L. Hench, "The story of Bioglass® ", *J. Mater. Sci. Mater. Med.,* vol. 17, no. 11, pp. 967-978, 2006.
[http://dx.doi.org/10.1007/s10856-006-0432-z] [PMID: 17122907]

[23] M.N. Rahaman, D.E. Day, B. Sonny Bal, Q. Fu, S.B. Jung, L.F. Bonewald, and A.P. Tomsia, "Bioactive glass in tissue engineering", *Acta Biomater.,* vol. 7, no. 6, pp. 2355-2373, 2011.
[http://dx.doi.org/10.1016/j.actbio.2011.03.016] [PMID: 21421084]

[24] K. Rezwan, Q.Z. Chen, J.J. Blaker, and A.R. Boccaccini, "Biodegradable and bioactive porous polymer/inorganic composite scaffolds for bone tissue engineering", *Biomaterials,* vol. 27, no. 18, pp. 3413-3431, 2006.
[http://dx.doi.org/10.1016/j.biomaterials.2006.01.039] [PMID: 16504284]

[25] A. Dolcimascolo, G. Calabrese, S. Conoci, and R. Parenti, "Innovative biomaterials for tissue engineering", In: *Biomaterial-supported Tissue Reconstruction or Regeneration.* IntechOpen, 2019.
[http://dx.doi.org/10.5772/intechopen.83839]

[26] D. Deponti, A.D. Giancamillo, F. Gervaso, M. Domenicucci, C. Domeneghini, A. Sannino, and G.M. Peretti, "Collagen scaffold for cartilage tissue engineering: the benefit of fibrin glue and the proper culture time in an infant cartilage model", *Tissue Eng. Part A,* vol. 20, no. 5-6, pp. 1113-1126, 2014.
[http://dx.doi.org/10.1089/ten.tea.2013.0171] [PMID: 24152291]

[27] A. Aravamudhan, D.M. Ramos, J. Nip, M.D. Harmon, R. James, M. Deng, C.T. Laurencin, X. Yu, and S.G. Kumbar, "Cellulose and collagen derived micro-nano structured scaffolds for bone tissue engineering", *J. Biomed. Nanotechnol.,* vol. 9, no. 4, pp. 719-731, 2013.
[http://dx.doi.org/10.1166/jbn.2013.1574] [PMID: 23621034]

[28] R.K. Schneider, A. Puellen, R. Kramann, K. Raupach, J. Bornemann, R. Knuechel, A. Pérez-Bouza, and S. Neuss, "The osteogenic differentiation of adult bone marrow and perinatal umbilical mesenchymal stem cells and matrix remodelling in three-dimensional collagen scaffolds", *Biomaterials,* vol. 31, no. 3, pp. 467-480, 2010.
[http://dx.doi.org/10.1016/j.biomaterials.2009.09.059] [PMID: 19815272]

[29] H. Lu, N. Kawazoe, T. Kitajima, Y. Myoken, M. Tomita, A. Umezawa, G. Chen, and Y. Ito, "Spatial immobilization of bone morphogenetic protein-4 in a collagen-PLGA hybrid scaffold for enhanced osteoinductivity", *Biomaterials,* vol. 33, no. 26, pp. 6140-6146, 2012.
[http://dx.doi.org/10.1016/j.biomaterials.2012.05.038] [PMID: 22698726]

[30] A.R. Costa-Pinto, V.M. Correlo, P.C. Sol, M. Bhattacharya, P. Charbord, B. Delorme, R.L. Reis, and N.M. Neves, "Osteogenic differentiation of human bone marrow mesenchymal stem cells seeded on melt based chitosan scaffolds for bone tissue engineering applications", *Biomacromolecules,* vol. 10, no. 8, pp. 2067-2073, 2009.
[http://dx.doi.org/10.1021/bm9000102] [PMID: 19621927]

[31] L. Bi, W. Cheng, H. Fan, and G. Pei, "Reconstruction of goat tibial defects using an injectable tricalcium phosphate/chitosan in combination with autologous platelet-rich plasma", *Biomaterials,* vol. 31, no. 12, pp. 3201-3211, 2010.
[http://dx.doi.org/10.1016/j.biomaterials.2010.01.038] [PMID: 20116844]

[32] S. Eğri, and N. Eczacıoğlu, "Sequential VEGF and BMP-2 releasing PLA-PEG-PLA scaffolds for bone tissue engineering: I. Design and *in vitro* tests"., *Artif. Cells Nanomed. Biotechnol.,* vol. 45, no. 2, pp. 321-329, 2017.
[http://dx.doi.org/10.3109/21691401.2016.1147454] [PMID: 26912262]

[33] H. Liu, E.B. Slamovich, and T.J. Webster, "Less harmful acidic degradation of poly(lactic-co-glycolic

acid) bone tissue engineering scaffolds through titania nanoparticle addition", *Int. J. Nanomedicine,* vol. 1, no. 4, pp. 541-545, 2006.
[http://dx.doi.org/10.2147/nano.2006.1.4.541] [PMID: 17722285]

[34]   A. Yusop, A. Bakir, N. Shaharom, M. Abdul Kadir, and H. Hermawan, "Porous biodegradable metals for hard tissue scaffolds: a review", *International journal of biomaterials,* vol. 2012, 2012.
[http://dx.doi.org/10.1155/2012/641430]

[35]   J.C. Wohlfahrt, M. Monjo, H.J. RÃ¸nold, A.M. Aass, J.E. Ellingsen, and S.P. Lyngstadaas, "Porous titanium granules promote bone healing and growth in rabbit tibia peri-implant osseous defects", *Clin. Oral Implants Res.,* vol. 21, no. 2, pp. 165-173, 2010.
[http://dx.doi.org/10.1111/j.1600-0501.2009.01813.x] [PMID: 19912270]

[36]   J. Zuchuat, M. Berli, Y. Maldonado, and O. Decco, "Influence of Chromium-Cobalt-Molybdenum Alloy (ASTM F75) on Bone Ingrowth in an Experimental Animal Model", *J. Funct. Biomater.,* vol. 9, no. 1, p. 2, 2017.
[http://dx.doi.org/10.3390/jfb9010002] [PMID: 29278372]

[37]   J. Qian, W. Xu, X. Yong, X. Jin, and W. Zhang, "Fabrication and *in vitro* biocompatibility of biomorphic PLGA/nHA composite scaffolds for bone tissue engineering"., *Mater. Sci. Eng. C,* vol. 36, pp. 95-101, 2014.
[http://dx.doi.org/10.1016/j.msec.2013.11.047] [PMID: 24433891]

[38]   S. Oughlis, S. Lessim, S. Changotade, F. Bollotte, F. Poirier, G. Helary, J.J. Lataillade, V. Migonney, and D. Lutomski, "Development of proteomic tools to study protein adsorption on a biomaterial, titanium grafted with poly(sodium styrene sulfonate)", *J. Chromatogr. B Analyt. Technol. Biomed. Life Sci.,* vol. 879, no. 31, pp. 3681-3687, 2011.
[http://dx.doi.org/10.1016/j.jchromb.2011.10.006] [PMID: 22036657]

[39]   O. Wichterle, and D. Lím, "Hydrophilic gels in biologic use", *Nature,* vol. 185, no. 4706, pp. 117-118, 1960.
[http://dx.doi.org/10.1038/185117a0]

[40]   J.L. Drury, and D.J. Mooney, "Hydrogels for tissue engineering: scaffold design variables and applications", *Biomaterials,* vol. 24, no. 24, pp. 4337-4351, 2003.
[http://dx.doi.org/10.1016/S0142-9612(03)00340-5] [PMID: 12922147]

[41]   D. Pasqui, P. Torricelli, M. De Cagna, M. Fini, and R. Barbucci, "Carboxymethyl cellulose-hydroxyapatite hybrid hydrogel as a composite material for bone tissue engineering applications", *J. Biomed. Mater. Res. A,* vol. 102, no. 5, pp. 1568-1579, 2014.
[http://dx.doi.org/10.1002/jbm.a.34810] [PMID: 23720392]

[42]   L.A. Kinard, R.L. Dahlin, J. Lam, S. Lu, E.J. Lee, F.K. Kasper, and A.G. Mikos, "Synthetic biodegradable hydrogel delivery of demineralized bone matrix for bone augmentation in a rat model", *Acta Biomater.,* vol. 10, no. 11, pp. 4574-4582, 2014.
[http://dx.doi.org/10.1016/j.actbio.2014.07.011] [PMID: 25046637]

[43]   B.P. Chan, and K.W. Leong, "Scaffolding in tissue engineering: general approaches and tissue-specific considerations", *Eur. Spine J.,* vol. 17, no. S4, suppl. Suppl. 4, pp. 467-479, 2008.
[http://dx.doi.org/10.1007/s00586-008-0745-3] [PMID: 19005702]

[44]   J.P. Vacanti, and R. Langer, "Tissue engineering: the design and fabrication of living replacement devices for surgical reconstruction and transplantation", *Lancet,* vol. 354, suppl. Suppl. 1, pp. S32-S34, 1999.
[http://dx.doi.org/10.1016/S0140-6736(99)90247-7] [PMID: 10437854]

[45]   R.S. Langer, and J.P. Vacanti, "Tissue engineering: the challenges ahead", *Sci. Am.,* vol. 280, no. 4, pp. 86-89, 1999.
[http://dx.doi.org/10.1038/scientificamerican0499-86] [PMID: 10201120]

[46]   J.A. Hubbell, "Biomaterials in tissue engineering", *Biotechnology (N. Y.),* vol. 13, no. 6, pp. 565-576, 1995.

[PMID: 9634795]

[47]   Y.S. Nam, J.J. Yoon, and T.G. Park, "A novel fabrication method of macroporous biodegradable polymer scaffolds using gas foaming salt as a porogen additive", *J. Biomed. Mater. Res.,* vol. 53, no. 1, pp. 1-7, 2000.
[http://dx.doi.org/10.1002/(SICI)1097-4636(2000)53:1<1::AID-JBM1>3.0.CO;2-R]   [PMID: 10634946]

[48]   A. Prasad, M.R. Sankar, and V. Katiyar, "State of art on solvent casting particulate leaching method for orthopedic scaffoldsfabrication", *Mater. Today Proc.,* vol. 4, no. 2, pp. 898-907, 2017.
[http://dx.doi.org/10.1016/j.matpr.2017.01.101]

[49]   S.B. Lee, Y.H. Kim, M.S. Chong, S.H. Hong, and Y.M. Lee, "Study of gelatin-containing artificial skin V: fabrication of gelatin scaffolds using a salt-leaching method", *Biomaterials,* vol. 26, no. 14, pp. 1961-1968, 2005.
[http://dx.doi.org/10.1016/j.biomaterials.2004.06.032] [PMID: 15576170]

[50]   Q.P. Pham, U. Sharma, and A.G. Mikos, "Electrospinning of polymeric nanofibers for tissue engineering applications: a review", *Tissue Eng.,* vol. 12, no. 5, pp. 1197-1211, 2006.
[http://dx.doi.org/10.1089/ten.2006.12.1197] [PMID: 16771634]

[51]   S.F. Badylak, "Xenogeneic extracellular matrix as a scaffold for tissue reconstruction", *Transpl. Immunol.,* vol. 12, no. 3-4, pp. 367-377, 2004.
[http://dx.doi.org/10.1016/j.trim.2003.12.016] [PMID: 15157928]

[52]   J.H. Ingram, S. Korossis, G. Howling, J. Fisher, and E. Ingham, "The use of ultrasonication to aid recellularization of acellular natural tissue scaffolds for use in anterior cruciate ligament reconstruction", *Tissue Eng.,* vol. 13, no. 7, pp. 1561-1572, 2007.
[http://dx.doi.org/10.1089/ten.2006.0362] [PMID: 17518726]

[53]   T.W. Gilbert, T.L. Sellaro, and S.F. Badylak, "Decellularization of tissues and organs", *Biomaterials,* vol. 27, no. 19, pp. 3675-3683, 2006.
[PMID: 16519932]

[54]   G.H. Borschel, Y.C. Huang, S. Calve, E.M. Arruda, J.B. Lynch, D.E. Dow, W.M. Kuzon, R.G. Dennis, and D.L. Brown, "Tissue engineering of recellularized small-diameter vascular grafts", *Tissue Eng.,* vol. 11, no. 5-6, pp. 778-786, 2005.
[http://dx.doi.org/10.1089/ten.2005.11.778] [PMID: 15998218]

[55]   L. Ansaloni, F. Catena, S. Gagliardi, F. Gazzotti, L. D'Alessandro, and A.D. Pinna, "Hernia repair with porcine small-intestinal submucosa", *Hernia,* vol. 11, no. 4, pp. 321-326, 2007.
[http://dx.doi.org/10.1007/s10029-007-0225-4] [PMID: 17443270]

[56]   R. Fiala, A. Vidlar, R. Vrtal, K. Belej, and V. Student, "Porcine small intestinal submucosa graft for repair of anterior urethral strictures", *Eur. Urol.,* vol. 51, no. 6, pp. 1702-1708, 2007.
[http://dx.doi.org/10.1016/j.eururo.2007.01.099] [PMID: 17306922]

[57]   W.A. Morrison, and A.J. Hussey, "Extracellular matrix as a bioactive material for soft tissue reconstruction", *ANZ J. Surg.,* vol. 76, no. 12, pp. 1047-1047, 2006.
[http://dx.doi.org/10.1111/j.1445-2197.2006.03970.x] [PMID: 17199686]

[58]   C.N. Manning, A.G. Schwartz, W. Liu, J. Xie, N. Havlioglu, S.E. Sakiyama-Elbert, M.J. Silva, Y. Xia, R.H. Gelberman, and S. Thomopoulos, "Controlled delivery of mesenchymal stem cells and growth factors using a nanofiber scaffold for tendon repair", *Acta Biomater.,* vol. 9, no. 6, pp. 6905-6914, 2013.
[http://dx.doi.org/10.1016/j.actbio.2013.02.008] [PMID: 23416576]

[59]   M.H. Zheng, J. Chen, Y. Kirilak, C. Willers, J. Xu, and D. Wood, "Porcine small intestine submucosa (SIS) is not an acellular collagenous matrix and contains porcine DNA: Possible implications in human implantation", *J. Biomed. Mater. Res. B Appl. Biomater.,* vol. 73B, no. 1, pp. 61-67, 2005.
[http://dx.doi.org/10.1002/jbm.b.30170] [PMID: 15736287]

[60]   X. Yu, X. Tang, S.V. Gohil, and C.T. Laurencin, "Biomaterials for bone regenerative engineering", *Adv. Healthc. Mater.,* vol. 4, no. 9, pp. 1268-1285, 2015.
[http://dx.doi.org/10.1002/adhm.201400760] [PMID: 25846250]

[61]   Y. Tabata, "Biomaterial technology for tissue engineering applications", *Journal of the Royal Society interface,* vol. vol.6, no. no. suppl_3, pp. 311-324, 2009.
[http://dx.doi.org/10.1098/rsif.2008.0448.focus]

[62]   B.D. Ratner, A.S. Hoffman, F.J. Schoen, and J.E. Lemons, "Biomaterials science: an introduction to materials in medicine", *MRS Bull.,* vol. 31, p. 59, 2006.

[63]   T. Ghassemi, A. Shahroodi, M.H. Ebrahimzadeh, A. Mousavian, J. Movaffagh, and A. Moradi, "Current concepts in scaffolding for bone tissue engineering", *Arch. Bone Jt. Surg.,* vol. 6, no. 2, pp. 90-99, 2018.
[PMID: 29600260]

[64]   E. Hadjipanayi, V. Mudera, and R.A. Brown, "Close dependence of fibroblast proliferation on collagen scaffold matrix stiffness", *J. Tissue Eng. Regen. Med.,* vol. 3, no. 2, pp. 77-84, 2009.
[http://dx.doi.org/10.1002/term.136] [PMID: 19051218]

[65]   F. Anjum, N.A. Agabalyan, H.D. Sparks, N.L. Rosin, M.S. Kallos, and J. Biernaskie, "Biocomposite nanofiber matrices to support ECM remodeling by human dermal progenitors and enhanced wound closure", *Sci. Rep.,* vol. 7, no. 1, p. 10291, 2017.
[http://dx.doi.org/10.1038/s41598-017-10735-x] [PMID: 28860484]

[66]   V.L. Tsang, and S.N. Bhatia, "Fabrication of three-dimensional tissues", *Adv. Biochem. Eng. Biotechnol.,* vol. 103, pp. 189-205, 2007.
[http://dx.doi.org/10.1007/10_010] [PMID: 17195464]

[67]   O.A. Abdelaal, and S.M. Darwish, "Review of rapid prototyping techniques for tissue engineering scaffolds fabrication", In: *Characterization and Development of Biosystems and Biomaterials.* Springer, 2013, pp. 33-54.
[http://dx.doi.org/10.1007/978-3-642-31470-4_3]

[68]   H.N. Chia, and B.M. Wu, "Recent advances in 3D printing of biomaterials", *J. Biol. Eng.,* vol. 9, no. 1, p. 4, 2015.
[http://dx.doi.org/10.1186/s13036-015-0001-4] [PMID: 25866560]

[69]   R.C. Dutta, M. Dey, A.K. Dutta, and B. Basu, "Competent processing techniques for scaffolds in tissue engineering", *Biotechnol. Adv.,* vol. 35, no. 2, pp. 240-250, 2017.
[http://dx.doi.org/10.1016/j.biotechadv.2017.01.001] [PMID: 28095322]

[70]   H. Qu, H. Fu, Z. Han, and Y. Sun, "Biomaterials for bone tissue engineering scaffolds: a review", *RSC Advances,* vol. 9, no. 45, pp. 26252-26262, 2019.
[http://dx.doi.org/10.1039/C9RA05214C] [PMID: 35531040]

[71]   Z. Li, M.B. Xie, Y. Li, Y. Ma, J.S. Li, and F.Y. Dai, "Recent progress in tissue engineering and regenerative medicine", *J. Biomater. Tissue Eng.,* vol. 6, no. 10, pp. 755-766, 2016.
[http://dx.doi.org/10.1166/jbt.2016.1510]

[72]   J. Sanzherrera, J. Garcíaaznar, and M. Doblaré, "On scaffold designing for bone regeneration: A computational multiscale approach", *Acta Biomater.,* vol. 5, no. 1, pp. 219-229, 2009.
[http://dx.doi.org/10.1016/j.actbio.2008.06.021] [PMID: 18725187]

[73]   D.T.M. Thanh, P.T.T. Trang, N.T. Thom, N.T. Phuong, P.T. Nam, N.T.T. Trang, J. Seo-Park, and T. Hoang, "Effects of porogen on structure and properties of poly lactic acid/hydroxyapatite nanocomposites (PLA/HAp)", *J. Nanosci. Nanotechnol.,* vol. 16, no. 9, pp. 9450-9459, 2016.
[http://dx.doi.org/10.1166/jnn.2016.12032]

[74]   E. Jahed, M.A. Khaledabad, H. Almasi, and R. Hasanzadeh, "Physicochemical properties of Carum copticum essential oil loaded chitosan films containing organic nanoreinforcements", *Carbohydr. Polym.,* vol. 164, pp. 325-338, 2017.

[http://dx.doi.org/10.1016/j.carbpol.2017.02.022] [PMID: 28325333]

[75]   P. Agrawal, Sonali, R.P. Singh, G. Sharma, A.K. Mehata, S. Singh, C.V. Rajesh, B.L. Pandey, B. Koch, and M.S. Muthu, "Bioadhesive micelles of d -α-tocopherol polyethylene glycol succinate 1000: Synergism of chitosan and transferrin in targeted drug delivery", *Colloids Surf. B Biointerfaces,* vol. 152, pp. 277-288, 2017.
[http://dx.doi.org/10.1016/j.colsurfb.2017.01.021] [PMID: 28122295]

[76]   S. Preethi Soundarya, V. Sanjay, A. Haritha Menon, S. Dhivya, and N. Selvamurugan, "Effects of flavonoids incorporated biological macromolecules based scaffolds in bone tissue engineering", *Int. J. Biol. Macromol.,* vol. 110, pp. 74-87, 2018.
[http://dx.doi.org/10.1016/j.ijbiomac.2017.09.014] [PMID: 28893682]

[77]   L. Roseti, V. Parisi, M. Petretta, C. Cavallo, G. Desando, I. Bartolotti, and B. Grigolo, "Scaffolds for Bone Tissue Engineering: State of the art and new perspectives", *Mater. Sci. Eng. C,* vol. 78, pp. 1246-1262, 2017.
[http://dx.doi.org/10.1016/j.msec.2017.05.017] [PMID: 28575964]

[78]   I. Aranaz, M. Gutiérrez, M. Ferrer, and F. del Monte, "Preparation of chitosan nanocomposites with a macroporous structure by unidirectional freezing and subsequent freeze-drying", *Mar. Drugs,* vol. 12, no. 11, pp. 5619-5642, 2014.
[http://dx.doi.org/10.3390/md12115619] [PMID: 25421320]

[79]   A. Kumar, K. Madhusudana Rao, A. Haider, S.S. Han, T.W. Son, J.H. Kim, and T.H. Oh, "Fabrication and characterization of multicomponent polysaccharide/nanohydroxyapatite composite scaffolds", *Polym. Plast. Technol. Eng.,* vol. 56, no. 9, pp. 983-991, 2017.
[http://dx.doi.org/10.1080/03602559.2016.1247279]

[80]   B. Nasiri, and S. Mashayekhan, "Fabrication of porous scaffolds with decellularized cartilage matrix for tissue engineering application", *Biologicals,* vol. 48, pp. 39-46, 2017.
[http://dx.doi.org/10.1016/j.biologicals.2017.05.008] [PMID: 28602577]

[81]   M.Y. Kim, and J. Lee, "Chitosan fibrous 3D networks prepared by freeze drying", *Carbohydr. Polym.,* vol. 84, no. 4, pp. 1329-1336, 2011.
[http://dx.doi.org/10.1016/j.carbpol.2011.01.029]

[82]   J. Venkatesan, I. Bhatnagar, and S.K. Kim, "Chitosan-alginate biocomposite containing fucoidan for bone tissue engineering", *Mar. Drugs,* vol. 12, no. 1, pp. 300-316, 2014.
[http://dx.doi.org/10.3390/md12010300] [PMID: 24441614]

[83]   Z. Karimi, M. Ghorbani, B. Hashemibeni, and H. Bahramian, "Evaluation of the proliferation and viability rates of nucleus pulposus cells of human intervertebral disk in fabricated chitosan-gelatin scaffolds by freeze drying and freeze gelation methods", *Advanced biomedical research,* vol. 4, 2015.

[84]   A.G. Guex, J.L. Puetzer, A. Armgarth, E. Littmann, E. Stavrinidou, E.P. Giannelis, G.G. Malliaras, and M.M. Stevens, "Highly porous scaffolds of PEDOT:PSS for bone tissue engineering", *Acta Biomater.,* vol. 62, pp. 91-101, 2017.
[http://dx.doi.org/10.1016/j.actbio.2017.08.045] [PMID: 28865991]

[85]   E. Torino, R. Aruta, T. Sibillano, C. Giannini, and P.A. Netti, "Synthesis of semicrystalline nanocapsular structures obtained by Thermally Induced Phase Separation in nanoconfinement", *Sci. Rep.,* vol. 6, no. 1, p. 32727, 2016.
[http://dx.doi.org/10.1038/srep32727] [PMID: 27604818]

[86]   F.C. Pavia, V. La Carrubba, S. Piccarolo, and V. Brucato, "Polymeric scaffolds prepared *via* thermally induced phase separation: Tuning of structure and morphology"., *J. Biomed. Mater. Res. A,* vol. 86A, no. 2, pp. 459-466, 2008.
[http://dx.doi.org/10.1002/jbm.a.31621] [PMID: 17975822]

[87]   J.J. Blaker, J.C. Knowles, and R.M. Day, "Novel fabrication techniques to produce microspheres by thermally induced phase separation for tissue engineering and drug delivery", *Acta Biomater.,* vol. 4, no. 2, pp. 264-272, 2008.

[http://dx.doi.org/10.1016/j.actbio.2007.09.011] [PMID: 18032120]

[88]   I.O. Smith, X.H. Liu, L.A. Smith, and P.X. Ma, "Nanostructured polymer scaffolds for tissue engineering and regenerative medicine", *Wiley Interdiscip. Rev. Nanomed. Nanobiotechnol.,* vol. 1, no. 2, pp. 226-236, 2009.
[http://dx.doi.org/10.1002/wnan.26] [PMID: 20049793]

[89]   L. D. Harris, B. S. Kim, and D. J. Mooney, "Open pore biodegradable matrices formed with gas foaming", *Journal of Biomedical Materials Research: An Official Journal of The Society for Biomaterials, The Japanese Society for Biomaterials, and the Australian Society for Biomaterials,* vol. 42, no. 3, pp. 396-402, 1998.
[http://dx.doi.org/10.1002/(SICI)1097-4636(19981205)42:3<396::AID-JBM7>3.0.CO;2-E]

[90]   A. Haider, K.C. Gupta, and I.K. Kang, "PLGA/nHA hybrid nanofiber scaffold as a nanocargo carrier of insulin for accelerating bone tissue regeneration", *Nanoscale Res. Lett.,* vol. 9, no. 1, p. 314, 2014.
[http://dx.doi.org/10.1186/1556-276X-9-314] [PMID: 25024679]

[91]   B. Zhu, W. Li, R.V. Lewis, C.U. Segre, and R. Wang, "E-spun composite fibers of collagen and dragline silk protein: fiber mechanics, biocompatibility, and application in stem cell differentiation", *Biomacromolecules,* vol. 16, no. 1, pp. 202-213, 2015.
[http://dx.doi.org/10.1021/bm501403f] [PMID: 25405355]

[92]   W.A. Sarhan, H.M.E. Azzazy, and I.M. El-Sherbiny, "Honey/chitosan nanofiber wound dressing enriched with Allium sativum and Cleome droserifolia: enhanced antimicrobial and wound healing activity", *ACS Appl. Mater. Interfaces,* vol. 8, no. 10, pp. 6379-6390, 2016.
[http://dx.doi.org/10.1021/acsami.6b00739] [PMID: 26909753]

[93]   A. K. HPS, "A review on chitosan-cellulose blends and nanocellulose reinforced chitosan biocomposites: Properties and their applications", *Carbohydrate polymers,* vol. 150, pp. 216-226, 2016.

[94]   G. Zhou, S. Liu, Y. Ma, W. Xu, W. Meng, X. Lin, W. Wang, S. Wang, and J. Zhang, "Innovative biodegradable poly(L-lactide)/collagen/hydroxyapatite composite fibrous scaffolds promote osteoblastic proliferation and differentiation", *Int. J. Nanomedicine,* vol. 12, pp. 7577-7588, 2017.
[http://dx.doi.org/10.2147/IJN.S146679] [PMID: 29075116]

[95]   D.W. Hutmacher, M. Sittinger, and M.V. Risbud, "Scaffold-based tissue engineering: rationale for computer-aided design and solid free-form fabrication systems", *Trends Biotechnol.,* vol. 22, no. 7, pp. 354-362, 2004.
[http://dx.doi.org/10.1016/j.tibtech.2004.05.005] [PMID: 15245908]

[96]   D.J. Yoo, "Recent trends and challenges in computer-aided design of additive manufacturing-based biomimetic scaffolds and bioartificial organs", *Int. J. Precis. Eng. Manuf.,* vol. 15, no. 10, pp. 2205-2217, 2014.
[http://dx.doi.org/10.1007/s12541-014-0583-7]

[97]   A. Eltom, G. Zhong, and A. Muhammad, "Scaffold techniques and designs in tissue engineering functions and purposes: a review", *Advances in Materials Science and Engineering,* vol. 2019, 2019.

[98]   R. Gauvin, Y.C. Chen, J.W. Lee, P. Soman, P. Zorlutuna, J.W. Nichol, H. Bae, S. Chen, and A. Khademhosseini, "Microfabrication of complex porous tissue engineering scaffolds using 3D projection stereolithography", *Biomaterials,* vol. 33, no. 15, pp. 3824-3834, 2012.
[http://dx.doi.org/10.1016/j.biomaterials.2012.01.048] [PMID: 22365811]

[99]   S. Xiao, T. Zhao, J. Wang, C. Wang, J. Du, L. Ying, J. Lin, C. Zhang, W. Hu, L. Wang, and K. Xu, "Gelatin methacrylate (GelMA)-based hydrogels for cell transplantation: an effective strategy for tissue engineering", *Stem Cell Rev. Rep.,* vol. 15, no. 5, pp. 664-679, 2019.
[http://dx.doi.org/10.1007/s12015-019-09893-4] [PMID: 31154619]

[100] P. Bajaj, R.M. Schweller, A. Khademhosseini, J.L. West, and R. Bashir, "3D biofabrication strategies for tissue engineering and regenerative medicine", *Annu. Rev. Biomed. Eng.,* vol. 16, no. 1, pp. 247-276, 2014.

[http://dx.doi.org/10.1146/annurev-bioeng-071813-105155] [PMID: 24905875]

[101] I. Zein, D.W. Hutmacher, K.C. Tan, and S.H. Teoh, "Fused deposition modeling of novel scaffold architectures for tissue engineering applications", *Biomaterials,* vol. 23, no. 4, pp. 1169-1185, 2002.
[http://dx.doi.org/10.1016/S0142-9612(01)00232-0] [PMID: 11791921]

[102] Z. Xiong, Y. Yan, S. Wang, R. Zhang, and C. Zhang, "Fabrication of porous scaffolds for bone tissue engineering *via* low-temperature deposition", *Scr. Mater.,* vol. 46, no. 11, pp. 771-776, 2002.
[http://dx.doi.org/10.1016/S1359-6462(02)00071-4]

[103] K.H. Tan, C.K. Chua, K.F. Leong, C.M. Cheah, P. Cheang, M.S. Abu Bakar, and S.W. Cha, "Scaffold development using selective laser sintering of polyetheretherketone–hydroxyapatite biocomposite blends", *Biomaterials,* vol. 24, no. 18, pp. 3115-3123, 2003.
[http://dx.doi.org/10.1016/S0142-9612(03)00131-5] [PMID: 12895584]

[104] W.Y. Zhou, S.H. Lee, M. Wang, W.L. Cheung, and W.Y. Ip, "Selective laser sintering of porous tissue engineering scaffolds from poly(l-lactide)/carbonated hydroxyapatite nanocomposite microspheres", *J. Mater. Sci. Mater. Med.,* vol. 19, no. 7, pp. 2535-2540, 2008.
[http://dx.doi.org/10.1007/s10856-007-3089-3] [PMID: 17619975]

[105] N. Sultana, and M. Wang, "PHBV/PLLA-based composite scaffolds fabricated using an emulsion freezing/freeze-drying technique for bone tissue engineering: surface modification and *in vitro* biological evaluation"., *Biofabrication,* vol. 4, no. 1, p. 015003, 2012.
[http://dx.doi.org/10.1088/1758-5082/4/1/015003] [PMID: 22258057]

[106] M. Wang, P. Favi, X. Cheng, N.H. Golshan, K.S. Ziemer, M. Keidar, and T.J. Webster, "Cold atmospheric plasma (CAP) surface nanomodified 3D printed polylactic acid (PLA) scaffolds for bone regeneration", *Acta Biomater.,* vol. 46, pp. 256-265, 2016.
[http://dx.doi.org/10.1016/j.actbio.2016.09.030] [PMID: 27667017]

[107] C.T. Kao, C.C. Lin, Y.W. Chen, C.H. Yeh, H.Y. Fang, and M.Y. Shie, "Poly(dopamine) coating of 3D printed poly(lactic acid) scaffolds for bone tissue engineering", *Mater. Sci. Eng. C,* vol. 56, pp. 165-173, 2015.
[http://dx.doi.org/10.1016/j.msec.2015.06.028] [PMID: 26249577]

[108] G. Zhong, M. Vaezi, P. Liu, L. Pan, and S. Yang, "Characterization approach on the extrusion process of bioceramics for the 3D printing of bone tissue engineering scaffolds", *Ceram. Int.,* vol. 43, no. 16, pp. 13860-13868, 2017.
[http://dx.doi.org/10.1016/j.ceramint.2017.07.109]

[109] M. Vaezi, and S. Yang, "Extrusion-based additive manufacturing of PEEK for biomedical applications", *Virtual Phys. Prototyp.,* vol. 10, no. 3, pp. 123-135, 2015.
[http://dx.doi.org/10.1080/17452759.2015.1097053]

[110] M. Vaezi, C. Black, D. Gibbs, R. Oreffo, M. Brady, M. Moshrefi-Torbati, and S. Yang, "Characterization of new PEEK/HA composites with 3D HA network fabricated by extrusion freeforming", *Molecules,* vol. 21, no. 6, p. 687, 2016.
[http://dx.doi.org/10.3390/molecules21060687] [PMID: 27240326]

[111] S.V. Murphy, and A. Atala, "3D bioprinting of tissues and organs", *Nat. Biotechnol.,* vol. 32, no. 8, pp. 773-785, 2014.
[http://dx.doi.org/10.1038/nbt.2958] [PMID: 25093879]

[112] M.S. Begam, and V. Dharani, "3D Bioprinting of Tissues and Organs", *Biomaterials,* vol. 30, pp. 2164-2174, 2014.

[113] A. Hacıoglu, H. Yılmazer, and C.B. Ustundag, "3D printing for tissue engineering applications", *Politeknik Dergisi,* vol. 21, no. 1, pp. 221-227, 2018.

[114] G. Ratheesh, C. Vaquette, P. Sonar, and Y. Xiao, "Strategies of 3D bioprinting and parameters that determine cell interaction with the scaffold-A review", *Regenerated Organs,* pp. 81-95, 2021.
[http://dx.doi.org/10.1016/B978-0-12-821085-7.00004-X]

[115] S. Adepu, N. Dhiman, A. Laha, C.S. Sharma, S. Ramakrishna, and M. Khandelwal, "Three-dimensional bioprinting for bone tissue regeneration", *Curr. Opin. Biomed. Eng.,* vol. 2, pp. 22-28, 2017.
[http://dx.doi.org/10.1016/j.cobme.2017.03.005]

[116] C. Kasper, D. Egger, and A. Lavrentieva, *Basic Concepts on 3D Cell Culture.* Springer, 2021.
[http://dx.doi.org/10.1007/978-3-030-66749-8]

[117] Z. Gu, J. Fu, H. Lin, and Y. He, "Development of 3D bioprinting: From printing methods to biomedical applications", *Asian Journal of Pharmaceutical Sciences,* vol. 15, no. 5, pp. 529-557, 2020.
[http://dx.doi.org/10.1016/j.ajps.2019.11.003] [PMID: 33193859]

[118] P. Apelgren, M. Amoroso, A. Lindahl, C. Brantsing, N. Rotter, P. Gatenholm, and L. Kölby, "Chondrocytes and stem cells in 3D-bioprinted structures create human cartilage *in vivo*", *PLoS One,* vol. 12, no. 12, p. e0189428, 2017.
[http://dx.doi.org/10.1371/journal.pone.0189428] [PMID: 29236765]

[119] N. Cubo, M. Garcia, J.F. del Cañizo, D. Velasco, and J.L. Jorcano, "3D bioprinting of functional human skin: production and *in vivo* analysis"., *Biofabrication,* vol. 9, no. 1, p. 015006, 2016.
[http://dx.doi.org/10.1088/1758-5090/9/1/015006] [PMID: 27917823]

CHAPTER 2

Biocomposites for Tissue Engineering

Amjad Khan[1,*] and **Naeem Khan**[2]

[1] *Department of Pharmacy, Kohat University of Science and Technology, Kohat, Pakistan*

[2] *Department of Chemistry, Kohat University of Science and Technology, Kohat, Pakistan*

Abstract: The goal of tissue engineering is to restore damaged tissue by combining cells with biomimetic material to initiate the growth of new tissue. Biomimetic material plays a crucial role in tissue engineering as it serves as a template and is responsible for providing a suitable environment for tissue development, which includes adhesion of cells, their proliferation and deposition of extracellular matrix. Biocomposites are composite materials, consisting of one or more multiphase materials of biological origin. In this chapter, the biocomposites used for tissue engineering are described in detail. The chapter also highlights the scaffolds and their mechanical properties. This chapter also includes various materials used for scaffold fabrication.

Keywords: Biocomposites, Ceramics, Polymers, Scaffold, Tissue Engineering.

INTRODUCTION

Biocomposites are composite materials composed of single- or multiphase material derived from natural sources, such as plant fibers, flax, cotton, or fibers from wood, waste paper, or food crop byproducts [1 - 5]. The criteria for selecting suitable fibers are determined by the required values of tensile strength, stiffness, elongation at break, adhesion of fiber and matrix, thermal stability, dynamic and long-term behavior of a composite, and processing cost [6]. Composite materials can be classified into (1) Particle reinforced composites, (2) Fiber reinforced composites, and (3) Structural composites. These materials have been used as scaffolds for tissue engineering. The aim of tissue engineering is to restore damaged tissue based on the combination of cells with biomimetic material. The biomimetic material should serve as a template for tissue regeneration and provide a suitable environment for tissue growth [7]. According to the National Science Foundation (1988), tissue engineering was defined as "the understanding of the relationship between structure and function of mammalian tissues under physiological and pathological conditions and their restoration, maintenance, or

---

* **Corresponding author Amjad Khan:** Department of Pharmacy, Kohat University of Science and Technology (KUST), Kohat, Pakistan; Tel: +92-3339334017; E-mail: dr.amjad@kust.edu.pk

improvement of function through the development of biological substitutes based on fundamental principles and procedures of engineering and biological sciences" [8]. Langer and Vacanti defined tissue engineering as "an interdisciplinary field involving the application of principles of engineering and biological sciences to the development of biological substitutes for the restoration, maintenance, and improvement of the function of a tissue" [9]. The basis of tissue engineering is the use of biomimetic material that provides a suitable environment for the development of tissues and serves as a template for cell adhesion, their proliferation and the development of an extracellular matrix until the complete restoration of tissues. Tissue engineering is based on various scientific principles, such as clinical medicine, material science, mechanical engineering and biological sciences [10 - 14]. The combination of scaffold, cells and growth factors (signaling molecules) forms the basis for tissue engineering [15]. Fig. (**1**) shows a schematic representation of the role of the scaffold in bone tissue regeneration.

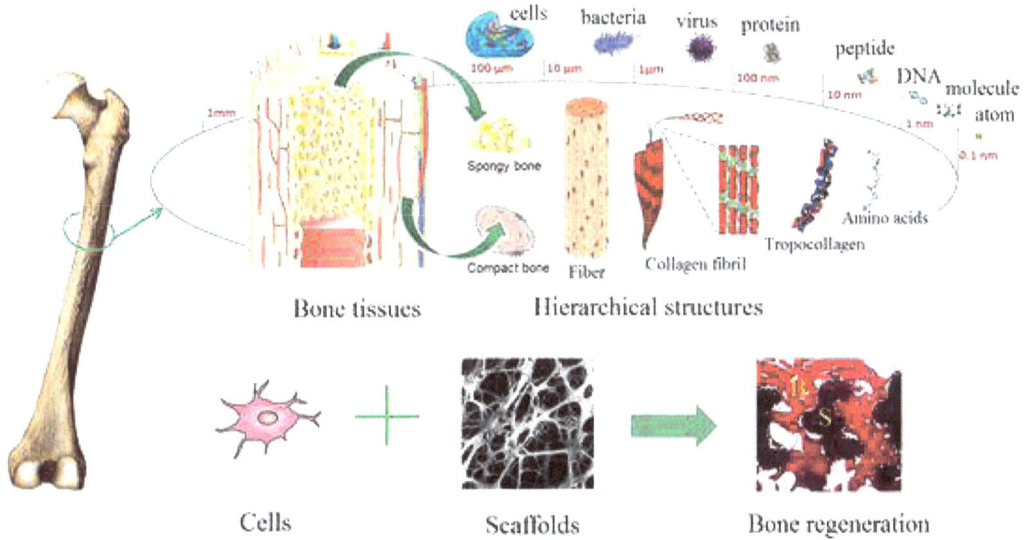

Fig. (1). Schematic presentation of the role of the scaffold in bone tissue regeneration [16].

BIOMATERIALS FOR TISSUE ENGINEERING

Biomaterials are "natural or synthetic substances (not drugs by nature) or their combination that can be used as part of a biological system to treat, support, or replace a tissue or organ" [17]. Since ancient times, natural materials of both animal and plant origin have been sought in nature for wound healing and maintenance and restoration of bodily functions. Plant fibers were used by the Egyptians and Romans to suture skin wounds and were capable of sculpting wooden prosthetic limbs [18]. Over time, various synthetic materials, including metallic and polymeric materials, were used to make medical devices. These materials had need-based properties and were suitable for use in medical devices. In the modern era, regenerative medicine and tissue engineering are based on biomaterials derived from both natural and synthetic sources. Biomaterials of different types such as polymers (natural and synthetic), ceramics, metal, composites, and hydrogels, have been used to fabricate scaffolds that are used in tissue engineering [19]. To be suitable for scaffold fabrication, any material should have basic properties such as biocompatibility, bioactivity and biodegradability.

Biocompatibility is the basic requirement for any biomaterial to be used for scaffold fabrication, and its compatibility with the biological system [20]. Any biomaterial to be used for tissue engineering should not induce an immune response or inflammatory reaction that may lead to rejection or interfere with wound healing after implantation into the living system. Rather, it should promote cell adhesion, cell proliferation and surface migration [21, 22]. The next is bioactivity, which is the ability of a biomaterial to interact with tissue and ensure that cell adhesion, proliferation, and differentiation occur [23]. The bioactivity of a biomaterial is high when the composition of the biomaterial is similar to the target tissue and capable of inducing the cellular responses required for tissue growth. Bioactivity can be increased by surface modification of the biomaterial by adding macromolecules from the extracellular matrix such as collagens, fibronectins and laminins. These macromolecules create an environment similar to the host tissue that modulates the cellular response [24]. The other important property is biodegradability, which is the breakdown of biomaterials by the living system into non-toxic products that can be easily excreted from the body without adverse effects on other body tissues. This is one of the fundamental properties of biomaterials used in tissue engineering, as the scaffolds only serve to support tissue repair and growth and should not remain in the body forever [25]. The *in vivo* degradation kinetics of any biomaterial should be accurately determined as it controls the rate of its elimination from the body. If the biodegradation rate of a biomaterial is high, the scaffold will not be able to support cell growth for a sufficient period of time. In the case of slow biodegradation, the scaffold remains in the body longer and may cause inflammation and necrosis [26].

SCAFFOLDS FOR TISSUE ENGINEERING

Scaffolds are intended to be implanted in an anatomical location in the body, and their structure should be suitable for the intended site of implantation. Scaffolds should have mechanical strength suitable for the anatomical site and be strong enough to withstand surgical manipulations during the implantation process [27]. The structural properties of a scaffold include macrostructural properties and microstructural properties. The macrostructural properties of the scaffold include its three-dimensional structure. It is of utmost importance as it simulates the extracellular matrix and helps the cells to maintain their differentiated physical shape, and the microstructural properties of the scaffolds include their porosity, the shape of the pores, and the size of the pores and their interconnectivity. Fig. (**2**) shows a schematic representation of the biological role of the microstructural properties of a scaffold.

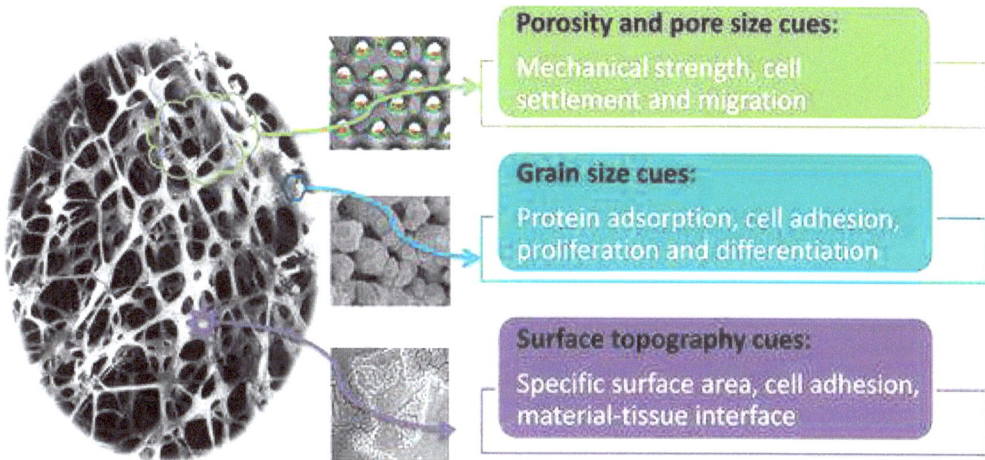

Scaffold structural cues

Fig. (2). Schematic presentation of the biological roles of micro-environmental structure of the scaffold [16].

The structure of a scaffold (both micro- and macro structure) controls all intended functions, including cell survival, surface adhesion properties, cell proliferation, cell differentiation, vascularization, and gene expression [28].

The scaffolds should have the sufficient mechanical strength to withstand body loading and manipulation during surgical implantation. Mechanical strength of a scaffold is controlled by the nature of bonds that hold atoms together and gives it a specific structure. It prevents deformation of solid structure of the scaffold because of the cellular burden on scaffold or its handling. The stiffness of the scaffold surface is measured by Young's Modulus. Stiffness of the scaffold has a significant effect on the proliferation and differentiation of cells, and various mechanisms have been reported to explain the cell response to scaffolds stiffness like ion channel activation and protein unfolding. A high rate of proliferation of human dermal fibroblast has been reported by increasing stiffness due to the free-floating collagen matrix [29].

In addition, they should have a porous structure to prevent cellular colonization. There should be a balance between the mechanical strength of the scaffold and its porous structure, as an increase in porosity will decrease the mechanical strength. Porosity, size of pores and interconnectivity of pores are fundamental parameters to be considered in the fabrication of scaffolds. These properties control cell penetration, and vascularization and ensure cell viability by regulating the supply of oxygen and nutrients to the cells and the newly formed extracellular matrix [30 - 34]. The efficiency of the scaffold depends mainly on the size of the pores. The pores should be large enough to allow cell penetration and migration within the scaffold, but also small enough to avoid cell colonization. The size of the pores in the scaffold is classified based on their dimensions as follows:

- Micro pores (0.1-2 nm)
- Meso-pores (2-50 nm)
- Macro pores (> 50 nm)

The type of host tissue determines the pore size of the scaffold used for tissue engineering. Normally, scaffolds have pores of macro size (> 50 nm), *i.e.*, they have a macroporous structure. For example, pores of about 20 µm are required for fibroblast and hepatocyte growth, and about 20 - 150 µm are required for soft tissue healing [35]. For bone tissue development, a pore size of 200 - 400 µm is generally recommended. Fig. (**3**) shows an SEM image of a scaffold that has an open pore structure, collagen sheets and collagen fibers.

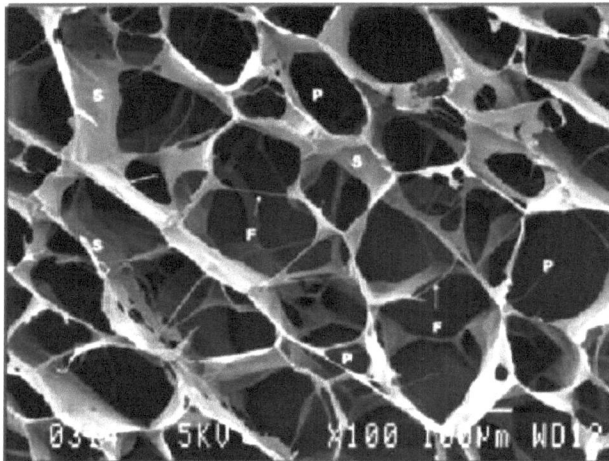

Fig. (3). SEM picture of a scaffold, presenting open pore structure (P), collagen sheets (S) and collagen fibers (F) [36].

BIOMATERIALS FOR SCAFFOLD PREPARATION

The successful replacement or regeneration of a tissue depends on the nature of the biomaterial used in tissue engineering, as it interacts with the site of implantation and directly affects the process of tissue regeneration. Biomaterials of different types such as metals, ceramics, polymers, composites, and hydrogels are used to create scaffolds. Some of the most commonly used biomaterials are:

Ceramics

Ceramics are inorganic biomaterials that contain metallic or nonmetallic elements and are derived from natural or synthetic sources [37]. Ceramics consist of polycrystalline solid substances, which can also be monocrystalline. They may have an amorphous structure. Ceramics have a hard surface with high mechanical stiffness, low elasticity and low expansion when heated. The properties of ceramics [38] depend upon:

- the composition of the particle size of the starting material
- their method of preparation or extraction

Scaffolds based on ceramics are osteoinductive and can recruit cells from the host tissue and support osteogenic differentiation. They are used for bone regeneration because they:

- High biocompatibility
- Low immunogenicity
- Infrequent formation of fibrous tissue surrounding the scaffold.

Despite their advantages, their fragile slow degradation rate limits the use of ceramic scaffolds in tissue engineering [39]. Based on their properties, ceramics are classified into the following categories [40] Bioinert ceramics; these ceramics are non-reactive and remain inert in the biological environment, Resorbable ceramics; these ceramics are degraded *in vivo* and phagocytosed or dissolved in biological fluids, and Bioactive ceramics; these ceramics interact chemically with the cell surface. Some of the common ceramics for tissue engineering include CaP, including hydroxyapatite (HA) ($Ca_{10}[PO_4]_6[OH]_2$), beta-tricalcium phosphate (BTF) ($Ca_3[PO_4]_2$) and bi-phasic calcium phosphate (mixture of hydroxyapatite and beta tricalcium phosphate).

Biomaterials based on CaP have a similar composition to bone and are commonly used for bone grafting. Beta-tricalcium phosphate (BTF) was first used for bone repair in 1920 by Albee and Morrinsen [41]. Coral or bovine bones are the main sources of natural hydroxyapatite, which may contain trace amounts of metals such as magnesium (Mg), fluorine (F), and sodium (Na), as well as carbonates (CO_3). HA can be produced synthetically by the sintering technique and can be in the form of dense or macroporous granules and blocks [42]. Ray and Ward used synthetic HA to regenerate canine long bones and iliac wings [43] and confirmed its biocompatibility and biomimicry. In another study, a two-layer type 1 composite of collagen, HAand Mg, was investigated for osteochondral regeneration [44].

Bio Glass

Hench developed bio glass for the first time for bio-medical use [45]. Bio glass contains SiO_2 (45%), CaO (24.5%), Na_2O (24.5%) and P_2O_5 (6%) on the basis of weight. Bio glass is prepared by different techniques, such as:

• Polymer Foam Replications
• Fusion of particles and fiber by Heating
• Sol-gel Process

Bio glass has certain advantages such as:

• Higher osteo-inductivity
• Controlled degradation rate
• Good bio-activity

The main limitation of bio glass is its lower mechanical strength [46].

Alumina (Al_2O_3)

The structure of alumina (Al_2O_3) is crystalline, and it is inherently fragile. However, its tribological properties are relatively better and resistant to wear [47]. The mechanical strength of alumina can be increased by reducing its particle size and porosity.

Zirconia oxide (ZrO_2)

Zirconia has a polymorphic structure, and it is commonly used in prosthesis and bone grafting [48] because of its:

• High bio-compatibility
• Higher breaking load
• Hard surface
• Lower conduction of heat
• Higher co-efficient of thermal expansion

Polymers

Polymers of biological origin exhibit high biocompatibility and bioactivity, and their surface is well suited for cell adhesion and growth. Their major limitations are low mechanical strength and a higher rate of biodegradation. Scaffolds based on biological polymers, such as collagen, alginates, proteoglycan, chitin and chitosan are widely used for tissue engineering [49]. Some of the polymers used in scaffold preparation are discussed in the following sections.

Collagen and its Derivatives

Scaffolds based on collagen and its derivatives show higher bioactivity and excellent adhesion of cells to their surface. Scaffolds containing collagen have low mechanical strength and are usually used in combination with other substances. The extracellular matrix of tendons and ligaments is composed of collagen fibers (type-1), so collagen-containing scaffolds are most suitable for the reconstruction of these tissues [50]. Micro and nano-structured scaffolds based on cellulose and collagen have been reported with higher mechanical strength [51] and are used for tissue engineering due to better surface adhesion of human osteoblast cells. In another study, controlled micro-nanostructured scaffolds made up of polyester were compared with the calcium deposition method [52].

Polysaccharides

Scaffold based polysaccharides such as chitin, chitosan and alginates are used for tissue engineering of both hard and soft tissues. Chitosan has a positive charge on its surface and interacts with glycosaminoglycans and proteoglycans, which ensures better cell adhesion. Chitosan-based scaffolds with better porosity and higher pores inter-connectivity have been prepared by the freeze-drying technique [53]. Costa-Pinto *et al.* reported that MSCs were grown from human bone marrow on chitosan-based scaffolds [54].

Synthetic Polymers

Synthetic polymers have a high molecular weight and are composed of many smaller units (monomers). Synthetic polymers can be in the form of fibers, films, bar, or viscous liquids. The advantage of synthetic polymers is that their mechanical properties and rate of degradation can be modulated by their constituents and method of preparation. Some of the synthetic polymers are less biocompatible and have lower mechanical strength. Some polymers have been found to have *in vivo* toxicity due to ions and monomeric units. Synthetic polymers can be classified based on several characteristics [55]. Based on their structure, synthetic polymers can be classified into the following categories:

- Linear Polymers
- Branched Polymers
- Cross liked Polymers

Based on their thermos mechanical characteristics:

- Thermo-plastic
- Thermo-setting

Biodegradable synthetic polymers are best suited for scaffold fabrication because their degradation products are non-toxic and have lower molecular weight. Various synthetic polymers used for the preparation of scaffolds [56] include:

- Polystyrenes
- Polylactic acids (PLA)
- Poly glycolic acids (PGA)
- Polylactic co glycolic acid (PLGA)
- Poly caprolactones

The morphology of different polymeric composites is presented in Fig. (**4**).

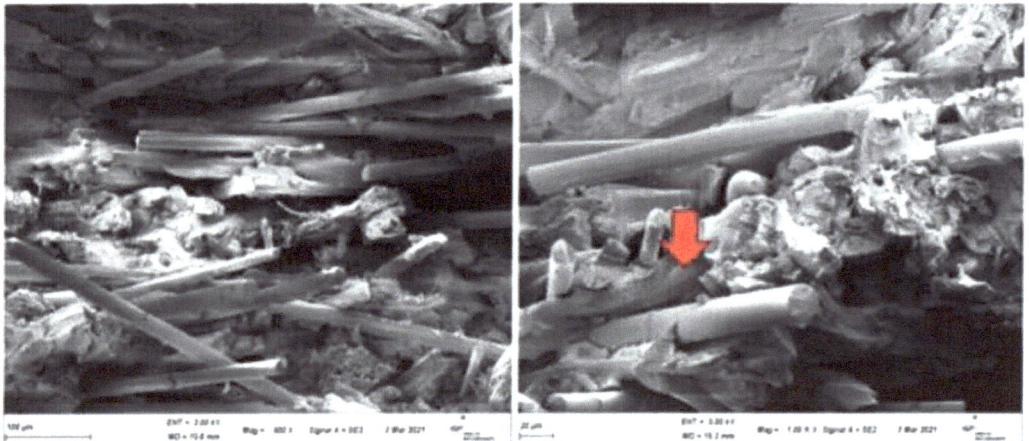

Fig. (4). Morphology of D/23BA/KEFL composite [57].

Metals

Metals have high mechanical strength and are best suited for scaffold preparation [58, 59] because of their:

- Higher elastic modulus
- Higher yield strength
- Good ductility
- Capability of bearing heavy loads without deforming

Limitations for the use of metals in scaffold preparation include:

- Lower cell adhesion to the surface of metals
- Leaching of metal ions and particles which can cause toxicity
- Corrosion caused by surrounding biological fluid

Some of the metals commonly used in scaffolds preparation are:

Stainless Steel

Stainless steel is an alloy of iron, chromium, and carbon. Iron and chromium are the major components of stainless steel, while carbon is present only in small amounts. Carbon is used to increase mechanical strength, but it makes scaffolds susceptible to corrosion because it forms carbides in biological fluids.

Alloy of Cobalt

Cobalt-based alloy is composed of different combinations like:

- Cobalt-chromium-molybdenum
- Cobalt-nickel-chromium-molybdenum

The mechanical strength of cobalt alloys is tailored by changing the quantity of chromium and molybdenum.

Titanium Alloys

Titanium alloy is classified as:

Alpha Alloy: Alpha alloy has higher mechanical strength, high weld-ability and is resistant to sliding. Alpha alloy contains alpha stabilizers like aluminum and gallium.

Beta Alloy: Beta alloy is highly ductile. Stabilizers used in beta alloys are vanadium and niobium molybdenum.

Alpha-Beta Bi-Phasic Alloy: This type of alloy is best suited for the preparation of scaffolds with biomedical applications as it is highly ductile, *e.g.*, Ti_6Al_4V.

Composite

Composite scaffolds are combinations of different biomaterials, such as polymers, ceramics and metals [60]. Both natural and synthetic polymers are used to make composites. Composites are gaining importance in tissue engineering due to their good biocompatibility, biodegradability and higher mechanical strength. The properties of composite scaffolds can be tailored to mimic the host tissue composition, which increases their applicability for tissue regeneration. There are a number of studies on the efficacy of composite scaffolds [61]. Different material used in preparation of scaffold for tissue engineering and their clinical applications are presented in Table **1**.

Hydrogel

Hydrogel is defined as hydrophilic polymeric material containing carboxyl, amide, amino and hydroxyl groups linked either by chemical bonds or physical molecular interactions [62]. Hydrogels swell upon absorption of biological fluids without dissolving. Based on their source, hydrogels are classified into different groups as follows:

Natural Hydrogels

These are composed of polypeptide and polysaccharides.

Synthetic Hydrogels

These are synthesized by polymerization of different monomers.

Semi-Synthetic Hydrogels

These are modified forms of natural hydrogels.

Physically hydrogels are amorphous or semi-crystalline in nature and may be cationic, anionic, neutral or ampholytic. On the basis of their response to biological fluids, hydrogels may be:

Durable Hydrogels

These hydrogels remain stable in a biological system without any physical or chemical modifications.

Biodegradable Hydrogels

These hydrogels are degraded by the biological system to oligomers and can be easily excreted.

Recently, research has focused on the development of smart hydrogels with tailored properties, such as mechanical strength and response to stimuli of the biological system (pH- and temperature-dependent hydrogels). Hydrogels are used for scaffold fabrication because they are biocompatible and degrade in a controlled manner. Hydrogels can be made suitable for cell growth and proliferation by altering their porosity, the size of the pores, and interconnectivity of the pores by crosslinking their molecules [63]. Cell adherence and differentiation process are enhanced by binding peptides or growth factors to the surface of hydrogels. In general, hydrogels derived from natural sources are safe and well tolerated by biological systems. A hybrid hydrogel of natural cellulose and hydroxyapatite is commonly used for bone tissue engineering. Synthetic hydrogels have the advantage that their mechanical strength and biodegradability can be modulated according to the requirements of the system. The major disadvantage of synthetic hydrogels is their incompatibility with biological systems. Kinard and co-workers used a hydrogel of oligo [poly(ethylene glycol) fumarate] for the delivery of demineralized bone matrix (DBM) to rats with defective bones. The rate of degradation of the hydrogel was controlled by the amount of DBM (the higher the amount of DBM, the higher the biodegradation) [64]. Moreover, the mechanical strength of the hydrogel was found to be related to the amount of DBM [65]. Fig. (5) shows SEM image of the surface and cross-section of the hydrogels.

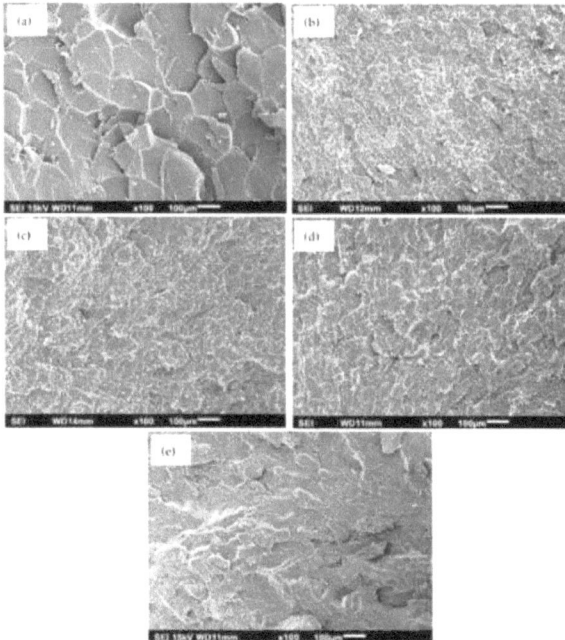

Fig. (5). SEM image of the surface of pure bio-epoxy and bio-epoxy composites with graphene oxide [66].

Table 1. Comparison of different materials used for tissue engineering and their clinical applications.

Material	Advantage	Limitations	Use
Ceramics	Hard Surface High mechanical strength Biocompatibility	Brittleness Slow rate of degradation Processing Difficulties	Hip Prosthesis Dental Prosthesis Bone and Cartilages
Natural Polymers	Biocompatibility Bioactivity	Poor mechanical strength Fast Biodegradation rate	Bone and Cartilages Tendon and ligament
Synthetic Polymers	Possibility of modulating porosity and mechanical properties during the synthesis process	Reduced cell adhesion to their surfaces Possible corrosion mediated by biological fluids	Sutures Catheters Bone cements Cardiac prosthesis
Metals	Good mechanical strength High elastic module, yield strength and high ductility	Reduced cell adhesion to their surfaces Possible erosion mediated by biological fluids	Dentistry and orthopedic prosthesis
Composites	Biocompatibility Good mechanical properties	Processing difficulties	Hard and Soft tissues
Hydrogel	Biocompatibility Controlled *in vivo* biodegradation Possibility to modulate their properties by cross-linking	-	Hard and soft tissues

TECHNIQUES USED FOR THE PREPARATION OF SCAFFOLDS

The properties of scaffolds depend on the type of biomaterial and the process used to produce it. Once a biomaterial with certain properties has been selected, the next step is to choose an appropriate preparation process to obtain a scaffold with controlled structural (micro and macro) properties. Essential features [67] of a process for the preparation of a scaffold are:

- Preparation process should be accurate and reproducible.
- It will result in a scaffold with controlled porosity and pores interconnectivity, and the scaffold will be of a regular shape.
- Upon repetition, there should be no changes in physico-chemical characteristics of the scaffolds.
- The preparation method will not affect the characteristics of the constituent biomaterials (*e.g.*, mechanical strength).
- The process should be capable of producing scaffolds free of residual solvents as some toxic solvents can affect biomedical applications of the scaffolds.

Different techniques used for preparation of scaffolds are summarized as follows:

Solvent Casting / Particulate Leaching

Mikos *et al.* [68], pioneered the solvent casting/particle leaching technique, which has proven successful in the fabrication of porous scaffolds for bone tissue engineering. In this method, the porosifying agent is dispersed in a suitable solvent, followed by a casting or freeze-drying process. This technique has been used to produce thin membranes with a porosity of 20-50% and a pore size of 30-300 μm. The time required (solvent evaporation is a slow process) and the use of organic solvents are the main limitations of this technique.

Melt Molding / Particulate Leaching

In the melt casting /particle leaching method, porogenic material and unrefined thermoplastic polymers are mixed. The mixture is placed in a mold of suitable shape and heated to a temperature above the glass transition temperature of the polymers. The Pore making agent(porogen) is dissolved by immersing the prepared solid in a suitable solvent. This method is advantageous because the porosity and size of the pores can be controlled by the amount of poring agent. The porosity of the scaffolds prepared by this method can be in the range of 80-84% [69]. Gomes and his co-workers [70] modified this technique by replacing the porosifying purifying agents with blowing agents based on citric acid. When heated, the decomposition of the blowing agents produces carbon dioxide, resulting in well-crosslinked pores or suitable shapes.

Gas Foaming

Mooney *et al.* pioneered the gas foam technique, which uses high pressure and no organic solvents [71]. Mooney *et al.* prepared a sponge of poly(D,L-lactic-co-glycolic acid) without organic solvents. This method exposes a solid polymer disk to CO_2 (5.5 MPa) under high pressure at ambient temperature. The pressure of the gas is reduced, resulting in a decrease in the solubility of the gas in the polymer. Due to the decreased solubility, the CO_2 leaves the polymer, resulting in interconnected pores in good shape.

Phase Inversion / Particulate Leaching

In the phase inversion technique, a polymer solution is prepared in a suitable solvent. The polymer solution is mixed with water, which leads to the precipitation of the polymer [72]. It is the most suitable technique for the preparation of polymer-based scaffolds and the properties of the scaffolds can be modulated by changing the polymer content and temperature. This method has been used to fabricate PLGA-based scaffolds for bone tissue engineering, whose structure mimics osseous trabecular [73].

Fiber Bonding

The fiber bonding technique is used to fabricate scaffolds with a high-density frame of fiber material that forms a three-dimensional structure with high porosity. In this method, PGA fibers are aligned in a specific way; the fibers are covered with a PLLA-methylene chloride solution and heated above the melting point of both polymers. PLLA is separated by dissolving with a suitable solvent, and a membrane of PGA fibers is obtained [74].

Freeze Drying Method

Freeze drying method involves the preparation of polymeric solution in a suitable solvent, freezing of polymeric solution below 0 °C and application of vacuum to remove the solvent by sublimation [75]. This method has been used to prepare scaffold based on natural and synthetic materials.

Solid Free Form Fabrication

Solid free form fabrication involves the application of information from a computer-aided designing system or computer-based medical imaging modalities in the preparation of a layered three-dimensional scaffold. It is a time-saving technique with precise modulation of the structure of the scaffold [76]. The most common methods based on solid free-form fabrication are:

• Three-Dimensional Printing (3D Printing)
• Fused Deposition Modeling (FDM)

Three Dimensional Printing

3D printing is based on solid free-form fabrication and has been successfully applied for tissue engineering. It involves the generation of shape by a CAD system and placement of a liquid binder on the surface of thin powder layer by a printer head to generate shape according to the instructions from CAD system. This technique has been applied for preparation of PLGA scaffold [77] and poly (1-lactic acid) disk [78] with different porosity and pore size distribution.

Fused Deposition Modeling

In fused deposition modeling, a three dimensional scaffold is prepared by heating a film of thermoplastic substance in a hot liquefier head, forcing it out with the help of an extruder and deposition on a platform [79]. Porosity, pore size and inter connectivity in a scaffold can be modulated by changing the deposition pattern of layers. Hutmacher *et al.* [80], prepared a poly-caprolactone scaffold by this technique, with a honeycomb-like structure of high porosity (61%).

CONCLUSION

Biomaterials are becoming increasingly important in tissue engineering due to their biomimetic nature, modifiable properties, and broad applicability. Both synthetic and natural biomaterials are the focus of research to better understand their physicochemical properties, biocompatibility, and fabrication techniques. Biomaterials have been used to fabricate three-dimensional (3D) bioactive scaffolds with enhanced compatibility and controlled degeneration in living tissues. Biomaterials used for scaffold fabrication include ceramics, natural and synthetic polymers, metals, composites, and hydrogels. Extensive studies have been conducted both *in vitro* and *in vivo* to explore safe and effective biomaterials, their fabrication methods, and their applications in regenerative medicine. The capabilities and limitations of biomaterials have been investigated with emphasis on their safety profile. More recently, research has been conducted on the development of biomaterials (natural, synthetic or semi-synthetic) that can promote cell proliferation and differentiation in tissue to restore the normal structure of the ECM. In conclusion, tissue engineering strategies, especially bone tissue regeneration, are an effective alternative. However, their success depends on a deeper knowledge of the properties of biomaterials and their possible combinations.

CONSENT FOR PUBLICATION

Not applicable.

CONFLICT OF INTEREST

The authors declare no conflict of interest, financial or otherwise.

ACKNOWLEDGEMENT

Declared none.

REFERENCES

[1] S.V. Dorozhkin, "Biocomposites and Hybrid Biomaterials Based on CaPO 4", *Calcium Orthophosphate-Based Bioceramics and Biocomposite,* no. Mar, pp. 287-374, 2016.
[http://dx.doi.org/10.1002/9783527699315.ch20]

[2] A. Partanen, and M. Carus, "Biocomposites, find the real alternative to plastic – An examination of biocomposites in the market", *Reinf. Plast.,* vol. 63, no. 6, pp. 317-321, 2019.
[http://dx.doi.org/10.1016/j.repl.2019.04.065]

[3] P. Olsén, N. Herrera, and L.A. Berglund, "Toward Biocomposites Recycling: Localized Interphase Degradation in PCL-Cellulose Biocomposites and its Mitigation", *Biomacromolecules,* vol. 21, no. 5,

pp. 1795-1801, 2020.
[http://dx.doi.org/10.1021/acs.biomac.9b01704] [PMID: 31958232]

[4]   L. Avérous, "Nano- and Biocomposites", *Mater. Today,* vol. 13, no. 4, p. 57, 2010.
[http://dx.doi.org/10.1016/S1369-7021(10)70063-8]

[5]   N. Reddy, and Y. Yang, "Biocomposites developed using water-plasticized wheat gluten as matrix and jute fibers as reinforcement", *Polym. Int.,* vol. 60, no. 4, pp. 711-716, 2011.
[http://dx.doi.org/10.1002/pi.3014]

[6]   M. Birsan, E. Andronescu, C. Birsan, C. Ghiţulica, and E.M. Stefan, "Ferromagnetic Biocomposites with Improved Bioactivity", *Key Eng. Mater.,* vol. 264-268, pp. 2043-2046, 2004.
[http://dx.doi.org/10.4028/www.scientific.net/KEM.264-268.2043]

[7]   "A review of chitosan-, alginate-, and gelatin-based biocomposites for bone tissue engineering", *Biomaterials and Tissue Engineering Bulletin,* vol. 5, no. 3–4, pp. 97-109, 2018.

[8]   S.M. Nainar, S. Begum, M.N.M. Ansari, and H. Anuar, "Tensile Properties and Morphological Studies on HA/PLA Biocomposites for Tissue Engineering Scaffolds", *Int. J. Engine Res.,* vol. 3, no. 3, pp. 186-189, 2014.
[http://dx.doi.org/10.17950/ijer/v3s3/312]

[9]   X. Li, Y. Yang, Y. Fan, Q. Feng, F. Cui, and F. Watari, "Biocomposites reinforced by fibers or tubes as scaffolds for tissue engineering or regenerative medicine", *J. Biomed. Mater. Res. A,* vol. 102, no. 5, pp. 1580-1594, 2014.
[http://dx.doi.org/10.1002/jbm.a.34801] [PMID: 23681610]

[10] I. J. Macha, S. Cazalbou, R. Shimmon, B. Ben-Nissan, and B. Milthorpe, "Development and dissolution studies of bisphosphonate (clodronate)-containing hydroxyapatite-polylactic acid biocomposites for slow drug delivery", *Tissue Eng Regenerat. Med.,* vol. 11, no. 6, pp. 1723-1731, 2017.

[11] D. Li, and H. J. Kaner, "A Unique Polymer Nano structure for Versatile Applications", *Acc. Ch em. Res,* vol. 42, pp. 135-145, 2014.

[12] E. Eskandari, M. Kosari, M.H.D. Farahani, N.D. Khiavi, M. Saeedikhani, R. Katal, and M. Zarinejad, "A review on polyaniline based materials applications in heavy metals removal and catalytic processes, Sep", *Pur. Technol. ,* vol. 231, p. 115, 2020.

[13] N.R. Tanguy, M. Arjmand, and N. Yan, "Nano composite of Nitrogen-Doped Graphene/ Polyaniline for enhanced Ammonia Gas Detection", *Adv. Mate r. Int erfaces,* vol. 6, p. 190, 2019 .

[14] G. Koronis, A. Silva, and M. Fontul, "Green composites: A review of adequate materials for automotive applications", *Compos., Part B Eng.,* vol. 44, no. 1, pp. 120-127, 2013.
[http://dx.doi.org/10.1016/j.compositesb.2012.07.004]

[15] J. Frketic, T. Dickens, and S. Ramakrishnan, "Automated manufacturing and processing of fiber-reinforced polymer (FRP) composites: An additive review of contemporary and modern techniques for advanced materials manufacturing", *Addit. Manuf.,* vol. 14, pp. 69-86, 2017.
[http://dx.doi.org/10.1016/j.addma.2017.01.003]

[16] X. Chen, H. Fan, X. Deng, L. Wu, T. Yi, L. Gu, C. Zhou, Y. Fan, and X. Zhang, "Scaffold structural microenvironmental cues to guide tissue regeneration in bone tissue applications", *Nanomaterials (Basel),* vol. 8, no. 11, p. 960, 2018.
[http://dx.doi.org/10.3390/nano8110960] [PMID: 30469378]

[17] R.A. Ilyas, S.M. Sapuan, R. Ibrahim, H. Abral, M.R. Ishak, E.S. Zainudin, M.S.N. Atikah, N. Mohd Nurazzi, A. Atiqah, M.N.M. Ansari, E. Syafri, M. Asrofi, N.H. Sari, and R. Jumaidin, "Effect of sugar palm nanofibrillated cellulose concentrations on morphological, mechanical and physical properties of biodegradable films based on agro-waste sugar palm (Arenga pinnata (Wurmb.) Merr) starch", *J. Mater. Res. Technol.,* vol. 8, no. 5, pp. 4819-4830, 2019.
[http://dx.doi.org/10.1016/j.jmrt.2019.08.028]

[18]   F.J. O'Brien, "Biomaterials & scaffolds for tissue engineering", *Mater. Today,* vol. 14, no. 3, pp. 88-95, 2011.
[http://dx.doi.org/10.1016/S1369-7021(11)70058-X]

[19]   A.J. Salgado, O.P. Coutinho, R.L. Reis, and R.L. Reis, "Bone tissue engineering: state of the art and future trends", *Macromol. Biosci.,* vol. 4, no. 8, pp. 743-765, 2004.
[http://dx.doi.org/10.1002/mabi.200400026] [PMID: 15468269]

[20]   D.W. Hutmacher, "Scaffolds in tissue engineering bone and cartilage", *Biomaterials,* vol. 21, no. 24, pp. 2529-2543, 2000.
[http://dx.doi.org/10.1016/S0142-9612(00)00121-6] [PMID: 11071603]

[21]   S. M R, S. Siengchin, J. Parameswaranpillai, M. Jawaid, C.I. Pruncu, and A. Khan, "A comprehensive review of techniques for natural fibers as reinforcement in composites: Preparation, processing and characterization", *Carbohydr. Polym.,* vol. 207, pp. 108-121, 2019.
[http://dx.doi.org/10.1016/j.carbpol.2018.11.083] [PMID: 30599990]

[22]   N. Nurazzi, K. Khalina, S. Sapuan, and R.A. Ilyas, "Mechanical properties of sugar palm yarn/woven glass fiber reinforced unsaturated polyester composites: effect of fiber loadings and alkaline treatment", *Polimery,* vol. 64, no. 10, pp. 665-675, 2019.
[http://dx.doi.org/10.14314/polimery.2019.10.3]

[23]   L.K. Kian, N. Saba, M. Jawaid, and M.T.H. Sultan, "A review on processing techniques of bast fibers nanocellulose and its polylactic acid (PLA) nanocomposites", *Int. J. Biol. Macromol.,* vol. 121, pp. 1314-1328, 2019.
[http://dx.doi.org/10.1016/j.ijbiomac.2018.09.040] [PMID: 30208300]

[24]   M. Irimia-Valdu, and J.W. Fergus, "Suitability of emeraldine base polyaniline PVA composite film for carbon dioxide sensing, Synth", *Met,* vol. 156, pp. 1401-1407, 2006.

[25]   N.R. Tanguy, M. Arjmand, and N. Yan, "Nanocompossite of nitrogen doped Graphene/polyaniline for enhanced ammonia gas detection", *Adv. Mate r. Int erfaces,* vol. 6, p. 190, 2019.

[26]   I. Sapurina, and J. Stejskal, "The mechanism of the oxidative polymerization of aniline and the formation of supramolecular polyaniline structures", *Polym. Int.,* vol. 57, no. 12, pp. 1295-1325, 2008.
[http://dx.doi.org/10.1002/pi.2476]

[27]   A. Pud, N. Ogurtsov, A. Korzhenko, and G. Shapoval, "Some aspects of preparation methods and properties of polyaniline blends and composites with organic polymers", *Prog. Polym. Sci.,* vol. 28, no. 12, pp. 1701-1753, 2003.
[http://dx.doi.org/10.1016/j.progpolymsci.2003.08.001]

[28]   S. Sedaghat, "Synthesis and Evaluation of Chitosan-polyaniline copolymer in presence of ammonium persulphate as initiater", *J. Appl. Chem. Res.,* vol. 8, pp. 47-54, 2014.

[29]   X. Yu, X. Tang, S.V. Gohil, and C.T. Laurencin, "Biomaterials for bone regenerative engineering", *Adv. Healthc. Mater.,* vol. 4, no. 9, pp. 1268-1285, 2015.
[http://dx.doi.org/10.1002/adhm.201400760] [PMID: 25846250]

[30]   R.A. Ilyas, S.M. Sapuan, A. Atiqah, R. Ibrahim, H. Abral, M.R. Ishak, E.S. Zainudin, N.M. Nurazzi, M.S.N. Atikah, M.N.M. Ansari, M.R.M. Asyraf, A.B.M. Supian, and H. Ya, "Sugar palm (Arenga pinnata [Wurmb.] Merr) starch films containing sugar palm nanofibrillated cellulose as reinforcement: Water barrier properties", *Polym. Compos.,* vol. 41, no. 2, pp. 459-467, 2020.
[http://dx.doi.org/10.1002/pc.25379]

[31]   R.A. Ilyas, S.M. Sapuan, M.L. Sanyang, M.R. Ishak, and E.S. Zainudin, "Nanocrystalline cellulose as reinforcement for polymeric matrix nanocomposites and its potential applications: A Review", *Curr. Anal. Chem.,* vol. 14, no. 3, pp. 203-225, 2018.
[http://dx.doi.org/10.2174/1573411013666171003155624]

[32]   E. Barrios, D. Fox, Y.Y. Li Sip, R. Catarata, J.E. Calderon, N. Azim, S. Afrin, Z. Zhang, and L. Zhai, "Nanomaterials in advanced, high-performance aerogel composites: A review", *Polymers (Basel),* vol.

11, no. 4, p. 726, 2019.
[http://dx.doi.org/10.3390/polym11040726] [PMID: 31010008]

[33]   A. Atiqah, M. Jawaid, S.M. Sapuan, M.R. Ishak, M.N.M. Ansari, and R.A. Ilyas, "Physical and thermal properties of treated sugar palm/glass fibre reinforced thermoplastic polyurethane hybrid composites", *J. Mater. Res. Technol.,* vol. 8, no. 5, pp. 3726-3732, 2019.
[http://dx.doi.org/10.1016/j.jmrt.2019.06.032]

[34]   M.N. Collins, M. Nechifor, F. Tanasă, M. Zănoagă, A. McLoughlin, M.A. Stróżyk, M. Culebras, and C.A. Teacă, "Valorization of lignin in polymer and composite systems for advanced engineering applications – A review", *Int. J. Biol. Macromol.,* vol. 131, pp. 828-849, 2019.
[http://dx.doi.org/10.1016/j.ijbiomac.2019.03.069] [PMID: 30872049]

[35]   R.A. Ilyas, S.M. Sapuan, M.N. Norizan, M.S.N. Atikah, M.R.M. Huzaifah, and A.M. Radzi, "Potential of natural fibre composites for transport industry: A review. ", In: *In: Prosiding Seminar Enau Kebangsaan,* 2019.

[36]   Q.B. Wessels, and E. Pretorius, "Enhanced stabilization of collagen-based dermal regeneration scaffolds through the combination of physical and chemical crosslinking", *S. Afr. J. Sci.,* vol. 104, no. 11/12, pp. 11-12, 2008.
[http://dx.doi.org/10.1590/S0038-23532008000600030]

[37]   M. Tirrell, E. Kokkoli, and M. Biesalski, "The role of surface science in bioengineered materials", *Surf. Sci.,* vol. 500, no. 1-3, pp. 61-83, 2002.
[http://dx.doi.org/10.1016/S0039-6028(01)01548-5]

[38]   J.E. Babensee, J.M. Anderson, L.V. McIntire, and A.G. Mikos, "Host response to tissue engineered devices", *Adv. Drug Deliv. Rev.,* vol. 33, no. 1-2, pp. 111-139, 1998.
[http://dx.doi.org/10.1016/S0169-409X(98)00023-4] [PMID: 10837656]

[39]   B. Ratner, A. Hoffman, F. Schoen, and J.J. Lemons, "Biomaterials Science", In: *An Introduction to Materials in Medicine.* 3rd Edition. Academic Press, 2012.

[40]   K.F. Leong, C.M. Cheah, and C.K. Chua, "Solid freeform fabrication of three-dimensional scaffolds for engineering replacement tissues and organs", *Biomaterials,* vol. 24, no. 13, pp. 2363-2378, 2003.
[http://dx.doi.org/10.1016/S0142-9612(03)00030-9] [PMID: 12699674]

[41]   E. Hadjipanayi, V. Mudera, and R.A. Brown, "Close dependence of fibroblast proliferation on collagen scaffold matrix stiffness", *J. Tissue Eng. Regen. Med.,* vol. 3, no. 2, pp. 77-84, 2009.
[http://dx.doi.org/10.1002/term.136] [PMID: 19051218]

[42]   P. Lichte, H.C. Pape, T. Pufe, P. Kobbe, and H. Fischer, "Scaffolds for bone healing: Concepts, materials and evidence", *Injury,* vol. 42, no. 6, pp. 569-573, 2011.
[http://dx.doi.org/10.1016/j.injury.2011.03.033] [PMID: 21489531]

[43]   M.S.N. Atikah, R.A. Ilyas, S.M. Sapuan, M.R. Ishak, E.S. Zainudin, R. Ibrahim, A. Atiqah, M.N.M. Ansari, and R. Jumaidin, "Degradation and physical properties of sugar palm starch/sugar palm nanofibrillated cellulose bionanocomposite", *Polimery,* vol. 64, no. 10, pp. 680-689, 2019.
[http://dx.doi.org/10.14314/polimery.2019.10.5]

[44]   H. Abral, J. Ariksa, M. Mahardika, D. Handayani, I. Aminah, N. Sandrawati, A.B. Pratama, N. Fajri, S.M. Sapuan, and R.A. Ilyas, "Transparent and antimicrobial cellulose film from ginger nanofiber", *Food Hydrocoll.,* vol. 98, p. 105266, 2020.
[http://dx.doi.org/10.1016/j.foodhyd.2019.105266]

[45]   M.R.M. Asyraf, M.R. Ishak, S.M. Sapuan, N. Yidris, and R.A. Ilyas, "Woods and composites cantilever beam: A comprehensive review of experimental and numerical creep methodologies", *J. Mater. Res. Technol.,* vol. 9, no. 3, pp. 6759-6776, 2020.
[http://dx.doi.org/10.1016/j.jmrt.2020.01.013]

[46]   E. Syafri, Sudirman, Mashadi, E. Yulianti, Deswita, M. Asrofi, H. Abral, S.M. Sapuan, R.A. Ilyas, and A. Fudholi, "Effect of sonication time on the thermal stability, moisture absorption, and

biodegradation of water hyacinth (Eichhornia crassipes) nanocellulose-filled bengkuang (Pachyrhizus erosus) starch biocomposites", *J. Mater. Res. Technol.,* vol. 8, no. 6, pp. 6223-6231, 2019.
[http://dx.doi.org/10.1016/j.jmrt.2019.10.016]

[47]   S. Bose, M. Roy, and A. Bandyopadhyay, "Recent advances in bone tissue engineering scaffolds", *Trends Biotechnol.,* vol. 30, no. 10, pp. 546-554, 2012.
[http://dx.doi.org/10.1016/j.tibtech.2012.07.005] [PMID: 22939815]

[48]   S.M. Best, A.E. Porter, F.S. Thian, and J. Huang, "Bioceramics: Past, present and for the future", *J. Eur. Ceram. Soc.,* vol. 28, no. 7, pp. 1319-1327, 2008.
[http://dx.doi.org/10.1016/j.jeurceramsoc.2007.12.001]

[49]   F.H. Albee, and H.F. Morrison, "Studies in bone growth", *Ann. Surg.,* vol. 71, no. 1, pp. 32-39, 1920.
[http://dx.doi.org/10.1097/00000658-192001000-00006] [PMID: 17864220]

[50]   R.Z. LeGeros, "Properties of osteoconductive biomaterials: calcium phosphates", *Clin. Orthop. Relat. Res.,* vol. 395, no. 395, pp. 81-98, 2002.
[http://dx.doi.org/10.1097/00003086-200202000-00009] [PMID: 11937868]

[51]   R.D. Ray, and A.A. Ward, "A preliminary report on studies of basic calcium phosphate in bone replacement", In: *Surgical Forum, American College of Surgeons, 1951.* WB Saunders Co: Philadelphia, 1952, pp. 429-434.

[52]   G. Calabrese, R. Giuffrida, C. Fabbi, E. Figallo, D. Lo Furno, R. Gulino, C. Colarossi, F. Fullone, R. Giuffrida, R. Parenti, L. Memeo, and S. Forte, "Collagen-hydroxyapatite scaffolds induce human adipose derived stem cells osteogenic differentiation *in vitro*"., *PLoS One,* vol. 11, no. 3, p. e0151181, 2016.
[http://dx.doi.org/10.1371/journal.pone.0151181] [PMID: 26982592]

[53]   L. Mathew, K.U. Joseph, and R. Joseph, "Isora fibres and their composites with natural rubber", *Prog. Rubber Plast. Recycl. Technol.,* vol. 20, no. 4, pp. 337-349, 2004.
[http://dx.doi.org/10.1177/147776060402000404]

[54]   S. Matkó, A. Toldy, S. Keszei, P. Anna, G. Bertalan, and G. Marosi, "Flame retardancy of biodegradable polymers and biocomposites", *Polym. Degrad. Stabil.,* vol. 88, no. 1, pp. 138-145, 2005.
[http://dx.doi.org/10.1016/j.polymdegradstab.2004.02.023]

[55]   J. Mirbagheri, M. Tajvidi, J.C. Hermanson, and I. Ghasemi, "Tensile properties of wood flour/kenaf fiber polypropylene hybrid composites", *J. Appl. Polym. Sci.,* vol. 105, no. 5, pp. 3054-3059, 2007.
[http://dx.doi.org/10.1002/app.26363]

[56]   S. Mishra, A.K. Mohanty, L.T. Drzal, M. Misra, S. Parija, S.K. Nayak, and S.S. Tripathy, "Studies on mechanical performance of biofibre/glass reinforced polyester hybrid composites", *Compos. Sci. Technol.,* vol. 63, no. 10, pp. 1377-1385, 2003.
[http://dx.doi.org/10.1016/S0266-3538(03)00084-8]

[57]   A. Saleem, L. Medina, and M. Skrifvars, "Improvement of performance profile of acrylic based polyester bio-composites by bast/basalt fibers hybridization for automotive applications", *Journal of Composites Science,* vol. 5, no. 4, p. 100, 2021.
[http://dx.doi.org/10.3390/jcs5040100]

[58]   A.K. Mohanty, L.T. Drzal, and M. Misra, "Engineered natural fiber reinforced polypropylene composites: influence of surface modifications and novel powder impregnation processing", *J. Adhes. Sci. Technol.,* vol. 16, no. 8, pp. 999-1015, 2002.
[http://dx.doi.org/10.1163/156856102760146129]

[59]   J. Rout, M. Misra, S.S. Tripathy, S.K. Nayak, and A.K. Mohanty, "The influence of fibre treatment on the performance of coir-polyester composites", *Compos. Sci. Technol.,* vol. 61, no. 9, pp. 1303-1310, 2001.
[http://dx.doi.org/10.1016/S0266-3538(01)00021-5]

[60] M.A.S. Azizi Samir, F. Alloin, and A. Dufresne, "Review of recent research into cellulosic whiskers, their properties and their application in nanocomposite field", *Biomacromolecules,* vol. 6, no. 2, pp. 612-626, 2005.
[http://dx.doi.org/10.1021/bm0493685] [PMID: 15762621]

[61] G. Calabrese, R. Giuffrida, S. Forte, L. Salvatorelli, C. Fabbi, E. Figallo, M. Gulisano, R. Parenti, G. Magro, C. Colarossi, L. Memeo, and R. Gulino, "Bone augmentation after ectopic implantation of a cell-free collagen-hydroxyapatite scaffold in the mouse", *Sci. Rep.,* vol. 6, no. 1, p. 36399, 2016.
[http://dx.doi.org/10.1038/srep36399] [PMID: 27821853]

[62] G. Calabrese, R. Giuffrida, S. Forte, C. Fabbi, E. Figallo, L. Salvatorelli, L. Memeo, R. Parenti, M. Gulisano, and R. Gulino, "Human adipose-derived mesenchymal stem cells seeded into a collagen-hydroxyapatite scaffold promote bone augmentation after implantation in the mouse", *Sci. Rep.,* vol. 7, no. 1, p. 7110, 2017.
[http://dx.doi.org/10.1038/s41598-017-07672-0] [PMID: 28769083]

[63] G. Calabrese, S. Forte, R. Gulino, F. Cefalì, E. Figallo, and L. Salvatorelli, "Combination of collagenbased scaffold and bioactive factors induces adipose-derived mesenchymal stem cells chondrogenic differentiation *in vitro*", *Front. Physiol.,* vol. 8, no. 50, 2017.

[64] G. Calabrese, R. Gulino, R. Giuffrida, S. Forte, E. Figallo, C. Fabbi, L. Salvatorelli, L. Memeo, M. Gulisano, and R. Parenti, *in vivo* evaluation of biocompatibility and chondrogenic potential of a cellfree collagen-based scaffold., *Front. Physiol.,* vol. 8, p. 984, 2017.
[http://dx.doi.org/10.3389/fphys.2017.00984] [PMID: 29238307]

[65] L.L. Hench, "The story of Bioglass® ", *J. Mater. Sci. Mater. Med.,* vol. 17, no. 11, pp. 967-978, 2006.
[http://dx.doi.org/10.1007/s10856-006-0432-z] [PMID: 17122907]

[66] A. Loeffen, D.E. Cree, M. Sabzevari, and L.D. Wilson, "Effect of graphene oxide as a reinforcement in a bio-epoxy composite", *Journal of Composites Science,* vol. 5, no. 3, p. 91, 2021.
[http://dx.doi.org/10.3390/jcs5030091]

[67] A. Aravamudhan, D.M. Ramos, J. Nip, M.D. Harmon, R. James, M. Deng, C.T. Laurencin, X. Yu, and S.G. Kumbar, "Cellulose and collagen derived micro-nano structured scaffolds for bone tissue engineering", *J. Biomed. Nanotechnol.,* vol. 9, no. 4, pp. 719-731, 2013.
[http://dx.doi.org/10.1166/jbn.2013.1574] [PMID: 23621034]

[68] N. Reddy, and Y. Yang, "Properties and potential applications of natural cellulose fibers from cornhusks", *Green Chem.,* vol. 7, no. 4, pp. 190-195, 2005.
[http://dx.doi.org/10.1039/b415102j]

[69] K.J. Seal, and L.H.G. Morton, "Chemical materials", *Biotechnology (N. Y.),* vol. 8, pp. 583-590, 1996.

[70] M.M. Nassar, E.A. Ashour, and S.S. Wahid, "Thermal characteristics of bagasse", *J. Appl. Polym. Sci.,* vol. 61, no. 6, pp. 885-890, 1996.
[http://dx.doi.org/10.1002/(SICI)1097-4628(19960808)61:6<885::AID-APP1>3.0.CO;2-D]

[71] P. Pan, B. Zhu, W. Kai, S. Serizawa, M. Iji, and Y. Inoue, "Crystallization behavior and mechanical properties of bio-based green composites based on poly(L-lactide) and kenaf fiber", *J. Appl. Polym. Sci.,* vol. 105, no. 3, pp. 1511-1520, 2007.
[http://dx.doi.org/10.1002/app.26407]

[72] Y. Shih, "A study of the fiber obtained from the water bamboo husks", *Bioresour. Technol.,* vol. 98, no. 4, pp. 819-828, 2007.
[http://dx.doi.org/10.1016/j.biortech.2006.03.025] [PMID: 16759852]

[73] M.S. Sreekala, J. George, M.G. Kumaran, and S. Thomas, "The mechanical performance of hybrid phenol-formaldehyde-based composites reinforced with glass and oil palm fibres", *Compos. Sci. Technol.,* vol. 62, no. 3, pp. 339-353, 2002.
[http://dx.doi.org/10.1016/S0266-3538(01)00219-6]

[74] X. Tang, J.D. Whitcomb, Y. Li, and H.J. Sue, "Micromechanics modeling of moisture diffusion in

woven composites", *Compos. Sci. Technol.,* vol. 65, no. 6, pp. 817-826, 2005.
[http://dx.doi.org/10.1016/j.compscitech.2004.01.015]

[75]   P. Pan, B. Zhu, W. Kai, S. Serizawa, M. Iji, and Y. Inoue, "Crystallization behavior and mechanical
       properties of bio-based green composites based on poly(L-lactide) and kenaf fiber", *J. Appl. Polym.
       Sci.,* vol. 105, no. 3, pp. 1511-1520, 2007.
       [http://dx.doi.org/10.1002/app.26407]

[76]   S.A. Theron, E. Zussman, and A.L. Yarin, "Experimental investigation of the governing parameters in
       the electrospinning of polymer solutions", *Polymer (Guildf.),* vol. 45, no. 6, pp. 2017-2030, 2004.
       [http://dx.doi.org/10.1016/j.polymer.2004.01.024]

[77]   N. Sekiya, S. Ichioka, D. Terada, S. Tsuchiya, and H. Kobayashi, "Efficacy of a poly glycolic acid
       (PGA)/collagen composite nanofibre scaffold on cell migration and neovascularisation *in vivo* skin
       defect model"., *J. Plast. Surg. Hand Surg.,* vol. 47, no. 6, pp. 498-502, 2013.
       [PMID: 23596989]

[78]   C. Wu, Y. Zhou, M. Xu, P. Han, L. Chen, J. Chang, and Y. Xiao, "Copper-containing mesoporous
       bioactive glass scaffolds with multifunctional properties of angiogenesis capacity, osteostimulation
       and antibacterial activity", *Biomaterials,* vol. 34, no. 2, pp. 422-433, 2013.
       [http://dx.doi.org/10.1016/j.biomaterials.2012.09.066] [PMID: 23083929]

[79]   S. Oughlis, S. Lessim, S. Changotade, F. Bollotte, F. Poirier, G. Helary, J.J. Lataillade, V. Migonney,
       and D. Lutomski, "Development of proteomic tools to study protein adsorption on a biomaterial,
       titanium grafted with poly(sodium styrene sulfonate)", *J. Chromatogr. B Analyt. Technol. Biomed. Life
       Sci.,* vol. 879, no. 31, pp. 3681-3687, 2011.
       [http://dx.doi.org/10.1016/j.jchromb.2011.10.006] [PMID: 22036657]

[80]   C. Tokoh, K. Takabe, J. Sugiyama, and M. Fujita, "M. Cp/mas13 C NMR and electron diffraction
       study of bacterial cellulose structure affected by cell wall polysaccharides", *Cellulose,* vol. 9, no. 3/4,
       pp. 351-360, 2002.
       [http://dx.doi.org/10.1023/A:1021150520953]

CHAPTER 3

Freeze Drying: A Versatile Technique for Fabrication of Porous Biomaterials

Shaukat Khan[1,*], Muhammad Umar Aslam Khan[2,3,4] and Zahoor Ullah[5]

[1] *Materials Science Institute, the PCFM and GDHPRC Laboratory, School of Chemistry, Sun Yatsen University, Guangzhou 510275, PR China*

[2] *BioInspired Device and Tissue Engineering Research Group, School of Biomedical Engineering and Health Sciences, 81300 Skudai, Johor, Malaysia*

[3] *Institute for Personalized Medicine, School of Biomedical Engineering, Shanghai Jiao Tong University, Shanghai 200030, China*

[4] *Nanoscience and Technology Department (NS & TD), National Center for Physics, Islamabad 44000, Pakistan*

[5] *Department of Chemistry, Balochistan University of Information Technology, Engineering and Management Sciences (BUITEMS), Takatu campus, Quetta 87100, Pakistan*

Abstract: The freeze-drying process involves solvent sublimation under vacuum from pre-frozen solution resulting in porous materials. Pore volume, pore size, and density depend on several variables, including freezing temperature, solute and solvent type, solution concentration, and freezing direction. Researchers have investigated aqueous and organic solutions, supercritical CO_2 solutions, and colloidal solutions to produce various porous structures. A more recent process involves freeze-drying of emulsions, which leads to controlled pore volume and pore morphology, and porous organic nanomaterials. Directional and spray freezing are used to produce aligned porous materials and porous particles. In this chapter, we describe the basic principles of the freeze-drying process, the factors affecting the porosity of freeze-dried biomaterials, and their biomedical applications. The freeze-dried porous biomaterials are discussed in detail based on their morphology: porous structures, micro- nanowires, and micro-nanoparticles. We have summarised the current status and given some directions for future research in this field.

Keywords: Freeze drying, directional freezing, biomaterials, porous structure, microwires, nanowires, microparticles, nanoparticles.

---

* **Corresponding Author Shaukat Khan:** Materials Science Institute, the PCFM and GDHPRC Laboratory, School of Chemistry, Sun Yat-sen University, Guangzhou 510275, PR China; Tel: 8615625106973; E-mail: khans@mail.sysu.edu.cn

Adnan Haider & Sajjad Haider (Eds.)
All rights reserved-© 2022 Bentham Science Publishers

INTRODUCTION

In recent decades, researchers have shown great interest in the fabrication of three-dimensional (3D) scaffolds for various biomedical applications, including tissue engineering.Various fabrication methods are based on the transformation of liquid precursors (mainly polymers and their composites) to solid-state, including 3D printing, gas foaming, electrospinning, solvent casting/porogen leaching, and freeze-drying (FD) [1 - 8]. The FD method can produce 3D scaffolds with a porosity of 90% and a pore diameter in the range of 20 - 400 μm. The FD method was first used by Shackell in 1909 for freeze-drying biological materials. The first patent for FD was filed by Tival in 1927, while Flosdorf patented the use of a modern FD method to prevent degeneration of blood serum [9 - 13]. However, its application for 3D porous scaffolds started only recently. Nowadays, FD technologies are widely used in various industries, including food industry, pharmaceutical industry, nanotechnology, biomaterial development, *etc.* [14]. It is the method of choice for high-value materials or heat-sensitive products, or has special applications due to the direct sublimation of the solvent from ice to vapors at low pressure and temperature.Therefore, sensitive materials, including biological samples and drugs, are neither vaporized nor decomposed. Accordingly, only the solvent is removed from the freeze-dried final product, and the properties of the ingredient are retained. In addition to 3D scaffolds, the FD method has also been developed for the preparation of various other biological materials. For example, nanoparticles and porous materials have been obtained by combining emulsion and freezing techniques, nanofibers and microwires by controlled freezing of polymer solutions, and colloidal suspensions and microparticles by spray freeze-drying.

In this chapter, we ought to explain the basics of the freeze-drying process and then introduce the biomaterials obtained through this process, including porous scaffolds, nano/microwires, nanoparticles, and microparticles. Due to the significant amount of research on porous structures, we have discussed them based on the solution system applied for fabrication; aqueous solutions, organic solutions, emulsions, and colloidal suspensions. Although the conventional method involves the immersion of liquid samples in liquid nitrogen, recent strategies involve directional freezing to fabricate porous materials with layered or aligned pores. Herein, we have introduced the conventional porous materials and then compared them to the materials obtained by directional freezing in each preparation process.

THE FREEZE-DRYING PROCESS

A typical freeze dryer contains refrigeration, vacuum and control systems, a product chamber, and a condenser. The freeze-drying process involves four basic steps: (1) formulation or pretreatment, (2) freezing, (3) primary drying, and (4) secondary drying [14]. In the first step, the precursor is prepared for the process, which may involve mixing or functionalization, leading to better stability in the FD process, such as increased resistance to the low pressure or enhanced 3D porosity. The freezing step involves the precursor loading into specific molds placed in freeze dryer shells by freezing using mechanical refrigeration, liquid nitrogen, or dry ice in aqueous methanol. The main objective is to obtain the temperature lower than the solvent triple point, which is the lowest temperature at which all three solvent phases coexist. Sublimation will occur at temperatures lower than the solvent triple point rather than melting during drying (Fig. **1**) [14 - 17]. It is worth mentioning that larger solvent crystals sublimate easily. Large and more uniform ice crystals are obtained through sluggish freezing or annealing. However, large ice crystals usually lead to non-uniform 3D porosity and weak mechanical properties. Therefore, the solution is rapidly frozen to a temperature lower than the eutectic point, which usually lies between -40 to -80 °C to avoid the formation of giant crystals. However, amorphous materials do not have a eutectic point, so their critical point is considered for the freeze-drying process. In any case, it is necessary to prevent the starting materials from melting or collapsing during the freeze-drying process. Almost 95% of solvent (mostly water) in the frozen samples is sublimated in the primary drying step. It is a prolonged step and usually takes several hours or days to avoid temperature-induced physical damage. The secondary drying involves the evaporation of solvent molecules that remained unfrozen during the freezing process. For efficient desorption of surface solvent molecules, the temperature is raised to 0 °C, and the pressure is dropped further. After complete drying, the vacuum is broken by an inert gas [1, 18, 19].

CONTROLLED FREEZING

The freezing step determines the morphology of the porous materials produced. During this step, the frozen solvent crystals grow, excluding the solute particles, until the sample is completely frozen. Freezing conditions, such as solute and solvent, solution concentration, freezing temperature, and direction determine the pore structure and pore density of the prepared material. For example, freezing aqueous solutions in liquid nitrogen results in rapid freezing and smaller ice crystals. However, freezing at -20 °C results in large ice crystals due to slow nucleation leading to porous materials with large pores after freeze-drying.

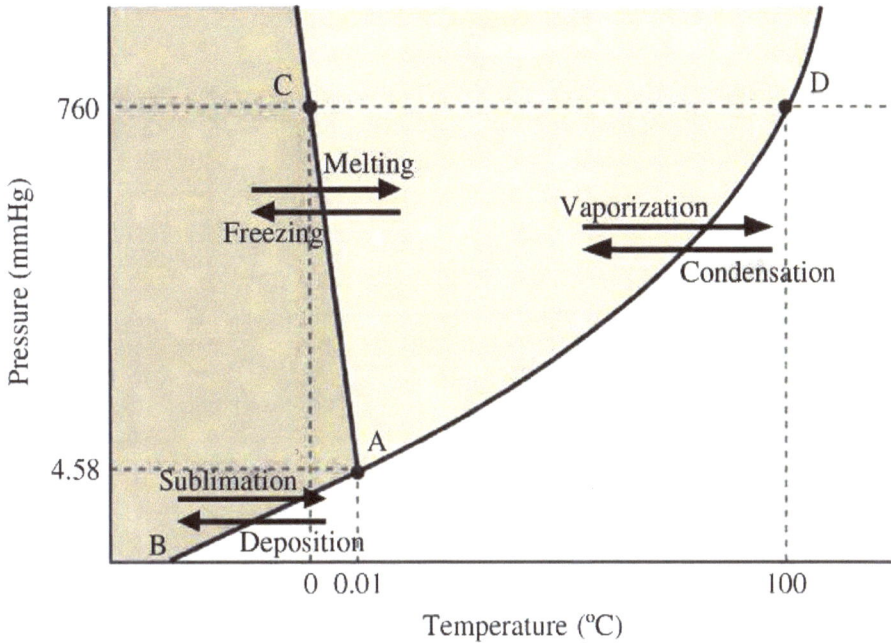

Fig. (1). Phase diagram of H_2O, the critical point occurs around T_c 5 373.946 C, p_c= 217.75 atm and ρ_c=356 kg/m^3 [17] (Reprinted with permission from Ref. 17, copyright Elsevier, 2018).

Directional Freezing

The process involves the ice crystal's growth unidirectionally as a result of controlled freezing in one direction. Ice crystals grow from the end at a lower temperature to the end at a higher temperature when a high-temperature difference is applied across the sample. Freeze drying results in pores that are unidirectional and aligned [20 - 22].

Spray-Freezing into Liquid (SFL)

A particle engineering method involves atomizing a feed solution in a cryogenic liquid (*e.g.*, liquid nitrogen) maintained at 5000 psi pressure by an HPLC pump (Fig. **2**) [23]. Upon contact with the cryogenic liquid, the feed solution is intensely atomized into micro-droplets which freeze instantly in the cryogen. Dry microparticles are obtained through separation and freeze-drying of the frozen microparticles from the cryogen.

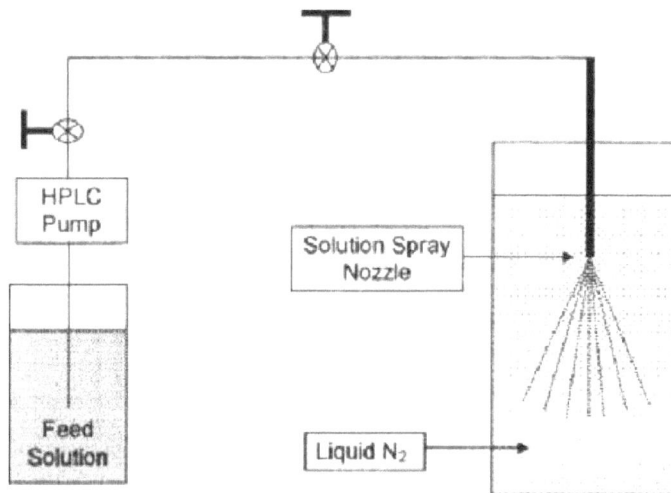

Fig. (2). Depiction of SFL method by using liquid nitrogen as the cryogenic medium [23] (Reprinted with permission from Ref. 23, copyright Elsevier Science 2002).

POROUS STRUCTURES

Aqueous Solutions

Water-based systems are the most intensively studied systems for the preparation of porous materials by freeze-drying. For example, porous poly(vinyl alcohol) (PVA) hydrogels have been prepared by freeze-drying the aqueous polymer solutions [24, 25]. Chitosan, on the other hand, is soluble in acidic water. Porous chitosan scaffolds were prepared by freeze-drying the chitosan solution in acetic acid (0.2 mol/L), which was frozen in glass tubes by immersion in liquid nitrogen [26]. It was found that the pore diameter increased with increasing freezing temperature. The growth and proliferation of mouse embryonic fibroblasts and endothelial cells were studied on the chitosan scaffolds [27]. Another study reported the covalent immobilization of glycosaminoglycan (an anionic polysaccharide derived from the extracellular matrix (ECM)) on the porous 3D chitosan scaffold. The composite scaffolds were used in perfusion cultures to study human cord blood stem cells [28]. An aqueous solution of sodium alginate and chitosan was mixed, and the mixture was freeze-dried to obtain a porous 3D chitosan-alginate scaffold [29]. Similarly, 3D scaffolds were also prepared from silk fibroin, and the effects of different freezing temperatures on the pore structure were investigated. The obtained scaffolds were examined for cell migration and proliferation [30].

When samples are frozen using a high-cold source, such as a freezer at -20 °C or dry ice at -78 °C or uncontrolled, the resulting scaffold exhibits random pores. Therefore, to produce scaffolds with highly organized and lined-up pores, the directional freezing method is used [20 - 22]. In this method, the vial containing the solution is immersed in liquid nitrogen at a controlled rate, with a purpose-built apparatus providing precise control.

PVA scaffolds with aligned pores were prepared by directional freezing of a 5 wt% aqueous solution. The distance between the aligned walls was adjusted between 12-50 μm by freezing rate from 10 to 100 μm/s [22]. PVA scaffolds with aligned pores were investigated for drug delivery, with the release of ciprofloxacin ranging from 10 minutes to several days [31]. Composite scaffolds of PVA and polypyrrole (PPy) were prepared by directional freeze-drying with prior vapor deposition polymerization of pyrrole on aligned porous PVA-FeCl3 [32]. The composite scaffold with 20 wt% PPy showed an electrical conductivity of 0.1 S/cm. A uniform mixture of PVA, polystyrene (PS), and graphene was unidirectionally frozen to obtain a polymer-graphene composite scaffold [33].

Most natural and hydrophilic polymers can be processed into both 3D and aligned porous scaffolds. The solution is prepared either at room temperature, at elevated temperature, or at various pH values. Gelatin scaffolds with aligned pores and controlled pore diameter were prepared by directional freeze-drying of their aqueous solution [34]. Cartilage cells grew well on these scaffolds, indicating that they are not cytotoxic. Scaffolds of starch with aligned pores were prepared and used as templates for the preparation of hierarchical porous silica [35]. Structured foams with aligned pores were prepared from nanosized chitin by directional freeze casting technique. The pore structure of the foam resembled the ice crystals formed during directional freezing (Fig. **3**). The obtained biofoams have been proposed for numerous applications, including scaffolding, drug carriers, filters, and especially as a template for the synthesis of multilayer composites [36].

Inorganic materials with aligned porous structures can also be prepared by directional freezing of their sol-gel precursors. Clear solutions of a commercial sodium silicate solution were obtained by diluting with water and adjusting the pH to 3, followed by unidirectional freezing. The solvent exchange was performed for the frozen samples followed by freeze drying, resulting in monolithic micro-honeycomb silica gel with a high surface area of 780 m2/g [37]. Directional freezing of silica hydrosols and hydrogels resulted in silica gels with different pore structures. It was found that the mobility of silica is the most important factor affecting the pore morphology of the silica gels obtained [38]. Resorcinol-formaldehyde hydrogels (RF) were prepared by unidirectional freezing of its aqueous solution followed by solvent exchange with t-butanol and freeze drying, resulting in RF cryogels with a micro-honeycomb structure. Porous carbon was obtained by pyrolysis of these cryogels [39].

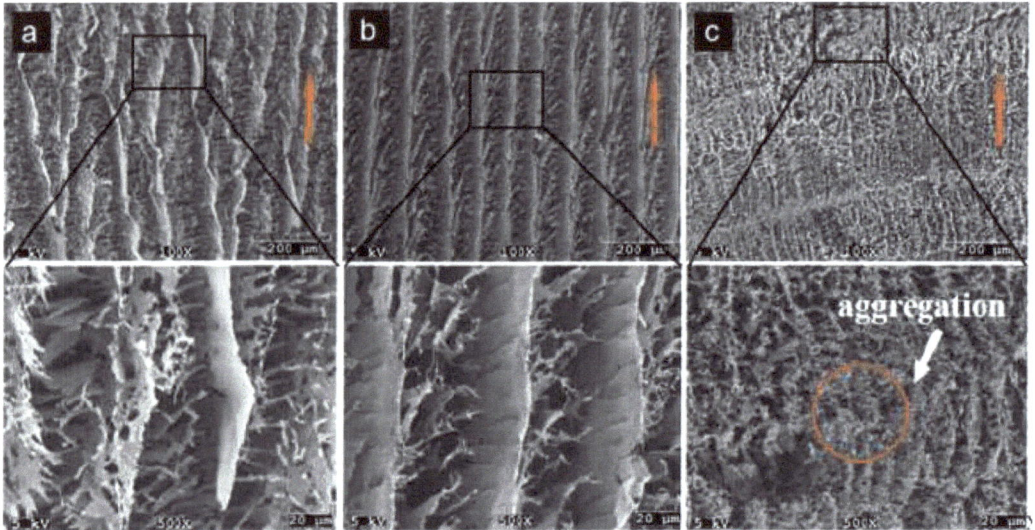

Fig. (3). SEM images of chitin foams made with the various chitin suspensions under directional freezing. The suspension concentrations are (a) 0.4 wt%, (b) 0.8 wt%, (c) 1.2 wt%. Red color arrows indicate the freezing direction [36]. Reprinted with permission from Ref. 36. Copyright Elsevier 2014.

The directional freezing method is also used to prepare hierarchical hybrid materials. A gel was prepared by mixing aqueous PVA, silica sol and pig liver esterase (an enzyme). The gel was cooled to the temperature of liquid nitrogen using the ice-segregation induced self-assembly (ISISA) method and then freeze-dried, resulting in a porous material with aligned and micrometer-sized pores [40]. The obtained biohybrid material preserved the structure and function of the enzyme along with its highly organized porous structure.

Aqueous Colloidal Suspensions

Polymer and inorganic suspensions are usually freeze-dried to obtain porous composites or porous inorganic materials. The morphology of the pores depends on the freezing method used. Usually, a dispersant or a steering reagent is used to stabilize the colloid suspension before freezing. For example, sodium carboxymethyl cellulose (SCMC) or PVA was added to a colloidal suspension of silica, resulting in pores with layered structures or parallel microchannels [22, 41]. In the absence of PVA, loose silica microparticles or microplates were obtained. PVA caused ice crystals to develop in an oriented pattern and confined the silica colloids in the porous product [22]. The channel size could be reduced to 530 nm by increasing the ice growth rate and temperature gradient during the freezing process, while the same size was reduced to 180 nm by adding dextran to the suspension before freezing [42].

Another important application of this research area is the preparation of advanced porous ceramics from their aqueous suspensions. Aqueous slurries of ceramic materials (20-50 wt%) have also been used to prepare porous ceramics. A polymeric dispersant is important to obtain a stable ceramic suspension and to keep the particles intact after freeze-drying. The freeze casting process is used for the production of porous ceramics [43]. Since alumina is a widely used ceramic material, its pore morphology and properties are extensively studied by freeze casting. Controlled slurry freezing followed by sublimation and sintering has resulted in alumina structures with regular patterns [44]. The porous architecture of alumina has been shown to depend on process variables, such as the composition of the slurry, the freezing rate and the pattern of the freezing surface [45]. Tallon *et al.* studied the effects of freeze casting on the morphology of porous alumina. The effects of various process parameters, such as freezing conditions (device and temperature), glycerol cryoprotectant and solid content were studied on the pore morphology and pore density of the sintered alumina [46]. Another study reported the use of impregnating freeze casting method to produce porous alumina. An aqueous alumina slurry was impregnated into a polyurethane sponge and then unidirectionally frozen and freeze-dried. Thanks to the porous template, local interruptions were introduced into the lamellar structure of the alumina [47]. Aligned porous alumina ceramics were prepared by unidirectional freezing of foamed alumina slurries, in which PVA served as a binder and mixing agent to stabilize air bubbles in aqueous alumina slurry [48]. The porous alumina was annealed at elevated temperatures after the initial freeze-drying and calcination process to obtain better mechanical properties. The hierarchical complexity in nature (structure of nacre) inspired the scientists to infiltrate porous lamellar alumina with polymethyl methacrylate (PMMA), resulting in a hybrid composite with high mechanical strength and toughness 300 times higher than its constituents [49, 50]. Chemical grafting at the PMMA-alumina interface resulted in improved initiation toughness and strength. This flexible approach can be used to fabricate various combinations of ceramics and polymers [51].

Hydroxyapatite (HA), an essential component of natural hard tissues, including bone, is widely used for tissue engineering of bone [52, 53]. Porous HA scaffolds with oriented architecture were prepared by unidirectional freezing of aqueous HA suspensions containing 10 vol% HA and 0.75 wt% dispersants. *In vitro* cell response to these scaffolds indicated their effective application for bone repair [54]. A binary solvent mixture (water-glycerol and water-dioxane) leads to dynamic changes in the porous morphology of HA scaffolds after freezing [55].

HA was also modified by the addition of SiO_2 nanoparticles during freeze-drying, resulting in a partial phase change from HA to β-tricalcium phosphate and improved structural stability due to minimal shrinkage during sintering. Porous ceramic scaffolds were investigated for the effects of their surface roughness on adhesion, growth, proliferation and differentiation of human osteoblasts [56]. A biohybrid composite mimicking natural nacre was prepared by filling a layered porous HA scaffold with epoxy. The resulting hybrid scaffold exhibited four times higher mechanical strength than currently used implantation materials for osseous tissue repair [57]. Highly porous composite scaffolds made of HA/biopolymer, including HA/collagen and HA/alginate, have shown their potential as suitable materials for bone tissue regeneration [58 - 60].

In addition to ceramic powders and silica suspensions, organic colloidal suspensions have also been used to prepare porous organic or composite materials. As with ceramic suspensions, a polymeric dispersant is used to stabilize the organic colloid suspension in water, *e.g.*, aqueous SCMC to stabilize polystyrene (PS) colloid. Directional freezing followed by freeze-drying resulted in the production of an aligned porous morphology, and the PS colloid concentration was found to influence the film thickness [41]. Graphene/ PS porous composites were prepared by freeze-drying graphene suspensions and PS colloids followed by compression molding. High electrical conductivity of 15 S/cm was measured at graphene loading of 1.6-2.0% [61]. Polymeric microgels with opposite charges were used to fabricate porous structures with aligned pores after being suspended in water at pH 7. The freeze-dried porous scaffolds are insoluble in water due to the microgels with opposite charges [62]. Another study reported the addition of poly(styrene-co-2-hydroxyethyl methacrylate) colloids (average diameter 385 nm) to silica sol (volume ratio of polymer/silica = 74.2/25.8). After freeze-drying, the obtained suspension gave honeycomb silica with cross-linked and highly ordered macroporous walls [63]. It was also reported that freeze-drying of an aqueous mixture of poly(vinyl laurate) latexes and silica nanoparticles resulted in composite foams [64]. When black carbon particles were added as a third colloid, the composites turned into conductive foams [64].

Carbon nanotubes (CNTs) have been incorporated into suitable materials for the synthesis of nanocomposites due to their advantageous properties. CNT-polymer composite scaffolds have been prepared by dispersion of CNTs in polymer-stabilized aqueous solution followed by ice templating. For example, chitosan-CNT scaffolds were prepared by dispersing 2.5 wt% multi-walled carbon nanotubes (MWCNTs) in chitosan solution [65]. Aligned and unaligned MWCNT scaffolds were prepared by flash freezing and directional freezing using silk fibroin as a binder. The aligned structure showed better electrical conductivity and thermal stability compared to the non-aligned one (Fig. **4**) [66]. Chitosan-

MWCNT scaffolds showed good compatibility with bacterial cells and have been proposed for the immobilization and proliferation of bacteria [67]. The chitosan-MWCNT scaffolds were tested *in vitro* for adhesion, growth and proliferation of mouse myoblastic cells (C_2C_{12}). The chamber structure of the scaffolds is shown in Fig. (**4**). These scaffolds promoted ectopic bone development in muscle tissue when incorporated with recombinant human bone morphogenetic protein-2 (rhBMP-2) [68]. These 3D scaffolds were also decorated with Pt nanoparticles, resulting in high conductivity and anode material for methanol fuel cells [69].

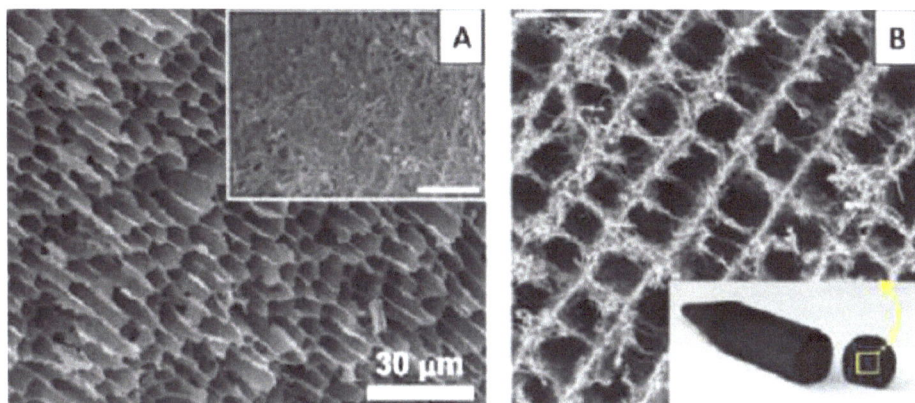

Fig. (4). (A) Inset scale bar 500 nm, aligned porous MWCNT with silk fibroin [66]. (Reprinted with permission from Ref. 66, copyright Elsevier 2009). (B) Macroporous morphology of MWCNT/chitosan with 20 um Scale bar [68]. (Reprinted with permission from Ref. 68 directional freeze–casting technique, copyright Elsevier 2008).

Porous ceramics have usually been prepared by aqueous slurries and colloids. However, camphene-based freeze-drying has emerged as an alternative technique. Freezing and subsequent sublimation of camphene at room temperature results in 3D porous networks. The study reports that camphene, zirconia and dispersant Texaphor 963 were mixed by ball milling, resulting in a slurry. The slurry was frozen in a mold at 15 °C, followed by sublimation of camphene at room temperature and sintering at 1450 °C, resulting in porous ZrO_2 foams with excellent compressive strength [70]. In a similar report, TiH_2 powder, camphene and oligomeric polyester dispersant were mixed by ball milling at 60 °C. The slurries containing 40, 25, and 10 vol% TiH_2 were poured into an aluminum mold in the order that the 40 vol% slurries were poured first and allowed to solidify for 5 min, followed by pouring the 25 and 10 vol% slurries at an interval of 5 min each. Sublimation of camphene, decomposition of TiH_2 at 400 °C and sintering at 1300 °C yielded titanium scaffolds with gradient porosity and pore size [71]. Directional freezing of camphene solutions has resulted in aligned porous materials, *e.g.*, polycarbosilane solutions with molten camphene were directionally frozen in polyethylene molds in liquid nitrogen or cool ethanol and then stored at -68 °C to increase their strength. Camphene was sublimed, and polycarbosilane was crosslinked by curing at 200 °C in air. Aligned porous SiC ceramics were obtained by pyrolysis at 1400 °C [72].

Organic Solutions

Freeze-drying of organic solutions of hydrophobic polymers results in porous scaffolds for tissue regeneration applications. A porous poly(caprolactone) (PCL) scaffold was prepared by freeze-drying a PCL-tetrahydrofuran solution at -80 °C. The effect of polymer concentration on the scaffold properties such as mechanical strength, morphology, biodegradability and porosity was investigated [73]. PCL and poly(D,L-lactide) scaffolds with controlled porosity were prepared by combined freeze-drying and porogen leaching method with 92 wt% sugar and salt particles in a polymer solution in 1,4-dioxane [74].

When applied to organic solutions, directional freezing has expanded considerably from its original application to aqueous solutions [20 - 22]. PCL scaffolds with aligned microchannels have been prepared by directional freezing of PCL-dichloroethane solutions followed by freeze-drying [22]. Directional freezing of poly(L-lactic acid) (PLLA) solutions in 1,4-dioxane also resulted in PLLA scaffolds with aligned pores [75]. The same group reported PLLA and poly(ethylene glycol) composite scaffolds (PEG) with aligned pores by directional freezing of the composite solution in 1,4-dioxane. The effects of molecular weight PEG and a weight ratio of PLLA to PEG on channel morphology and channel pores were studied [76]. Small pores in the channel walls are formed by the leaching of PEG in ethanol. The density and size of these pores depend on the concentration and molecular weight of PEG used (Fig. **5**) [76].

Emulsions

A heterogeneous mixture of immiscible liquid droplets in another is called an emulsion. A distinction is made between water-in-oil (W/O) and oil-in-water (O/W) emulsions. Emulsions are used as templates for the preparation of porous materials by polymerization of monomers in the continuous phase and subsequent removal of the solvent from the droplet phase [77]. An alternative approach is to freeze-fix the structure of the emulsion and then freeze-dry the continuous phase and droplet phase solvents to produce porous structures. Porous poly(lactic-co-glycolic acid) (PLGA) scaffolds were prepared by freeze-drying a water/PLGA-methylene chloride emulsion [78]. Porous scaffolds were prepared in the same way from horseradish peroxide, and bovine serum albumin (BSA) dispersed in PLG-methylene chloride emulsions and then freeze-dried. The protein-loaded scaffolds were used as a matrix for protein delivery [13, 79]. Porous PCL and

PLGA scaffolds were also prepared by freeze-drying over a water-in-oil emulsion containing the surfactant Span 80. The PLGA scaffolds were surface modified with heparin to allow better delivery of growth factors for soft tissue regeneration. Adhesion, growth and proliferation of human bladder stromal cells were studied on these scaffolds [80].

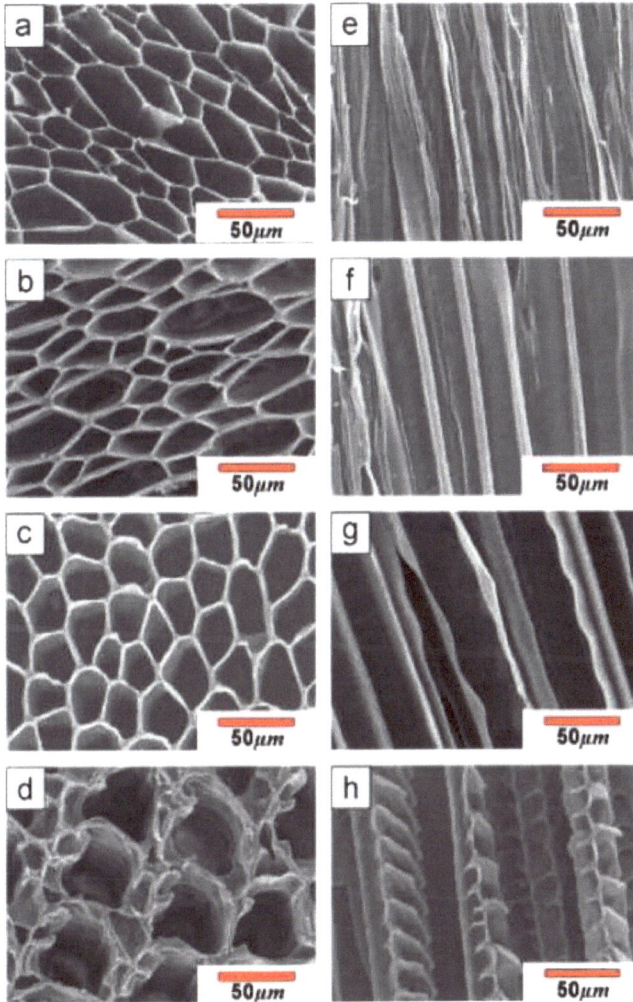

Fig. (5). SEM images of porous PLLA prepared from PLLA/ 1,4-dioxane solution. **(a)-(d)** Cross-sectional area perpendicular to freezing direction, **(e)-(h)** cross-sectional area parallel to freezing direction, **(a)** and **(e)** 3 wt%, **(b)** and **(f)** 5 wt%, **(c)** and **(g)** 7 wt% and **(d)** and **(h)** 10 wt% of PLLA [76] (Reprinted with permission from Ref. 76. Copyright Elsevier, 2008.

Freeze-drying and emulsion templating have been advantageously combined in the preparation of porous scaffolds. For example, rapid freezing locks the emulsion structure, which facilitates emulsion stability. In addition, the droplet phase ratio can be precisely controlled in the range of 10-95%, ultimately resulting in significant control over the porosity and pore morphology of the fabricated scaffolds. For example, porous scaffolds containing both ice-templated and emulsion-templated pores were prepared by emulsifying cyclohexane in an aqueous SCMC solution with SDS surfactant in volume ratios of 0:100, 20:80, 40:60, 60:40 and 75:25 followed by freeze-drying [41]. These scaffolds were used as templates for the preparation of porous zirconia with very controlled porosity and pore size.

MICRO- AND NANOWIRES

A fibrous PLA network scaffold was prepared by thermally induced gelation, solvent exchange and freeze-drying. The diameter of the nanofibers varied depending on variables, such as solution concentration, solvent exchange, and annealing and freezing temperatures [81, 82]. It has also been found that dilute aqueous polymer solutions form nanofibrous scaffolds with a fiber width of 200-600 nm when frozen in liquid nitrogen and freeze-dried [83]. This method can be used to prepare nanofibers from hydrophilic polymers such as alginate, SCMC, and PVA. These nanofiber scaffolds are also used as templates for the preparation of Fe_2O_3 nanofibers and hollow titanium dioxide microtubes [83]. Chitosan fibers with a diameter of 100-700 nm were also prepared by freeze-drying a 0.1 wt% acetic acid solution [84]. This method has the advantage over the electrospinning method because it does not require concentrated acetic acid and toxic organic solvents. The chitosan fibers showed good release properties for BSA and rhodamine B.

The preparation of nanofibrous polymeric scaffolds by freeze-drying is a recent accomplishment. However, the preparation of oxide fibers by directional freezing was reported decades ago. Directional freezing was first used to prepare silica fibers in 1980 [85]. Aqueous silica with a pH of 5.0 was gelled to produce a silica sol, followed by 30 min aging and directional freezing. Thawing resulted in parallel silica nanofibers with polygonal holes. After freeze-drying, directional freezing of a silicon nitride slurry in water leads to fibers with aligned pores [86]. To increase the strength of the fibers, titanium tetraisopropoxide polymerizes in a water/ethanol mixture to form a TiO_2 hydrogel, which is unidirectionally frozen in liquid nitrogen and stored at -30°C. After thawing, washing with t-butyl alcohol and freeze-drying the hydrogels, fibrous TiO_2 cryogels were obtained [87].

Recently, scientists have shown great interest in the preparation of microwires using aqueous colloidal suspensions. Gold microwires with a diameter of 700 nm were prepared by aggregating gold nanoparticles (diameter 15 nm) suspended in water. Simultaneously, metal oxide nanoparticles (20-40 nm average diameter) resulted in microrods, and PS colloid (450 nm diameter) also resulted in microwires [88]. Free-standing PS fibers were also prepared by directional freezing of an aqueous suspension of PS nanoparticles (80 nm average diameter) [89]. The brittleness of these fibers was overcome by using chemical vapor deposition (CVD). Fibers with ordered inorganic and organic phases comprising the arrangement of bricks and mortar were prepared by directional freezing of Silica@Poly (*N*-isopropylacrylamide) (PNIPAM) microgels [90]. Also, composite fibers were prepared from silica@polyacrylonitrile (PAN) colloids [91]. AN fibers were carbonized by pyrolysis at 800 ºC, while the silica core was washed out with concentrated alkali, leaving hollow carbon fibers. Carbon/metal oxide fibers are formed when another inorganic nanoparticle is added to the colloid (*e.g.*, $SiO_2@ZrO_2@$ PAN).

Micro- And Nanoparticles

Microparticles can be prepared by spray freezing in liquid (SFL), which makes it a suitable method to prepare microparticles of drugs with poor water solubility (*e.g.*, carbamazepine and danazol) either neat or with their excipients [23, 92, 93]. The dissolution rate is greatly improved by increasing the contact area by decreasing the particle size. For example, danazol can dissolve 95% in 2 min from danazol microparticles (with an effective strength of 91%), compared to only 30% dissolving in 2 min from danazol bulk. Enzyme and protein particles have also been prepared by spray freeze drying (SFD) and SFL methods [94 - 96]. In the SFD method, the solution is sprayed into a half-filled vessel containing a cryogenic liquid. The atomized droplets freeze first in the cold vapours and then completely in the cryogenic liquid. In the SFL process, little protein denaturation and enzyme activity loss were observed due to exposure at the air-water interface during atomization. Spray freezing can result in different particle morphologies by using different cooling rates and spray geometries [97], using a four-fluid nozzle [98], or a solid-in-oil suspension [99]. The spray freezing method has already been used to produce food [100], fat [101], and inorganic particles that can be used as a catalyst for the oxidation of methane and CO [102].

The emulsification/freeze-drying and spray/freeze-drying processes are used to produce hollow, biodegradable PLA particles with porous shell walls. The particles were first suspended in an aqueous solution of the drug and then plasticized with compressed CO_2 or dichloromethane, leading to pore closure and final encapsulation of the drug for controlled future release [103]. Alternatively, PS hollow particles with controlled surface pores were prepared by freeze-drying and swelling. PS colloids were suspended in a water/toluene mixture where the colloids were swollen by the entry of toluene, followed by dropwise addition of the above suspension into liquid nitrogen. The frozen suspension was carefully heated to evaporate the solvent and leave surface pores on the hollow PS particles. A variety of functional materials, including proteins, drugs, and minor colloids, can be encapsulated in these hollow PS colloids, with the pores closed by annealing above the glass transition temperature [104]. The surface pores can also be closed by treating the PS particles with a second swelling solvent after encapsulation [105].

Porous microparticles can also be prepared by directly freezing and processing an emulsion. For example, a PCL-xylene solution was emulsified in a PVA-sodium dodecyl sulphate solution (SDS), resulting in an O/W emulsion, followed by directional freezing and freeze-drying to form aligned porous particles that were implanted into an aligned porous matrix [106]. Another method for producing aligned porous microparticles involves the preparation of an aqueous solution of the composite material followed by filtration. The porous PCL microparticles showed good biocompatibility with mouse embryonic stem cells and were non-toxic over a culture period of 7 days [106]. Freeze-drying of emulsions is also used to prepare organic microparticles. Oil Red (OR), an organic dye, was dissolved in a W/O emulsion with PVA in the continuous aqueous phase. After freeze-drying, this emulsion, nanoparticles (average diameter 90 nm) embedded in a porous PVA matrix were obtained [107]. An antibacterial agent, triclosan, was dissolved in an O/W emulsion and then freeze-dried. Dissolution in water resulted in an aqueous suspension of triclosan nanoparticles, which showed better biocidal activity compared to a triclosan solution in a water/ethanol system [107]. The emulsion method is a general way to prepare an aqueous nanoparticle dispersion of organic materials with poor water solubility [108]. Many drugs have poor water solubility, which limits their biosorption, while this technique can improve their water solubility [92, 93]. Moreover, organic nanoparticles can also be prepared by a temperature-dependent porous crosslinked PNIPAM [109]. At room temperature, the nanoparticles in PNIPAM are kept intact, while at 32 °C (low critical dissolution temperature of PNIPAM), they are released in water [109].

Potential Application of Freeze Drying

Freeze-drying is mainly used in the food industry, pharmaceuticals and biology as a drying process [110]. It is also known as cryodesiccation or freeze drying. In this process, the materials are made more transportable [111]. The mechanism of freeze drying is based on sublimation, where frozen products are directly dehydrated [112]. Haider *et al.* [113] reported that in freeze-drying, the material is prepared in three steps: (a) the material solution is prepared, (b) the solution is moulded or poured, frozen and swelled at 70-80 °C, (c) finally, the material is dried at low pressure in a chamber. In the chamber, the ice is removed from the material by sublimation, while the non-frozen residual water is removed in a secondary drying process, the adsorp- tion process.

Freeze drying is often used in the pharmaceutical industry to extend the shelf life of products. By removing the water and enclosing the substance in a vial, the material can be easily stored, shipped, and then returned to its original form for injection. Another example from the pharmaceutical field is the use of freeze-drying to produce wafers or tablets, which have the advantage of containing fewer excipients and providing a dosage form that can be quickly absorbed and administered.

Freeze-drying is used to preserve food, and the final product is light. Freeze-dried ice cream, which is an example of astronaut cuisine, is a common example of this process. It is also commonly used to make substances or flours that are added to foods. Freeze-dried foods are also ideal and popular for hikers as they weigh very little. Dried foods can be easily transported as more compared to the weight of wet foods and have the advantage of having a longer shelf life than wet foods, which spoil quickly. The most common methods of drying coffee are evaporation by a stream of hot air or projection onto hot metallic surfaces. Similarly, freeze-dried fruits are used in some morning cereals or offered as snacks. They are a popular snack for dieters, preschoolers, and toddlers, and are also used by some pet owners as a treat for birds. Kitchen herbs are also freeze-dried, although air-dried herbs are far more common and less expensive.

The Document Preservation Laboratory of the U.S. National Archives and Records Administration (NARA) has conducted research on freeze-drying as a method of restoring books and papers damaged by water. When a document is made of several different materials with different absorption properties, the document expands unevenly, which can cause deformation.

Freeze-drying is used in bacteriology to preserve specific strains. In high-altitude environments, natural mummies are sometimes produced by freeze drying due to low pressure and temperature [111]. Silva *et al.* [114] used a freeze-drying technique to produce sponges based on Eri silk fibres for biomedical applications using genipin for chemical cross-linking. Haider *et al.* [113] reported the -drying technique widely used to fabricate 3D porous scaffolds. The scaffold fabricated using this technique has various applications in the biomedical field, especially in tissue engineering. They also reported that other researchers were working on porous 3D scaffolds fabricated by a freeze-drying technique using polysaccharides.

CONCLUSION AND PROSPECTS

This chapter is mainly devoted to the synthesis of porous and micro- and nanostructured biomaterials by freeze-drying. These materials can be easily synthesized by processing their solutions, colloids and emulsions. The freezing temperature, the type of material and solvent, the concentration and the freezing direction affect the porous morphology of the obtained materials. Among these methods, directional freezing leads to porous materials with aligned pores, while spray freezing leads to porous microparticles. The combination of freeze-drying and emulsion templating allows precise control of pore structure and volume, as well as the incorporation of pharmaceutical and organic nanoparticles into porous materials. We have studied various material morphologies, such as porous structures, micro nanowires, and micro nanoparticles, as well as various materials used to fabricate such porous materials, including polymers, metals and metal oxides, silica, pharmaceuticals, and proteins.

Freeze-drying should be investigated for its versatility and advantages of its myriad applications and the development of novel processes. The following research directions are expected to advance in the fabrication of porous and nanomaterials using freeze-drying: (i) novel morphologies, mechanical strength, and hierarchical porosity; (ii) large-scale fabrication with engineered defect-free structures; (iii) biodegradable and biocompatible porous scaffolds and particles with tunable mechanical strength and surface functionality for improved cell adhesion, growth, and tissue regeneration; (iv) functional polymeric and inorganic materials for energy storage and catalysis.

CONSENT FOR PUBLICATION

Not applicable.

CONFLICT OF INTEREST

The authors declare no conflict of interest, financial or otherwise.

ACKNOWLEDGEMENT

Declared none.

REFERENCES

[1]   S.M. Giannitelli, P. Mozetic, M. Trombetta, and A. Rainer, "Combined additive manufacturing approaches in tissue engineering", *Acta Biomater.*, vol. 24, pp. 1-11, 2015.
[http://dx.doi.org/10.1016/j.actbio.2015.06.032] [PMID: 26134665]

[2]   Z. Fereshteh, M. Fathi, A. Bagri, and A.R. Boccaccini, "Preparation and characterization of aligned porous PCL/zein scaffolds as drug delivery systems *via* improved unidirectional freeze-drying

method", *Mater. Sci. Eng. C,* vol. 68, pp. 613-622, 2016.
[http://dx.doi.org/10.1016/j.msec.2016.06.009] [PMID: 27524061]

[3]   Z. Fereshteh, M. Fathi, and R. Mozaffarinia, "Mg-doped fluorapatite nanoparticles-poly(-
        -caprolactone) electrospun nanocomposite: Microstructure and mechanical properties", *Superlattices Microstruct.,* vol. 75, pp. 208-221, 2014.
        [http://dx.doi.org/10.1016/j.spmi.2014.07.011]

[4]   Z. Fereshteh, P. Nooeaid, M. Fathi, A. Bagri, and A.R. Boccaccini, "Mechanical properties and drug
        release behavior of PCL/zein coated 45S5 bioactive glass scaffolds for bone tissue engineering
        application", *Data Brief,* vol. 4, pp. 524-528, 2015.
        [http://dx.doi.org/10.1016/j.dib.2015.07.013] [PMID: 26966716]

[5]   Z. Fereshteh, P. Nooeaid, M. Fathi, A. Bagri, and A.R. Boccaccini, "The effect of coating type on
        mechanical properties and controlled drug release of PCL/zein coated 45S5 bioactive glass scaffolds
        for bone tissue engineering", *Mater. Sci. Eng. C,* vol. 54, pp. 50-60, 2015.
        [http://dx.doi.org/10.1016/j.msec.2015.05.011] [PMID: 26046267]

[6]   T. Lu, Y. Li, and T. Chen, "Techniques for fabrication and construction of three-dimensional scaffolds
        for tissue engineering", *Int. J. Nanomedicine,* vol. 8, pp. 337-350, 2013.
        [http://dx.doi.org/10.2147/IJN.S38635] [PMID: 23345979]

[7]   Q.L. Loh, and C. Choong, "Three-dimensional scaffolds for tissue engineering applications: role of
        porosity and pore size", *Tissue Eng. Part B Rev.,* vol. 19, no. 6, pp. 485-502, 2013.
        [http://dx.doi.org/10.1089/ten.teb.2012.0437] [PMID: 23672709]

[8]   F. Sun, H. Zhou, and J. Lee, "Various preparation methods of highly porous hydroxyapatite/polymer
        nanoscale biocomposites for bone regeneration", *Acta Biomater.,* vol. 7, no. 11, pp. 3813-3828, 2011.
        [http://dx.doi.org/10.1016/j.actbio.2011.07.002] [PMID: 21784182]

[9]   E.W. Flosdorf, "Drying Penicillin by Sublimation", *BMJ,* vol. 1, no. 4389, pp. 216-218, 1945.
        [http://dx.doi.org/10.1136/bmj.1.4389.216] [PMID: 20785906]

[10]   W.J. Elser, R.A. Thomas, and G.I. Steffen, "The desiccation of sera and other biological products
        (including microörganisms) in the frozen state with the preservation of the original qualities of
        products so treated", *J. Immunol.,* vol. 28, pp. 433-473, 1935.

[11]   J.H. de Groot, A.J. Nijenhuis, P. Bruin, A.J. Pennings, R.P.H. Veth, J. Klompmaker, and H.W.B.
        Jansen, "Use of porous biodegradable polymer implants in meniscus reconstruction. 1) Preparation of
        porous biodegradable polyurethanes for the reconstruction of meniscus lesions", *Colloid Polym. Sci.,*
        vol. 268, no. 12, pp. 1073-1081, 1990.
        [http://dx.doi.org/10.1007/BF01410672]

[12]   H. Elema, J.H. de Groot, A.J. Nijenhuis, A.J. Pennings, R.P.H. Veth, J. Klompmaker, and H.W.B.
        Jansen, "Use of porous biodegradable polymer implants in meniscus reconstruction. 2) Biological
        evaluation of porous biodegradable polymer implants in menisci", *Colloid Polym. Sci.,* vol. 268, no.
        12, pp. 1082-1088, 1990.
        [http://dx.doi.org/10.1007/BF01410673]

[13]   K. Whang, C.H. Thomas, K.E. Healy, and G. Nuber, "A novel method to fabricate bioabsorbable
        scaffolds", *Polymer (Guildf.),* vol. 36, no. 4, pp. 837-842, 1995.
        [http://dx.doi.org/10.1016/0032-3861(95)93115-3]

[14]   A.R.V. Morais, É.N. Alencar, F.H. Xavier Júnior, C.M. Oliveira, H.R. Marcelino, G. Barratt, H. Fessi,
        E.S.T. Egito, and A. Elaissari, "Freeze-drying of emulsified systems: A review", *Int. J. Pharm.,* vol.
        503, no. 1-2, pp. 102-114, 2016.
        [http://dx.doi.org/10.1016/j.ijpharm.2016.02.047] [PMID: 26943974]

[15]   L. Garcia-Amezquita, J. Welti-Chanes, F. Vergara-Balderas, and D. Bermúdez-Aguirre, "Freeze-
        drying: the basic process", 2016.
        [http://dx.doi.org/10.1016/B978-0-12-384947-2.00328-7]

[16]   S.L. Nail, S. Jiang, S. Chongprasert, and S.A. Knopp, "Fundamentals of freeze-drying", In: *in Development and manufacture of protein pharmaceuticals.* Springer, 2002.
[http://dx.doi.org/10.1007/978-1-4615-0549-5_6]

[17]   Z. Fereshteh, "Freeze-drying technologies for 3D scaffold engineering", In: *Functional 3D tissue engineering scaffolds.* Elsevier, 2018.
[http://dx.doi.org/10.1016/B978-0-08-100979-6.00007-0]

[18]   R. Geidobler, and G. Winter, "Controlled ice nucleation in the field of freeze-drying: Fundamentals and technology review", *Eur. J. Pharm. Biopharm.,* vol. 85, no. 2, pp. 214-222, 2013.
[http://dx.doi.org/10.1016/j.ejpb.2013.04.014] [PMID: 23643793]

[19]   S. Wu, X. Liu, K.W.K. Yeung, C. Liu, and X. Yang, "Biomimetic porous scaffolds for bone tissue engineering", *Mater. Sci. Eng. Rep.,* vol. 80, pp. 1-36, 2014.
[http://dx.doi.org/10.1016/j.mser.2014.04.001]

[20]   H. Zhang, J. Long, and A.I. Cooper, "Aligned porous materials by directional freezing of solutions in liquid CO_2".. *J. Am. Chem. Soc.,* vol. 127, no. 39, pp. 13482-13483, 2005.
[http://dx.doi.org/10.1021/ja054353f] [PMID: 16190696]

[21]   M.C. Gutiérrez, M.L. Ferrer, and F. del Monte, "Ice-templated materials: Sophisticated structures exhibiting enhanced functionalities obtained after unidirectional freezing and ice-segregation-induced self-assembly", *Chem. Mater.,* vol. 20, no. 3, pp. 634-648, 2008.
[http://dx.doi.org/10.1021/cm702028z]

[22]   H. Zhang, I. Hussain, M. Brust, M.F. Butler, S.P. Rannard, and A.I. Cooper, "Aligned two- and three-dimensional structures by directional freezing of polymers and nanoparticles", *Nat. Mater.,* vol. 4, no. 10, pp. 787-793, 2005.
[http://dx.doi.org/10.1038/nmat1487] [PMID: 16184171]

[23]   T.L. Rogers, J. Hu, Z. Yu, K.P. Johnston, and R.O. Williams III, "A novel particle engineering technology: spray-freezing into liquid", *Int. J. Pharm.,* vol. 242, no. 1-2, pp. 93-100, 2002.
[http://dx.doi.org/10.1016/S0378-5173(02)00154-0] [PMID: 12176230]

[24]   F. Yokoyama, I. Masada, K. Shimamura, T. Ikawa, and K. Monobe, "Morphology and structure of highly elastic poly(vinyl alcohol) hydrogel prepared by repeated freezing-and-melting", *Colloid Polym. Sci.,* vol. 264, no. 7, pp. 595-601, 1986.
[http://dx.doi.org/10.1007/BF01412597]

[25]   H.H. Trieu, and S. Qutubuddin, "Polyvinyl alcohol hydrogels I. Microscopic structure by freeze-etching and critical point drying techniques", *Colloid Polym. Sci.,* vol. 272, no. 3, pp. 301-309, 1994.
[http://dx.doi.org/10.1007/BF00655501]

[26]   S.V. Madihally, and H.W.T. Matthew, "Porous chitosan scaffolds for tissue engineering", *Biomaterials,* vol. 20, no. 12, pp. 1133-1142, 1999.
[http://dx.doi.org/10.1016/S0142-9612(99)00011-3] [PMID: 10382829]

[27]   Y. Huang, M. Siewe, and S.V. Madihally, "Effect of spatial architecture on cellular colonization", *Biotechnol. Bioeng.,* vol. 93, no. 1, pp. 64-75, 2006.
[http://dx.doi.org/10.1002/bit.20703] [PMID: 16142800]

[28]   C.H. Cho, J.F. Eliason, and H.W.T. Matthew, "Application of porous glycosaminoglycan-based scaffolds for expansion of human cord blood stem cells in perfusion culture", *J. Biomed. Mater. Res. A,* vol. 86A, no. 1, pp. 98-107, 2008.
[http://dx.doi.org/10.1002/jbm.a.31614] [PMID: 17941019]

[29]   Z. Li, and M. Zhang, "Chitosan-alginate as scaffolding material for cartilage tissue engineering", *J. Biomed. Mater. Res. A,* vol. 75A, no. 2, pp. 485-493, 2005.
[http://dx.doi.org/10.1002/jbm.a.30449] [PMID: 16092113]

[30]   B.B. Mandal, and S.C. Kundu, "Cell proliferation and migration in silk fibroin 3D scaffolds", *Biomaterials,* vol. 30, no. 15, pp. 2956-2965, 2009.

[http://dx.doi.org/10.1016/j.biomaterials.2009.02.006] [PMID: 19249094]

[31]   M.C. Gutiérrez, Z.Y. García-Carvajal, M. Jobbágy, F. Rubio, L. Yuste, F. Rojo, M.L. Ferrer, and F. del Monte, "Poly (vinyl alcohol) scaffolds with tailored morphologies for drug delivery and controlled release", *Adv. Funct. Mater.,* vol. 17, no. 17, pp. 3505-3513, 2007.
[http://dx.doi.org/10.1002/adfm.200700093]

[32]   H. Bai, C. Li, F. Chen, and G. Shi, "Aligned three-dimensional microstructures of conducting polymer composites", *Polymer (Guildf.),* vol. 48, no. 18, pp. 5259-5267, 2007.
[http://dx.doi.org/10.1016/j.polymer.2007.06.071]

[33]   J.L. Vickery, A.J. Patil, and S. Mann, "fabrication of graphene–polymer nanocomposites with higher-order three-dimensional architectures", *Adv. Mater.,* vol. 21, no. 21, pp. 2180-2184, 2009.
[http://dx.doi.org/10.1002/adma.200803606]

[34]   X. Wu, Y. Liu, X. Li, P. Wen, Y. Zhang, Y. Long, X. Wang, Y. Guo, F. Xing, and J. Gao, "Preparation of aligned porous gelatin scaffolds by unidirectional freeze-drying method", *Acta Biomater.,* vol. 6, no. 3, pp. 1167-1177, 2010.
[http://dx.doi.org/10.1016/j.actbio.2009.08.041] [PMID: 19733699]

[35]   Y. Zhang, L. Hu, J. Han, Z. Jiang, and Y. Zhou, "Soluble starch scaffolds with uniaxial aligned channel structure for *in situ* synthesis of hierarchically porous silica ceramics"., *Microporous Mesoporous Mater.,* vol. 130, no. 1-3, pp. 327-332, 2010.
[http://dx.doi.org/10.1016/j.micromeso.2009.11.030]

[36]   Y. Zhou, S. Fu, Y. Pu, S. Pan, and A.J. Ragauskas, "Preparation of aligned porous chitin nanowhisker foams by directional freeze–casting technique", *Carbohydr. Polym.,* vol. 112, pp. 277-283, 2014.
[http://dx.doi.org/10.1016/j.carbpol.2014.05.062] [PMID: 25129745]

[37]   S.R. Mukai, H. Nishihara, and H. Tamon, "Formation of monolithic silica gel microhoneycombs (SMHs) using pseudosteady state growth of microstructural ice crystals", *Chem. Commun. (Camb.),* no. 7, pp. 874-875, 2004.
[http://dx.doi.org/10.1039/b316597c] [PMID: 15045107]

[38]   S.R. Mukai, H. Nishihara, and H. Tamon, "Morphology maps of ice-templated silica gels derived from silica hydrogels and hydrosols", *Microporous Mesoporous Mater.,* vol. 116, no. 1-3, pp. 166-170, 2008.
[http://dx.doi.org/10.1016/j.micromeso.2008.03.031]

[39]   H. Nishihara, S.R. Mukai, and H. Tamon, "Preparation of resorcinol–formaldehyde carbon cryogel microhoneycombs", *Carbon,* vol. 42, no. 4, pp. 899-901, 2004.
[http://dx.doi.org/10.1016/j.carbon.2004.01.075]

[40]   M.C. Gutiérrez, M. Jobbágy, N. Rapún, M.L. Ferrer, and F. del Monte, "A biocompatible bottom-up route for the preparation of hierarchical biohybrid materials", *Adv. Mater.,* vol. 18, no. 9, pp. 1137-1140, 2006.
[http://dx.doi.org/10.1002/adma.200502550]

[41]   L. Qian, A. Ahmed, A. Foster, S.P. Rannard, A.I. Cooper, and H. Zhang, "Systematic tuning of pore morphologies and pore volumes in macroporous materials by freezing", *J. Mater. Chem.,* vol. 19, no. 29, pp. 5212-5219, 2009.
[http://dx.doi.org/10.1039/b903461g]

[42]   H. Nishihara, S. Iwamura, and T. Kyotani, "Synthesis of silica-based porous monoliths with straight nanochannels using an ice-rod nanoarray as a template", *J. Mater. Chem.,* vol. 18, no. 31, pp. 3662-3670, 2008.
[http://dx.doi.org/10.1039/b806005c]

[43]   S. Deville, "Freeze-casting of porous ceramics: a review of current achievements and issues", *Adv. Eng. Mater.,* vol. 10, no. 3, pp. 155-169, 2008.
[http://dx.doi.org/10.1002/adem.200700270]

[44]   S. Deville, E. Saiz, and A.P. Tomsia, "Ice-templated porous alumina structures", *Acta Mater.,* vol. 55, no. 6, pp. 1965-1974, 2007.
        [http://dx.doi.org/10.1016/j.actamat.2006.11.003]

[45]   E. Munch, E. Saiz, A.P. Tomsia, and S. Deville, "Architectural control of freeze-cast ceramics through additives and templating", *J. Am. Ceram. Soc.,* vol. 92, no. 7, pp. 1534-1539, 2009.
        [http://dx.doi.org/10.1111/j.1551-2916.2009.03087.x]

[46]   C. Tallón, R. Moreno, and I.M. Nieto, "Shaping of porous alumina bodies by freeze casting", *Adv. Appl. Ceramics,* vol. 108, no. 5, pp. 307-313, 2009.
        [http://dx.doi.org/10.1179/174367608X369280]

[47]   J. Han, L. Hu, Y. Zhang, and Y. Zhou, "Fabrication of Ceramics with Complex Porous Structures by the Impregnate-Freeze-Casting Process", *J. Am. Ceram. Soc.,* vol. 92, no. 9, pp. 2165-2167, 2009.
        [http://dx.doi.org/10.1111/j.1551-2916.2009.03168.x]

[48]   H.J. Yoon, U.C. Kim, J.H. Kim, Y.H. Koh, W.Y. Choi, and H.E. Kim, "Macroporous alumina ceramics with aligned microporous walls by unidirectionally freezing foamed aqueous ceramic suspensions", *J. Am. Ceram. Soc.,* vol. 93, pp. 1580-1582, 2010.
        [http://dx.doi.org/10.1111/j.1551-2916.2010.03627.x]

[49]   E. Munch, M.E. Launey, D.H. Alsem, E. Saiz, A.P. Tomsia, and R.O. Ritchie, "Tough, bio-inspired hybrid materials", *Science,* vol. 322, no. 5907, pp. 1516-1520, 2008.
        [http://dx.doi.org/10.1126/science.1164865] [PMID: 19056979]

[50]   M.E. Launey, E. Munch, D.H. Alsem, H.B. Barth, E. Saiz, A.P. Tomsia, and R.O. Ritchie, "Designing highly toughened hybrid composites through nature-inspired hierarchical complexity", *Acta Mater.,* vol. 57, no. 10, pp. 2919-2932, 2009.
        [http://dx.doi.org/10.1016/j.actamat.2009.03.003]

[51]   M.E. Launey, E. Munch, D.H. Alsem, E. Saiz, A.P. Tomsia, and R.O. Ritchie, "A novel biomimetic approach to the design of high-performance ceramic–metal composites", *J. R. Soc. Interface,* vol. 7, no. 46, pp. 741-753, 2010.
        [http://dx.doi.org/10.1098/rsif.2009.0331] [PMID: 19828498]

[52]   M. Jarcho, J.F. Kay, K.I. Gumaer, R.H. Doremus, and H.P. Drobeck, "Tissue, cellular and subcellular events at a bone-ceramic hydroxylapatite interface", *J. Bioeng.,* vol. 1, no. 2, pp. 79-92, 1977.
        [PMID: 355244]

[53]   M. Matsusaki, K. Kadowaki, K. Tateishi, C. Higuchi, W. Ando, D.A. Hart, Y. Tanaka, Y. Take, M. Akashi, H. Yoshikawa, and N. Nakamura, "Scaffold-free tissue-engineered construct-hydroxyapatite composites generated by an alternate soaking process: potential for repair of bone defects", *Tissue Eng. Part A,* vol. 15, no. 1, pp. 55-63, 2009.
        [http://dx.doi.org/10.1089/ten.tea.2007.0424] [PMID: 18673091]

[54]   Q. Fu, M.N. Rahaman, B.S. Bal, and R.F. Brown, "*In vitro* cellular response to hydroxyapatite scaffolds with oriented pore architectures"., *Mater. Sci. Eng. C,* vol. 29, no. 7, pp. 2147-2153, 2009.
        [http://dx.doi.org/10.1016/j.msec.2009.04.016]

[55]   M.N. Rahaman, and Q. Fu, "Manipulation of porous bioceramic microstructures by freezing of suspensions containing binary mixtures of solvents", *J. Am. Ceram. Soc.,* vol. 91, no. 12, pp. 4137-4140, 2008.
        [http://dx.doi.org/10.1111/j.1551-2916.2008.02795.x]

[56]   S. Blindow, M. Pulkin, D. Koch, G. Grathwohl, and K. Rezwan, "Hydroxyapatite/SiO_2 composites *via* freeze casting for bone tissue engineering"., *Adv. Eng. Mater.,* vol. 11, no. 11, pp. 875-884, 2009.
        [http://dx.doi.org/10.1002/adem.200900208]

[57]   S. Deville, E. Saiz, R.K. Nalla, and A.P. Tomsia, "Freezing as a path to build complex composites", *Science,* vol. 311, no. 5760, pp. 515-518, 2006.
        [http://dx.doi.org/10.1126/science.1120937] [PMID: 16439659]

[58] T. Yoshida, M. Kikuchi, Y. Koyama, and K. Takakuda, "Osteogenic activity of MG63 cells on bone-like hydroxyapatite/collagen nanocomposite sponges", *J. Mater. Sci. Mater. Med.,* vol. 21, no. 4, pp. 1263-1272, 2010.
[http://dx.doi.org/10.1007/s10856-009-3938-3] [PMID: 19924517]

[59] J. Han, Z. Zhou, R. Yin, D. Yang, and J. Nie, "Alginate–chitosan/hydroxyapatite polyelectrolyte complex porous scaffolds: Preparation and characterization", *Int. J. Biol. Macromol.,* vol. 46, no. 2, pp. 199-205, 2010.
[http://dx.doi.org/10.1016/j.ijbiomac.2009.11.004] [PMID: 19941890]

[60] S.E. Kim, H.W. Choi, H.J. Lee, J.H. Chang, J. Choi, K.J. Kim, H.J. Lim, Y.J. Jun, and S.C. Lee, "Designing a highly bioactive 3D bone-regenerative scaffold by surface immobilization of nano-hydroxyapatite", *J. Mater. Chem.,* vol. 18, no. 41, pp. 4994-5001, 2008.
[http://dx.doi.org/10.1039/b810328c]

[61] E. Tkalya, M. Ghislandi, A. Alekseev, C. Koning, and J. Loos, "Latex-based concept for the preparation of graphene-based polymer nanocomposites", *J. Mater. Chem.,* vol. 20, no. 15, pp. 3035-3039, 2010.
[http://dx.doi.org/10.1039/b922604d]

[62] X. Yao, H. Yao, and Y. Li, "Hierarchically aligned porous scaffold by ice-segregation-induced self-assembly and thermally triggered electrostatic self-assembly of oppositely charged thermosensitive microgels", *J. Mater. Chem.,* vol. 19, no. 36, pp. 6516-6520, 2009.
[http://dx.doi.org/10.1039/b909059b]

[63] J.W. Kim, K. Tazumi, R. Okaji, and M. Ohshima, "Honeycomb monolith-structured silica with highly ordered, three-dimensionally interconnected macroporous walls", *Chem. Mater.,* vol. 21, no. 15, pp. 3476-3478, 2009.
[http://dx.doi.org/10.1021/cm901265y]

[64] C.A.L. Colard, R.A. Cave, N. Grossiord, J.A. Covington, and S.A.F. Bon, "Conducting nanocomposite polymer foams from ice-crystal-templated assembly of mixtures of colloids", *Adv. Mater.,* vol. 21, no. 28, pp. 2894-2898, 2009.
[http://dx.doi.org/10.1002/adma.200803007]

[65] C. Lau, M.J. Cooney, and P. Atanassov, "Conductive macroporous composite chitosan-carbon nanotube scaffolds", *Langmuir,* vol. 24, no. 13, pp. 7004-7010, 2008.
[http://dx.doi.org/10.1021/la8005597] [PMID: 18517231]

[66] S.M. Kwon, H.S. Kim, and H.J. Jin, "Multiwalled carbon nanotube cryogels with aligned and non-aligned porous structures", *Polymer (Guildf.),* vol. 50, no. 13, pp. 2786-2792, 2009.
[http://dx.doi.org/10.1016/j.polymer.2009.04.056]

[67] M.C. Gutiérrez, Z.Y. García-Carvajal, M.J. Hortigüela, L. Yuste, F. Rojo, M.L. Ferrer, and F. del Monte, "Biocompatible MWCNT scaffolds for immobilization and proliferation of E. coli", *J. Mater. Chem.,* vol. 17, no. 29, pp. 2992-2995, 2007.
[http://dx.doi.org/10.1039/B707504A]

[68] A. Abarrategi, M.C. Gutiérrez, C. Moreno-Vicente, M.J. Hortigüela, V. Ramos, J.L. López-Lacomba, M.L. Ferrer, and F. del Monte, "Multiwall carbon nanotube scaffolds for tissue engineering purposes", *Biomaterials,* vol. 29, no. 1, pp. 94-102, 2008.
[http://dx.doi.org/10.1016/j.biomaterials.2007.09.021] [PMID: 17928048]

[69] M.C. Gutiérrez, M.J. Hortigüela, J.M. Amarilla, R. Jiménez, M.L. Ferrer, and F. del Monte, "Macroporous 3D architectures of self-assembled MWCNT surface decorated with Pt nanoparticles as anodes for a direct methanol fuel cell", *J. Phys. Chem. C,* vol. 111, no. 15, pp. 5557-5560, 2007.
[http://dx.doi.org/10.1021/jp0714365]

[70] C. Hong, X. Zhang, J. Han, J. Du, and W. Zhang, "Camphene-based freeze-cast ZrO$_2$ foam with high compressive strength", *Mater. Chem. Phys.,* vol. 119, no. 3, pp. 359-362, 2010.
[http://dx.doi.org/10.1016/j.matchemphys.2009.10.031]

[71] H.D. Jung, S.W. Yook, H.E. Kim, and Y.H. Koh, "Fabrication of titanium scaffolds with porosity and pore size gradients by sequential freeze casting", *Mater. Lett.,* vol. 63, no. 17, pp. 1545-1547, 2009.
[http://dx.doi.org/10.1016/j.matlet.2009.04.012]

[72] B.H. Yoon, E.J. Lee, H.E. Kim, and Y.H. Koh, "Highly aligned porous silicon carbide ceramics by freezing polycarbosilane/camphene solution", *J. Am. Ceram. Soc.,* vol. 90, no. 6, pp. 1753-1759, 2007.
[http://dx.doi.org/10.1111/j.1551-2916.2007.01703.x]

[73] I. Gerçek, R.S. Tığlı, and M. Gümüşderelioğlu, "A novel scaffold based on formation and agglomeration of PCL microbeads by freeze-drying", *J. Biomed. Mater. Res. A,* vol. 86A, no. 4, pp. 1012-1022, 2008.
[http://dx.doi.org/10.1002/jbm.a.31723] [PMID: 18067167]

[74] Q. Hou, D.W. Grijpma, and J. Feijen, "Preparation of interconnected highly porous polymeric structures by a replication and freeze-drying process", *J. Biomed. Mater. Res.,* vol. 67B, no. 2, pp. 732-740, 2003.
[http://dx.doi.org/10.1002/jbm.b.10066] [PMID: 14598400]

[75] J.W. Kim, K. Taki, S. Nagamine, and M. Ohshima, "Preparation of poly(L-lactic acid) honeycomb monolith structure by unidirectional freezing and freeze-drying", *Chem. Eng. Sci.,* vol. 63, no. 15, pp. 3858-3863, 2008.
[http://dx.doi.org/10.1016/j.ces.2008.04.036]

[76] J.W. Kim, K. Taki, S. Nagamine, and M. Ohshima, "Preparation of porous poly(L-lactic acid) honeycomb monolith structure by phase separation and unidirectional freezing", *Langmuir,* vol. 25, no. 9, pp. 5304-5312, 2009.
[http://dx.doi.org/10.1021/la804057e] [PMID: 19290649]

[77] H. Zhang, and A.I. Cooper, "Synthesis and applications of emulsion-templated porous materials", *Soft Matter,* vol. 1, no. 2, pp. 107-113, 2005.
[http://dx.doi.org/10.1039/b502551f] [PMID: 32646082]

[78] K. Whang, T.K. Goldstick, and K.E. Healy, "A biodegradable polymer scaffold for delivery of osteotropic factors", *Biomaterials,* vol. 21, no. 24, pp. 2545-2551, 2000.
[http://dx.doi.org/10.1016/S0142-9612(00)00122-8] [PMID: 11071604]

[79] O. Grinberg, I. Binderman, H. Bahar, and M. Zilberman, "Highly porous bioresorbable scaffolds with controlled release of bioactive agents for tissue-regeneration applications", *Acta Biomater.,* vol. 6, no. 4, pp. 1278-1287, 2010.
[http://dx.doi.org/10.1016/j.actbio.2009.10.047] [PMID: 19887123]

[80] G. Rohman, S.C. Baker, J. Southgate, and N.R. Cameron, "Heparin functionalisation of porous PLGA scaffolds for controlled, biologically relevant delivery of growth factors for soft tissue engineering", *J. Mater. Chem.,* vol. 19, no. 48, pp. 9265-9273, 2009.
[http://dx.doi.org/10.1039/b911625g]

[81] P.X. Ma, and R. Zhang, "Synthetic nano-scale fibrous extracellular matrix", *J. Biomed. Mater. Res.,* vol. 46, no. 1, pp. 60-72, 1999.
[http://dx.doi.org/10.1002/(SICI)1097-4636(199907)46:1<60::AID-JBM7>3.0.CO;2-H] [PMID: 10357136]

[82] X. Liu, Y. Won, and P.X. Ma, "Porogen-induced surface modification of nano-fibrous poly(l-lactic acid) scaffolds for tissue engineering", *Biomaterials,* vol. 27, no. 21, pp. 3980-3987, 2006.
[http://dx.doi.org/10.1016/j.biomaterials.2006.03.008] [PMID: 16580063]

[83] L. Qian, E. Willneff, and H. Zhang, "A novel route to polymeric sub-micron fibers and their use as templates for inorganic structures", *Chem. Commun. (Camb.),* no. 26, pp. 3946-3948, 2009.
[http://dx.doi.org/10.1039/b905130a] [PMID: 19662261]

[84] L. Qian, and H. Zhang, "Green synthesis of chitosan-based nanofibers and their applications", *Green Chem.,* vol. 12, no. 7, pp. 1207-1214, 2010.

[http://dx.doi.org/10.1039/b927125b]

[85]   W. Mahler, and M.F. Bechtold, "Freeze-formed silica fibres", *Nature,* vol. 285, no. 5759, pp. 27-28, 1980.
[http://dx.doi.org/10.1038/285027a0]

[86]   T. Fukasawa, Z.Y. Deng, M. Ando, T. Ohji, and S. Kanzaki, "Synthesis of porous silicon nitride with unidirectionally aligned channels using freeze-drying process", *J. Am. Ceram. Soc.,* vol. 85, no. 9, pp. 2151-2155, 2002.
[http://dx.doi.org/10.1111/j.1151-2916.2002.tb00426.x]

[87]   S.R. Mukai, H. Nishihara, S. Shichi, and H. Tamon, Preparation of porous TiO$_2$ cryogel fibers through unidirectional freezing of hydrogel followed by freeze-drying., *Chem. Mater.,* vol. 16, no. 24, pp. 4987-4991, 2004.
[http://dx.doi.org/10.1021/cm0491328]

[88]   H. Zhang, J.Y. Lee, A. Ahmed, I. Hussain, and A.I. Cooper, "Freeze-align and heat-fuse: microwires and networks from nanoparticle suspensions", *Angew. Chem. Int. Ed.,* vol. 47, no. 24, pp. 4573-4576, 2008.
[http://dx.doi.org/10.1002/anie.200705512] [PMID: 18446918]

[89]   J. Yan, Z. Chen, J. Jiang, L. Tan, and X.C. Zeng, "Free-standing all-nanoparticle thin fibers: a novel nanostructure bridging zero-and one-dimensional nanoscale features", *Adv. Mater.,* vol. 21, no. 3, pp. 314-319, 2009.
[http://dx.doi.org/10.1002/adma.200801130]

[90]   Q. Shi, Z. An, C.K. Tsung, H. Liang, N. Zheng, C.J. Hawker, and G.D. Stucky, "Ice-Templating of Core/Shell Microgel Fibers through 'Bricks-and-Mortar' Assembly**", *Adv. Mater.,* vol. 19, no. 24, pp. 4539-4543, 2007.
[http://dx.doi.org/10.1002/adma.200700819]

[91]   Q. Shi, H. Liang, D. Feng, J. Wang, and G.D. Stucky, "Porous carbon and carbon/metal oxide microfibers with well-controlled pore structure and interface", *J. Am. Chem. Soc.,* vol. 130, no. 15, pp. 5034-5035, 2008.
[http://dx.doi.org/10.1021/ja800376t] [PMID: 18355006]

[92]   J. Hu, K.P. Johnston, and R.O. Williams III, "Rapid dissolving high potency danazol powders produced by spray freezing into liquid process", *Int. J. Pharm.,* vol. 271, no. 1-2, pp. 145-154, 2004.
[http://dx.doi.org/10.1016/j.ijpharm.2003.11.003] [PMID: 15129981]

[93]   T.L. Rogers, A.C. Nelsen, M. Sarkari, T.J. Young, K.P. Johnston, and R.O. Williams III, "Enhanced aqueous dissolution of a poorly water soluble drug by novel particle engineering technology: spray-freezing into liquid with atmospheric freeze-drying", *Pharm. Res.,* vol. 20, no. 3, pp. 485-493, 2003.
[http://dx.doi.org/10.1023/A:1022628826404] [PMID: 12669973]

[94]   Z. Yu, K.P. Johnston, and R.O. Williams III, "Spray freezing into liquid *versus* spray-freeze drying: Influence of atomization on protein aggregation and biological activity", *Eur. J. Pharm. Sci.,* vol. 27, no. 1, pp. 9-18, 2006.
[http://dx.doi.org/10.1016/j.ejps.2005.08.010] [PMID: 16188431]

[95]   J.D. Engstrom, D.T. Simpson, C. Cloonan, E.S. Lai, R.O. Williams III, G. Barrie Kitto, and K.P. Johnston, "Stable high surface area lactate dehydrogenase particles produced by spray freezing into liquid nitrogen", *Eur. J. Pharm. Biopharm.,* vol. 65, no. 2, pp. 163-174, 2007.
[http://dx.doi.org/10.1016/j.ejpb.2006.08.002] [PMID: 17027245]

[96]   Z. Yu, A.S. Garcia, K.P. Johnston, and R.O. Williams III, "Spray freezing into liquid nitrogen for highly stable protein nanostructured microparticles", *Eur. J. Pharm. Biopharm.,* vol. 58, no. 3, pp. 529-537, 2004.
[http://dx.doi.org/10.1016/j.ejpb.2004.04.018] [PMID: 15451527]

[97]   J.D. Engstrom, D.T. Simpson, E.S. Lai, R.O. Williams III, and K.P. Johnston, "Morphology of protein particles produced by spray freezing of concentrated solutions", *Eur. J. Pharm. Biopharm.,* vol. 65, no.

2, pp. 149-162, 2007.
[http://dx.doi.org/10.1016/j.ejpb.2006.08.005] [PMID: 17010582]

[98]   T. Niwa, H. Shimabara, M. Kondo, and K. Danjo, "Design of porous microparticles with single-micron size by novel spray freeze-drying technique using four-fluid nozzle", *Int. J. Pharm.,* vol. 382, no. 1-2, pp. 88-97, 2009.
[http://dx.doi.org/10.1016/j.ijpharm.2009.08.011] [PMID: 19686828]

[99]   W.T. Leach, D.T. Simpson, T.N. Val, E.C. Anuta, Z. Yu, R.O. Williams III, and K.P. Johnston, "Uniform encapsulation of stable protein nanoparticles produced by spray freezing for the reduction of burst release", *J. Pharm. Sci.,* vol. 94, no. 1, pp. 56-69, 2005.
[http://dx.doi.org/10.1002/jps.20209] [PMID: 15761930]

[100] J.P. Hindmarsh, A.B. Russell, and X.D. Chen, "Fundamentals of the spray freezing of foods—microstructure of frozen droplets", *J. Food Eng.,* vol. 78, no. 1, pp. 136-150, 2007.
[http://dx.doi.org/10.1016/j.jfoodeng.2005.09.011]

[101] C.G. Gwie, R.J. Griffiths, D.T. Cooney, M.L. Johns, and D.I. Wilson, "Microstructures formed by spray freezing of food fats", *J. Am. Oil Chem. Soc.,* vol. 83, no. 12, pp. 1053-1062, 2006.
[http://dx.doi.org/10.1007/s11746-006-5162-3]

[102] S. Lee, J. Lee, Y. Park, J. Wee, and K. Lee, "Complete oxidation of methane and CO at low temperature over $LaCoO_3$ prepared by spray-freezing/freeze-drying method", *Catal. Today,* vol. 117, no. 1-3, pp. 376-381, 2006.
[http://dx.doi.org/10.1016/j.cattod.2006.05.035]

[103] W. Yin, and M.Z. Yates, "Encapsulation and sustained release from biodegradable microcapsules made by emulsification/freeze drying and spray/freeze drying", *J. Colloid Interface Sci.,* vol. 336, no. 1, pp. 155-161, 2009.
[http://dx.doi.org/10.1016/j.jcis.2009.03.065] [PMID: 19423128]

[104] S. Hyuk Im, U. Jeong, and Y. Xia, "Polymer hollow particles with controllable holes in their surfaces", *Nat. Mater.,* vol. 4, no. 9, pp. 671-675, 2005.
[http://dx.doi.org/10.1038/nmat1448] [PMID: 16086022]

[105] W. Yin, and M.Z. Yates, "Effect of interfacial free energy on the formation of polymer microcapsules by emulsification/freeze-drying", *Langmuir,* vol. 24, no. 3, pp. 701-708, 2008.
[http://dx.doi.org/10.1021/la7022693] [PMID: 18173290]

[106] H. Zhang, D. Edgar, P. Murray, A. Rak-Raszewska, L. Glennon-Alty, and A.I. Cooper, "Synthesis of porous microparticles with aligned porosity", *Adv. Funct. Mater.,* vol. 18, no. 2, pp. 222-228, 2008.
[http://dx.doi.org/10.1002/adfm.200701309]

[107] H. Zhang, D. Wang, R. Butler, N.L. Campbell, J. Long, B. Tan, D.J. Duncalf, A.J. Foster, A. Hopkinson, D. Taylor, D. Angus, A.I. Cooper, and S.P. Rannard, "Formation and enhanced biocidal activity of water-dispersable organic nanoparticles", *Nat. Nanotechnol.,* vol. 3, no. 8, pp. 506-511, 2008.
[http://dx.doi.org/10.1038/nnano.2008.188] [PMID: 18685640]

[108] A. Ahmed, N. Grant, L. Qian, and H. Zhang, "Formation of Organic Nanoparticles by Freeze-Drying and Their Controlled Release", *Nanosci. Nanotechnol. Lett.,* vol. 1, no. 3, pp. 185-189, 2009.
[http://dx.doi.org/10.1166/nnl.2009.1036]

[109] H. Zhang, and A.I. Cooper, "Thermoresponsive "particle pumps": activated release of organic nanoparticles from open-cell macroporous polymers", *Adv. Mater.,* vol. 19, no. 18, pp. 2439-2444, 2007.
[http://dx.doi.org/10.1002/adma.200602794]

[110] H. Zhang, "Introduction to freeze-drying and ice templating", In: *Ice Templating and Freeze-Drying for Porous Materials and Their Applications.,* A. Jungbauer, Ed., Wiley-VCH: Weinheim, 2018, pp. 1-27.

[111] U. K. Das, R. Bordoloi, and S. Ganguly, "Freeze-drying technique and its wide application in biomedical and pharmaceutical sciences", *Res. J. Chem. Environ. Sci ,* 2014.

[112] T.M. Oyinloye, and W.B. Yoon, "Effect of freeze-drying on quality and grinding process of food produce: A review", *Processes (Basel),* vol. 8, no. 3, p. 354, 2020.
[http://dx.doi.org/10.3390/pr8030354]

[113] A. Haider, S. Haider, M. Rao Kummara, T. Kamal, A.A.A. Alghyamah, F. Jan Iftikhar, B. Bano, N. Khan, M. Amjid Afridi, S. Soo Han, A. Alrahlah, and R. Khan, "Advances in the scaffolds fabrication techniques using biocompatible polymers and their biomedical application: A technical and statistical review", *J. Saudi Chem. Soc.,* vol. 24, no. 2, pp. 186-215, 2020.
[http://dx.doi.org/10.1016/j.jscs.2020.01.002]

[114] S.S. Silva, N.M. Oliveira, M.B. Oliveira, D.P.S. da Costa, D. Naskar, J.F. Mano, S.C. Kundu, and R.L. Reis, "Fabrication and characterization of Eri silk fibers-based sponges for biomedical application", *Acta Biomater.,* vol. 32, pp. 178-189, 2016.
[http://dx.doi.org/10.1016/j.actbio.2016.01.003] [PMID: 26766632]

CHAPTER 4

Centrifugal and Solution Blow Spinning Techniques in Tissue Engineering

Muhammad Umar Aslam Khan[1,*], Saiful Izwan Abd. Razak[1,2], Rawaiz Khan[1,3], Sajjad Haider[4], Mohsin Ali Raza[5], Rashid Amin[6], Saqlain A. Shah[7] and Anwarul Hasan[8,9]

[1] *Department of Polymer Engineering, School of Chemical and Energy, Faculty of Engineering, Universiti Teknologi Malaysia, 81310 UTM Johor Bahru, Johor, Malaysia*

[2] *Centre for Advanced Composite Materials, Universiti Teknologi Malaysia, 81300 Skudai, Johor, Malaysia*

[3] *Department of Polymer Engineering, School of Chemical and Energy, Faculty of Engineering, Universiti Teknologi Malaysia, 81310 UTM Johor Bahru, Johor, Malaysia*

[4] *Department of Chemical Engineering, College of Engineering, King Saud University, P.O. Box 800, Riyadh 11421, KSA, Saudi Arabia*

[5] *Department of Metallurgy and Materials Engineering, University of the Punjab, Lahore, Pakistan*

[6] *Department of Biology, College of Sciences, University of Hafr Al Batin, Hafar Al-Batin 39524, Saudi Arabia*

[7] *Department of Physics, Forman Christian College (University) Lahore, Pakistan*

[8] *Department of Mechanical and Industrial Engineering, College of Engineering, Qatar University, Doha, Qatar*

[9] *Biomedical Research Center, Qatar University, Doha, Qatar*

Abstract: Nanofibers are a necessary source for fibrous materials and other useful applications such as tissue engineering, filtration, safety fabrics, batteries for the production of nanofibers so far. However, due to its low production rate, the wide commercial use of electrospinning is minimal. Almost all nanofiber fabrication techniques (*e.g.*, melt blowing, two-component processes, phase splitting, template synthesis, and self-assembly, *etc.*) are used to produce nanofibers from a limited number of polymeric materials. Centrifugal spinning (CS) and solution blow spinning (SBS) are advanced replacement processes to fabricate nanofibers with full performance from various low-cost raw materials. This chapter focuses on a comprehensive overview of CS and SBS as well as various other aspects of the fabrication of nanofibers.

* **Corresponding author Muhammad Umar Aslam Khan:** BioInspired Device and Tissue Engineering Research Group, School of Biomedical Engineering and Health Sciences, Faculty of Engineering, Universiti Teknologi Malaysia, 81300 Skudai, Johor, Malaysia; E-mail: umar007khan@gamil.com

Keywords: Centrifugal spinning, Nanofibers, Solution blow spinning, Tissue engineering.

INTRODUCTION

Electrospinning is a well-known technique for the production of nanofibers to prepare scaffolds for tissue engineering. Various polymers, including synthetic and natural polymers [1, 2], can be used to develop scaffolds for tissue engineering using different techniques. The specific surface area, porosity, biomimetic structure of the extracellular matrix (ECM), and improved biocompatibility are all advantages of scaffolds fabricated by electrospinning for tissue engineering. The ECM can associate, release and trigger signalling molecules and stimulate cell response [3, 4]. Scaffold nanofibers can be filled with various bioactive compounds such as proteins, peptides and small molecule drugs to functionalize the scaffolds and promote cell adherence, differentiation and proliferation. As a result, electrospun scaffolds offer significant advantages in biomimetic ECM processes and packaging of bioactive materials. Electrospun scaffolds are also used for drug delivery. In recent years, interest in submicron fibre mats for tissue engineering applications has increased. They provide a good surface area for cell adhesion and mimic the fibrillar structure of native ECM. The porosity of the mat favours the diffusion of nutrients, leading to rapid cell proliferation [5, 6]. The presence of fibres in the form of implants often makes them easy to handle during surgery. Submicron fibres for tissue engineering applications are currently being developed primarily using electrospinning technology, but the process has several limitations. Limitations of the process include low efficiency, limited protective features, and poor alignment and reproducibility of fibre morphology. In addition, the electrospinning process is an environmentally sensitive fibre production technique where even a small change in humidity affects fibre production and consistency. For tissue engineering applications, a new method that can overcome the above limitations is highly desirable [7].

Due to their high rate and easy production of fibres with different morphologies, fibres produced by centrifugal force have attracted the attention of scientists in recent years [8]. Polymer concentration, solvent selection and evaporation rate, spinneret rotation speed and collector unit distance from the spinneret are important parameters that contribute to the improvement of fibre quality [9]. By changing the spinneret configuration and the type of fibre collection, fibres with different morphologies can be produced. An aligned fibre mat can be easily obtained to develop biomaterials for biomedicine [10].

Micro/nanofibers are widely used in both nature and industry due to their exceptional properties and utility. These fibres are now being used in tissue bandages. These tissue bandages have high filtration efficiency, optical sensor, large surface area, rough surface and intense interfacial interaction [11]. Various techniques can be used to develop continuous microfibers, such as melt spinning, wet spinning, coaxial spinning, electrospinning and blow spinning. The biomaterial based on these fibre fabrication techniques has certain limitations and suffers from non-uniformity of shape and size [12]. Various hardware issues, dynamic configurations and low throughput are industrial obstacles. Consequently, it is a major limitation to fabricate continuous submicron/nanofibers with tunable and uniform morphology [13]. It is true that electrospinning is widely used in biomedical, energy, environmental, catalysis, *etc*. But as mentioned earlier, the process has its limitations when it comes to the use of high static voltages, safety and equipment [14] and the conductivity of the polymer solution. This limits the spinnability of the non-conductive polymers [15]. At the same time, the most important argument is that the nanofibers produced by electrospinning have poor yield. It is difficult to produce large quantities, which significantly hinders commercial production. CS and SBS have been proposed to overcome these limitations and eliminate the safety concerns associated with the electrospinning process. Therefore, there is a need to develop a new solution for nanofiber development that overcomes the limitations of the above approaches [16]. In this chapter, alternative methods using centrifugal spinning and solution blow spinning are discussed to economically fabricate nanofibers from various materials with maximum production. CS and SBS prevent high voltage as a simple and scalable method to fabricate nanofibers for various biomedical applications.

CONVENTIONAL FABRICATION TECHNIQUES

The fibrous material can be produced by a number of conventional techniques. In the late 19th century, Lord Rayleigh produced nanofibers through a technique known as electrospinning. This technique has the ability and potential to produce nanofibers with specific properties. Spun nanofibers have numerous advantages, including an extremely high surface-to-volume ratio, adjustable porosity, formability, pore size and shape, and the ability to control the morphology and size of the nanofiber to achieve desired properties. Nanofibers have unique advantages as they are used as basic structural building blocks in living organisms [17]. In addition to their use in tissue engineering, nanofibers prepared from biopolymers and synthetic polymers are also widely used in drug discovery [18, 19]. In the following, we will discuss some of the known conventional techniques (Fig. **1**) for the fabrication of nanofibers.

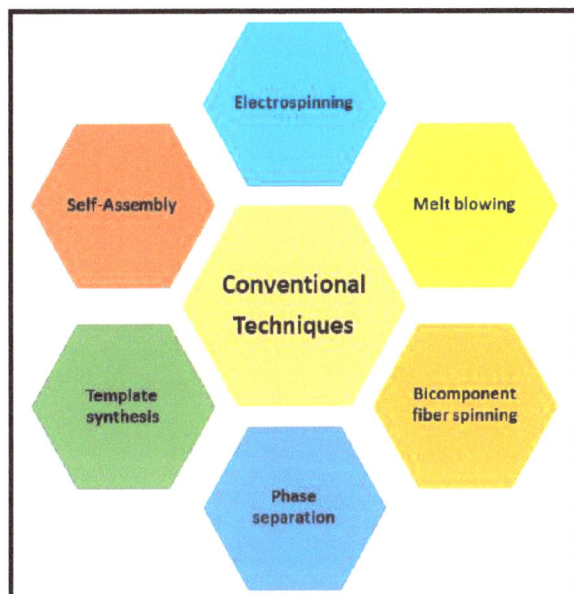

Fig. (1). The conventional technique for the synthesis of nanofibers.

Electrospinning Technique

Electrospinning is a common and simple method for producing nanofibers. A standard single-nozzle electrospinning system consists of a syringe, a metal nozzle, a high-voltage source (DC), and a collector. During electrospinning, a higher voltage difference is applied between the liquid-containing syringe and a metallic collector [20 - 22]. The high potential difference creates a conical droplet through the liquid. Then, the liquid jet is shaped to stretch after reaching the critical point of the voltage. A stretched and whipped electrically charged jet was produced. The diameter of the whipped electrically charged jet decreases from several hundred micrometres to a few nanometers. The whipped electrically charged jet is collected as a nanofiber on the grounded collector after evaporation of the solvent in flight between the needle and the collector [23 - 25]. The schematic diagram of electrospinning is shown in Fig. (**2**).

a)

b)

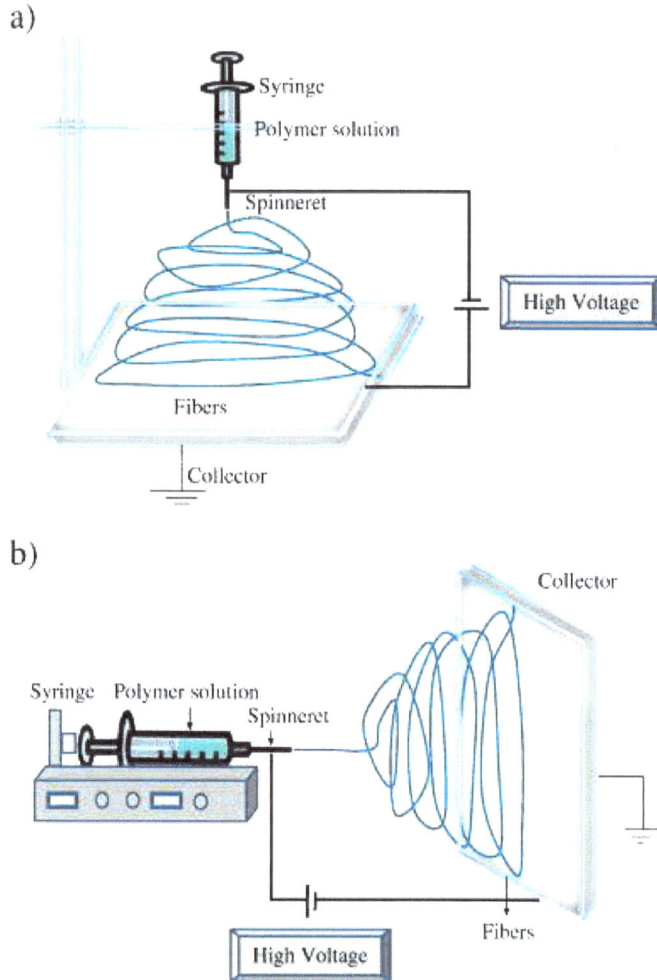

Fig. (2). Illustration of electrospinning reprinted with permission published by N. Bhardwaj *et al*. [26].

The morphology and diameter of the electrospun nanofibers depend on the fabrication parameters. These include: (i) Inherent fluid properties of the polymers such as viscosity, molecular weight and concentration of the polymer solution and (ii) Operating conditions, voltage potential difference, solution flow rate, nozzle diameter, nozzle to collector distance and collector movement.

Therefore, the desired nanofibers can be fabricated by optimising or adjusting the parameters with a controlled diameter of the fibre. The distance during electrospinning and the length of the nanofiber influence the deposition thickness.

Melt Blowing Technique

Melt blowing is typically one of the leading techniques for producing nonwoven fibres with a diameter greater than 1 μm [27]. The polymer is first melted, then polymer jets continuously generated through multiple nozzles. An optimized attenuated heated stream with high velocity and decreasing jet diameters ensures the production of fine fibres [28]. In different orientations, the fibres are gradually collected into a nonwoven web by selectively changing the parameters: Polymer melt, airflow velocities, polymer temperature, intrinsic properties of the polymer, nozzle geometry, *etc.* [29]. The scheme of melt blowing is shown in Fig. (3).

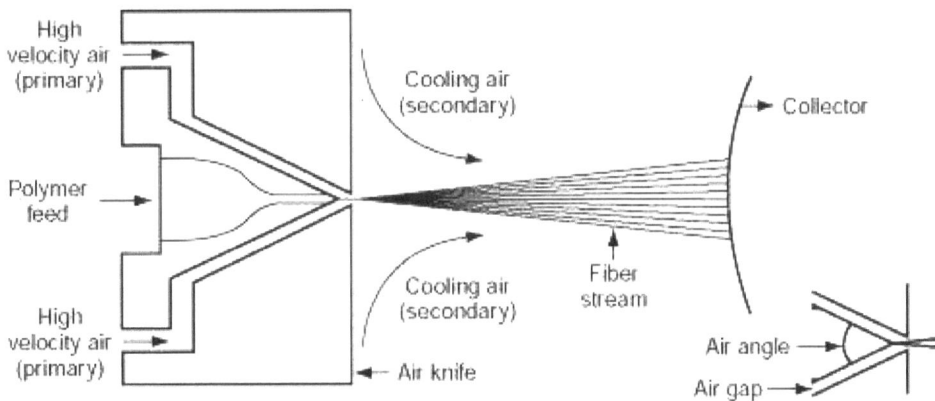

Fig. (3). Illustration of melt blowing under the necessary conditions and reprinted with permission published work of Irwin M. Hutten *et al.* [30].

Bicomponent Fibre Spinning

Bicomponent, hybrid, or multicomponent fibre spinning refers to the technology in which two or more different polymers are combined at the spinneret orifice so that each spun filament contains all the polymer components in separate parts of the cross-section [31, 32]. Bicomponent fibres are mainly sold as bonding components in thermally bonded nonwovens, as self-crimping fibres for the production of textured yarn, or as fibres with the surface functionality of special polymers and additives at a lower cost, depending on the properties of the different polymers. A bicomponent fibre is prepared by the segmented-pie technique from two incompatible polymers that adhere poorly and split into microfibers when subjected to mechanical stress. The segmented pie technique is mainly used to produce microfibers with smaller diameters than conventional melt spinning [33]. The polysaccharide fibres are produced from the melt by segmented pie with a suitable cross-sectional structure. After the bicomponent fibres are made, the components are separated, or one part is omitted to produce nanofibers [34, 35].

Phase Separation Technique

A phase separation mechanism also produces fiber. The common phase separation methods used to produce fibers are polymer dissolution, gelation, phase separation, solvent elimination, and drying. The polymer is first immersed at room temperature or a high temperature to produce a homogeneous solution [36, 37]. The solvent is kept at gelling temperature. After solvent extraction and drying of the matrix, the phase divides, and fibers are formed [38, 39]. The schematic diagram of the phase separation can be seen in Fig. (**4**).

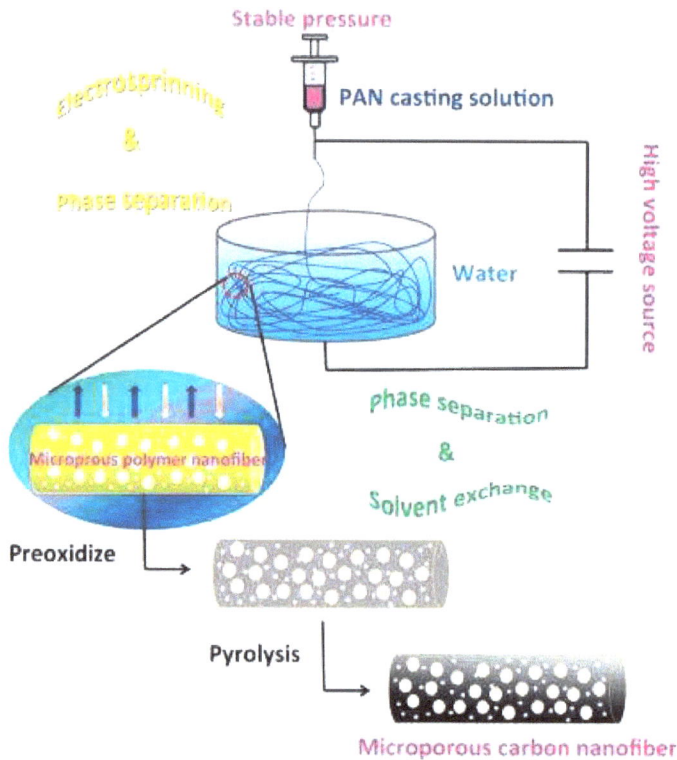

Fig. (4). Illustration of phase separation under the necessary conditions and reprinted with permission published work of Chang Liu *et al.* [40].

Template Synthesis Technique

Nanofibers are fabricated in hollow channels by incorporating porous ceramics into polymeric templates during template synthesis. In the hollow regions of the porous templates, the polymeric nanofibers are fabricated by monomers either chemically or electrochemically. Detached nanofibers are prepared by dissolution or etching after the mould is removed [41, 42]. Nanofibers with the hollow structure are also prepared by this technique. The synthesised polymer is deposited on the inner surfaces of hollow channels. Nanofibers can also be prepared directly from polymer solutions [43, 44]. In this step, polymer solutions enter the hollow channels instead of monomers and solidify into nanofibers after removing the solvent. The nanofibers prepared from polymer solutions usually have a larger diameter than those prepared from monomers. Due to the high viscosity of polymer solutions, it is difficult to use hollow channels to produce nanofibers with smaller diameters [45, 46]. The fabrication of nanofibers using template synthesis is shown in Fig. (**5**).

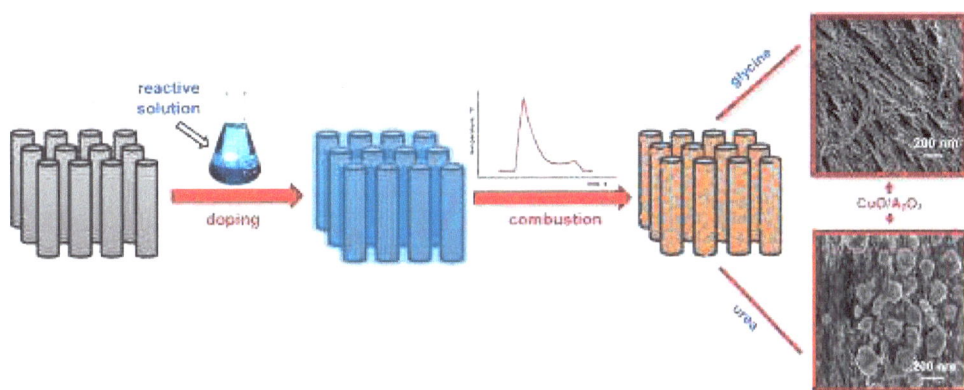

Fig. (5). Illustration of phase separation under the necessary conditions and reprinted with permission published work of Marina Aghayan *et al*. [47].

Self-Assembly Technique

In this technique, small molecules combine *via* intermolecular interactions to form nanofibers. Based on their chemical properties, the small molecules have been assembled into nanofibers by various mechanisms [48, 49]. The most widely used mechanisms are the preparation of hydrogels, which consist of two interpenetrating phases, namely the liquid phase and the solid phase. The hydrogel molecules are self-assembled to produce nanofibers. Elimination of the liquid phase from the hydrogel results in well-dried nanofibers - the hydrogelator molecules to nanofibers from the chemical structure of the hydrogelator [50, 51]. The nanofibers are formed by the self-assembly of weak intermolecular interactions and hydrogen bonds between the molecules of the hydrogelator [52].

Limitation of Conventional Techniques

After a brief comparison of the fabrication rate, material selection, protection, and economics of different techniques used to produce nanofibers, electrospinning was the most commonly used technique to produce nanofibers [53, 54]. However, the main obstacle to widespread industrial use is the low production frequency and the associated significant cost per gram of nanofibers produced. However, an electrospinning system with multiple nozzles can increase the production rate. However, these techniques are not sufficient to produce large quantities of nanofibers for various applications [55, 56]. The laboratory scale electrospinning system is easy to install, low cost and suitable for academic research. However, for large-scale production, the equipment is expensive. Electrospinning can be used to produce nanofibers from various polymeric materials [57].

In addition, few polymeric materials are available for other nanofiber fabrication techniques, including melt blowing, bicomponent, phase splitting, prototype synthesis, and self-assembly. Bicomponent fibre spinning requires two different nanoscale polymers to be distributed on the fibre cross-section. In this process, the rheological properties of the polymers and the choice of polymer materials must remain in balance [58, 59]. There is a significant difference between melt blowing and bicomponent spinning compared to electrospinning. The phase separation technique involves several steps such as polymer gelation, solvent extraction and freeze drying [60, 61]. Small molecule self-assembly leads to the formation of nanofibers from supramolecular hydrogels through Coulomb interactions such as hydrogen bonding and van der Waal forces [62, 63]. Moreover, the above approaches can be influenced by engineering parameters. Centrifugal spinning is also known as rotary spinning or rotary jet spinning. Due to its capacity and feasibility, it has recently attracted interest to produce a large amount of nanofibers with the desired structural morphology cheaply and quickly [64].

Centrifugal Spinning

This technique is not new to the industry and is commonly used to produce larger diameter glass fibres and is also known in the glass fibre industry as the glass wool process. The molten glass stream is passed through the rotating head, which is simply a metal container with a uniform number of holes distributed along the side wall [65]. The temperature of the spinning head is usually maintained at 900-1100°C to keep the glass in its liquid state. The rotating head spins at a high variable speed of 2000-3000 rpm to generate centrifugal forces. The excited molten glass flows through the holes in the side wall of the rotating head to create fine streams of glass. The glass streams exiting the sidewall are attenuated by high-velocity air (gas or steam) and broken down into fine glass fibres that are many centimetres long (> 1 μm) [66, 67]. Once these fibres are made, a binder is sprayed onto the glass fibres. The broken glass fibres enter a 'formation bonnet' located under the spinning head and are deposited in the form of a mat on a conveyor belt. The formation bonnet aims to evenly distribute the broken fibres and form a random arrangement around a glass fibre mat [68, 69]. These glass fibre mats are placed in a heating oven where the binder dries and cures, and they can be cut into desired lengths and widths for various applications. There is no information in the literature on specific processing parameters such as the rotor diameter, the radius of the spinning head, the viscosity of the glass, the airflow and the temperature [39, 70]. Similarly, production was found to be based on centrifugal spinning to produce various fibres. In contrast, centrifugal spinning has generally been used in the glass fibre industry to produce micron-sized glass fibres. Centrifugal spinning is relatively new in the development of polymer fibres, especially polymer nanofibers [71, 72]. The functional capabilities of CS are summarised in Table **1**.

Table 1. Discuss the functional possibilities of the CS and SBS.

Technique	Centrifugal Spinning	Solution Blow Spinning
Diameter of nanofiber	25 nm to 50 mm [73]	40 nm to several mm [74]
Rate of injection (Rate of production)	Up to 1 ml min^{-1} / nozzle [75]	20 mL min^{-2}
Influencing parameters	Viscosity Rotational speed Orifice diameter Evaporation rate Distance [75]	Geometry of nozzle Solution viscosity Solution feeding rate Gas pressure [76]
Commercialization	Yes	Yes
aligned nanofibers production	Yes [10, 77]	Yes [78]
Melt spinning	Yes [79, 80]	Yes [80]
Concentrated polymeric solutions for spinning	Yes [80]	Yes [80]
Voltage requirement	No	No
Composite fibre from polymers and ceramics	Yes [81, 82]	Yes [83]
Fabrication of core-shell nanofiber	Yes [84]	Yes [85]
Production of 3D nanofibrous	Yes [77, 86]	Yes [87]

Centrifugal Spinning Systems

For functional centrifugal spinning systems, the main components are as follows:

i. Rotating head
ii. Nanofiber collector

Rotating Head

The two similar syringes contain the same volume of rotating fluid and are attached to the rotor in the configuration. Two jets of liquid are produced by the centrifugal force of the needles at high speed of rotation. The rotating head is designed to stabilize the weights by spinning two fluid-filled syringes through the needles throughout the centrifugal spinning phase. The shape of the rotating head is cylindrical [52]. The diameter of the fiber is regulated by the diameter of the nozzle and the length. The cylindrical rotating head spins the nanofibers directly from a polymer solution. The heating components are required to melt the polymer and produce the spun nanofiber. Multiple nozzles are added to increase the production and output rate of the nanofibers. Production can be increased by optimizing the rotary head, spheroidal, oblate, or trapezoidal configurations [88, 89].

Nanofiber Collecting System

Different types of collectors are used to obtain nanofibers in different ways. In centrifugal spinning devices used for nanofibers, a spherical metal or plastic collector is applied to the inner wall. The most important parameter for regulating the structure of the resulting nanofibers is the diameter of the circular collector, which determines the size of the nozzle and the collector. The circular collector is ideal for batch production of nanofibers [90, 91]. Continuous production of nanofibers can be achieved by a rotating substrate mounted under a spinning head. The gravity of the nanofibers causes the nanofibers to settle on the substrate and a seamless nanofiber fabric is formed. Suction force or air jets can help in the continuous collection of nanofiber webs. These nanofiber nonwovens have been modified to support various applications such as porous substrates, textiles, paper, membranes and others [92]. These nanofibrous nonwovens should be collected at applied suction forces. The nanofiber materials are collected in a water tank. Then it is processed on a spinning roller to form a continuous nanofiber yarn [93, 94].

Types of Centrifugal Spinning Nanofibers

Nanofibers are produced from polymeric materials by centrifugal spinning. This technique can also be used to obtain nanofibers from other materials, such as carbon and ceramics, by polymer precursors for thermal treatment.

Polymer Nanofibers

Polymeric nanofibers were prepared from a molten polymer solution by centrifugal spinning. It has been reported that nanofibers can be prepared from various polymer solutions, including PEO, PLA, PS, polyacrylic acid (PAA), PVDF, PMMA, and polycaprolactone, which has been shown to be useful for the preparation of nanofibers (PCL) [52, 95 - 97]. These PEO nanofibers with flat morphology have a regular diameter of about 500 nm. The nanofibers can be aligned in the radial direction to the spinning head by the centrifugal spinning process. The resulting nanofibers have good mechanical strength and are suitable for various applications. Centrifugal spinning mainly focuses on the production of nanofibers with polymer solutions. This technique is more practical than melting polymers for the production of nanofibers. The centrifugal spinning technique and the cotton candy machine work by the same mechanism [98]. However, the spinning head of the cotton candy machine has a comparatively large nozzle. The cotton candy machine rotates at a low speed during the preparation of cotton candy in operation. The direct production of nanofibers from the polymer melt avoids harmful solvents and leads to more occupational safety for employees and the overall environment [99]. Centrifugal spinning technology is more attractive because no solvents are required in the production of nanofibers. It is much easier to report structures fabricated using a solvent-free process for commercial biomedical products [100].

Carbon Nanofibers

Spinning precursor polymers such as PAN, polyimide, pitch, *etc.* and then calcining the desired precursor fibres also produces carbon nanofibers (CNFs). The fabrication of nanofibers from carbon precursors is a well-established method for CNFs. There are two methods of heat treatment for fabrication: stabilisation in air and carbonization in an inert environment [101, 102]. The polymer PAN was transferred to carbon during the carbonization process, while PMMA was degraded to create porosity in the nanofiber structures. The produced nanofibers are lightweight, have a larger surface area, considerable electrical conductivity, exceptional adsorption capacity, and are suitable for use against pathogens [103, 104].

Ceramic Nanofibers

Ceramic nanofibers are prepared by calcination of centrifugally spun precursors. Liu and colleagues developed titanium tetrachloride precursor nanofibers containing TiO_2 [105]. Acetylacetone was used as a chelating agent prior to centrifugal spinning to convert titanium tetrachloride to titanium polyacetylacetonate to produce nanofibers. TiO_2 nanofibers were obtained after calcining the nanofibers at 700 °C. Liu and co-workers also prepared ZrO_2-based nanofibers by calcining zirconyl chloride precursors using the same technique [106].

Processing Centrifugal Spinning

In centrifugal spinning, the structural characteristics of nanofiber are based on different processing parameters:

i. Intangible properties of the spinning fluid such as viscosity, surface tension, molecular structure, molecular weight, solution concentration, solvent structure and additive structure.
ii. Operating parameters such as rotational speed, spinneret diameter, nozzle diameter and distance between nozzle and collector.

Fluid Properties

There are various fluid properties (*e.g.*, viscosity, surface tension, molecular structure, weight, and solution concentration, *etc.*) that can affect the structural properties of nanofibers when the polymer solution is used to make nanofibers [107, 108]. Not all of these properties are critical to the structure of the nanofiber. Fibre formation is dominated by viscosity and surface tension, while other variables affect the process by changing the solution properties. Similarly, melting properties such as viscosity, surface tension, molecular structure, molecular weight and additive have a synergistic effect on the process of fibre formation, with the first two factors being dominant [109, 110].

Viscosity

The viscosity of the polymer melt is usually optimised by specific adjustments of the heat treatment, molecular weight, and structure. In addition, the concentration of the polymer solution can be adjusted to the controlled viscosity [65, 111]. It is useful to apply the Berry number (Be), which refers to the viscosity of the polymer solution, to produce nanofibers with the desired diameter by controlling the concentration of the polymer solution. Berry number is a dimensionless term derived from inherent viscosity and solution concentration and is commonly used to regulate fibre morphology in electrospinning. Similarly, the Berry number can be used in centrifugal spinning to show how polymeric chain linkages affect the diameters of the resulting nanofibers. Be can exceed a critical value, Be*, if the concentration and viscosity are sufficiently high. The polymer chains are then too tightly knotted to produce nanofibers by centrifugal spinning. On the other hand, if Be* is higher than Be, the production of nanofibers becomes complex due to the overlap of the short chains. At a higher Be value, the fibres become larger in diameter, which is crucial for choosing an appropriate Be value. There are larger amounts of small diameter fibres to obtain bead-free nanofibers [52, 112, 113].

Surface Tension

Surface tension also plays an important role in the development of nanofibers. It is a driving force in the bead formation of a liquid jet into a sphere due to its limited surface area. The morphology of nanofibers mainly depends on three forces: Surface tension, centrifugal force and rheological forces, and the force during centrifugal spinning [114, 115]. Especially, the centrifugal force contributes to attracting the liquid jet and increasing the surface area. The rheological forces counteract sudden structural changes and enable the production of smooth fibers. Surface tension can be controlled by regulating and selecting the desired polymer solution with optimized solvent type, molecular structure and weight. The addition of fillers and a mixture of different solvents to obtain a detailed polymer solution is also a common practice [116, 117].

Operational Conditions

There are several governing factors in the operating conditions that can affect the morphology and structural behavior of nanofibers by centrifugal electrospinning. These include the rotational speed, the diameter of the spinning head, the diameter of the nozzle, the distance between the nozzle and the collector, *etc*. These operating conditions were described as follows:

Rotating Speed

The most important operating conditions are the rotational speed of the spinning head, which directly affects the centrifugal and air friction forces. During centrifugal spinning, these forces expand jets of fluid that contribute to the production of nanofibers. During centrifugal spinning, the forces exerted on the liquid (Fcentri) can be as follows:

$$Fcentri = m\omega 2D/2 \qquad (1)$$

Where m= fluid mass, ω = speed of rotating spinning head, D = diameter of spinning head.

The rotational speed must reach an optimal value to eject the liquid jet from the nozzle tip and generate the centrifugal force necessary to overcome the surface tension of the spinning liquid. Therefore, it is essential for the production of nanofibers to regulate the critical rotational speed of the spinning fluid. Different spinning fluids have different surface tensions and viscosities. Therefore, the proper rotational speed is required to generate the centrifugal force. The design and diameter of the rotating spinning heads to achieve maximum centrifugal forces [52, 115, 118]. The rotational speed used in centrifugal electrospinning of polymer nanofibers ranges from 3,000 to 12,000 rpm, as reported in the literature. The frictional force acting on the jet during ejection from the jet tip can be calculated using equation (1). However, the rotational speed of the jet should be modified, and D is the diameter of the jet path [52]. Alternatively, it is possible to calculate the air friction force (Ffri) as follows:

$$Ffri = \pi C\rho A\omega 2D2/2 \qquad (2)$$

C = drag coefficient, ρ = air density, A = cross-section area of jet, ω = rotating speed of jet, D = jet path diameter.

The fluid jet moves from the syringe to the collector with a gradually decreasing rotational speed. However, the higher rotational speed of the spinner head gives the jet a higher rotational speed after it has travelled a distance and increases the rotation. The centrifugal and frictional forces of the air are created by the spinning head. These forces cause the fluid jet to rise, and the higher elongation of the fluid jet causes a decrease in fibre diameter. The average fibre diameter decreases from 663-440 nm at a rotation speed of 2,000-4,000 rpm. Thus, it was observed that a higher rotation speed produces a thicker fibre due to the shorter flight time of the polymeric liquid jet to the collector. The shorter flight time does not provide enough time to stretch and elongate the polymer jet from the nozzle. In addition, a higher rotational speed results in a larger mass flow rate of the jet and thus a larger fibre diameter. Therefore, determining the optimum rotation speed that results in the smallest fibre diameter is critical [52, 118].

Diameter of the Spinning Head

The morphology of nanofibers can be optimized with the diameter of the spinning head, which plays a crucial role in the production of nanofibers according to equation (**1**). As the diameter of the spinning head increases, the centrifugal force increases and produces a larger jet [107]. Moreover, greater stretching and elongation of the jet results in finer nanofibers. The larger diameter of the spinning head cannot be achieved with a higher-speed motor. If the diameter of the spinning head is too large, it is not possible for such motors to achieve stable rotation to produce balanced nanofibers [52].

Diameter of Nozzle

Another way to control the nanofiber structure is to adjust the nozzle diameter. In this technique, the mass flow rate of the liquid jet is adjusted to change the nanofiber structure. A smaller nozzle diameter significantly restricts the mass bandwidth to produce finer nanofibers. A decrease in nanofiber diameter from 895-665 nm was observed when the nozzle diameter was changed from 1.0-0.4 mm. This indicates that a nozzle with a smaller diameter produces a fibre with a thinner diameter. A nozzle with a smaller diameter is not suitable for the production of nanofibers because the liquid jet cannot eject. In this way, the desired nanofibers are not developed. Therefore, the nozzle diameter is crucial for the production of nanofibers with optimised diameter [98, 115].

Distance of Nozzle from the Collector

The distance of the nozzle collector affects the time of flight of the liquid jet, and the distance is directly proportional to the time of flight of the liquid jet. When a solution is used as a centrifugal liquid, the distance from the nozzle to the collector should be as small as possible. Evaporation of the liquid jet was observed before it reached the collector. However, a liquid jet must travel a sufficient distance to reduce the fibre diameter. In addition, other operating parameters are more critical than the distance of the nozzle from the collector. A slight change in the distance from the nozzle to the collector (10-30 cm) resulted in an average fibre diameter (665-647 nm). Therefore, 10 cm is a reasonable distance for solvent evaporation, and a distance of the nozzle from the collector of more than 10 cm leads to a nominal change in fibre diameter [98, 119].

BRIEF INTRODUCTION OF SBS

Recently, nanomaterials have gained popularity due to their multifunctional behavior and advanced properties in the fields of environment, energy storage, nanomedicine, *etc.* They have been extensively explored, with nanomembranes, nanospheres, nanotubes and nanofibers receiving considerable attention. Among 1D nanostructures, nanofibers are the most researched [120, 121]. The electrospinning (e-spinning) techniques to produce nanofibers and these techniques to produce nanofibers are widely used in biomedicine, energy sector, environmental purification and catalysis. The high voltage of static electricity has been used for the fabrication of nanofibers, which is a major problem for reading reliability and energy consumption [122]. For the fabrication of nanofibers by the e-spinning technique, optimized conductivity of polymer solution is required, which limits the spinnability of some non-conductive polymers. The production of nanofibers by e-spinning is impossible for mass production because the yield is very low [20, 123, 124].

Centrifugal spinning and solution jet spinning have been proposed for the production of nanofibers to overcome safety issues and numerous technical problems in manufacturing and to increase the yield of nanofibers. SBS is an advanced technology for the fabrication of nanofibers compared to conventional techniques [125, 126]. SBS has a high yield, short preparation time and high utility compared to e-spinning technology. SBS provides a wide range of polymer solutions compared to a wide spinning solution for the production of nanofibers. In SBS, the conductivity of polymers is not required, and no high voltage static field is required during the solution spinning process. Therefore, the conductivity and safety issues are solved by SBS with higher equipment requirements. SBS has a wider range of available raw materials and is more suitable than the melt-blowing process. The various SBS parameters are shown in Fig. (6). However, since the process uses compressed air at normal temperatures, it can effectively prevent the thermal degradation of the polymer. In addition, the SBS process is increasingly attracting the attention of researchers due to the new spinning technology. Research studies are concerned with the feasibility of the process, fiber properties, and applications, and all aspects have been thoroughly investigated [122, 127, 128]. The functional capabilities of CS are summarized in Table **1**.

APPLICATION IN TISSUE ENGINEERING

Electrospinning offers an easier and more cost-effective way to fabricate scaffolds with contiguous pore size and submicron fibre diameters than self-assembly and phase separation. 3D scaffolds have been developed as advanced biomaterials in various fields of biomedical engineering with improved functionalities and properties. The use of electrospun scaffolds for tissue engineering is highlighted in the following sections.

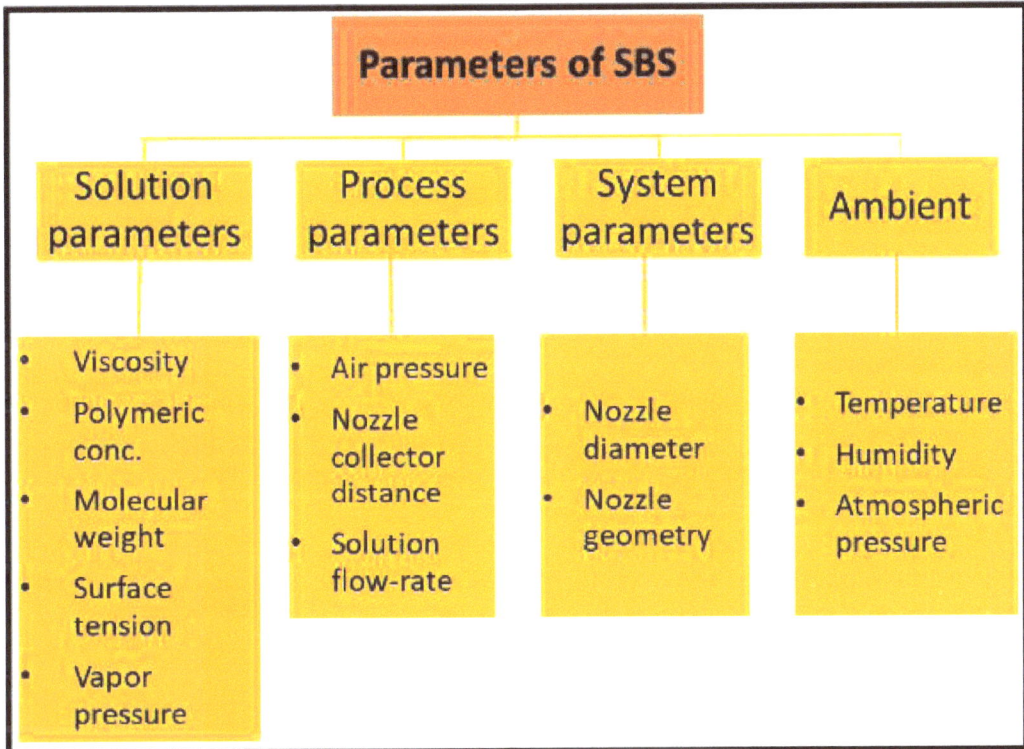

Fig. (6). Different parameters of SBS.

Synthetic Extracellular Matrix

As Langer and Vacanti defined it in 1993, tissue engineering is a multidisciplinary system that combines engineering and bioscience principles. Biomaterials have evolved as biological substitutes for tissues to restore, preserve, or enhance functions [129]. Cell behavior has been modulated for tissue engineering, and polymeric scaffolds with improved biomechanical properties compared to native extracellular matrix (ECM) have been developed. During *in vivo* analysis, cell lines come into contact with the ECM, which is a network protein [130]. It provides direct information signals to the complex spatial system that influences phenotypic and other cellular behaviors [131]. In bone, an organized collagen type I ECM is required for integrin binding [132]. Cellular activities, such as migration, proliferation, differentiation, gene expression and hormone secretion, and growth factors can influence cell-ECM interactions [133]. It creates a microenvironment similar to the natural microenvironment and can be customized with various properties (*e.g.*, chemical composition, surface morphology, function, *etc.*). The biocompatible materials have been successfully fabricated for tissue engineering [134, 135].

Until repair or regeneration, scaffolds for tissue engineering act as temporary ECMs. A nanofiber scaffold provides an *in vivo* 2-D structure to which cells can attach and develop [136]. The nanofiber structure can then be implanted into a defective tissue repair and regeneration site. Although the main features of a nanofiber scaffold vary slightly depending on the tissue to be replicated, the main advantages are desirable. First and foremost, the scaffold should be biocompatible, *i.e.*, it should integrate with the host tissue without eliciting a significant immune response [137]. To allow cell attachment, ingrowth, and exchange of nutrients during *in vivo* or *in vivo* culture, the nanofibrous scaffold should be porous and have a high surface-to-volume ratio [137]. Moreover, the porous nature of the nanofibrous scaffold with a large surface area enables angiogenesis when implanted at a defective site (in vascularized tissues). In addition, the nanofiber scaffold serves as temporary support for cell attachment and proliferation. It should architecturally and structurally mimic the native ECM [138]. Finally, a nanofiber scaffold for tissue engineering should be biodegradable so that the implant does not need to be removed by a second surgery. To mimic the rate of neo-tissue formation, the rate of degradation should be uniform or at least controllable [130, 137].

Biological Response to Nanofibrous

Nanoscale elements, such as collagen, are an essential aspect of native ECM. Nanofibrous scaffolds for tissue engineering should therefore contain nanophasic elements. Studies have shown that nanoscale elements can influence cell behaviour [139]. For example, the activity of osteoblasts and osteoclasts was increased on spherical nanophasic alumina particles that resemble the structure of hydroxyapatite crystals in bone [140]. Nanofiber architecture has also been shown to selectively promote osteoblast proliferation and differentiation in carbon nanofibers. Nanorippled surfaces can induce human corneal epithelial cells to contact encouragement, causing them to elongate and align their cytoskeleton with these topological features [141, 142]. Highly porous PLLA scaffolds with nanoscale pores generated by liquid-liquid phase separation have been used to culture neuronal electrospinning stem cells from polymeric nanofibers. They have been shown to have a beneficial effect on neurite growth. Since the fibre diameters of the nanofiber scaffold are orders of magnitude smaller than the cell size, the cells can organise or spread around the fibres and attach to adsorbed proteins at multiple sites [130, 143, 144].

CONCLUSION AND PROSPECTS

Considering the importance of nanofibers in current research and applications, we have designed this chapter to provide an overview of the different processes used to produce nanofibers. In addition, the types of fibres produced and their applications are also discussed. Fibre materials have various useful applications in tissue engineering, filtration, safety wipes, batteries, energy storage, *etc*. Electrospinning is a commonly used technique to produce nanofibers. However, due to the low production rate, the wide commercial use of electrospinning is minimal. Therefore, various other techniques have been used to produce nanofibers. These include melt blowing, phase separation, template synthesis and self-assembly. However, all these techniques had their own limitations. The need arose to search for a technique that either minimizes or overcomes the limitations of the above techniques. CS and SBS are advanced replacement approaches to produce nanofibers from various low-cost raw materials. This chapter is valuable for novices and experts who are looking for literature that could contribute to their understanding of fibre fabrication techniques.

CONSENT FOR PUBLICATION

Not applicable.

CONFLICT OF INTEREST

The authors declare no conflict of interest, financial or otherwise.

ACKNOWLEDGEMENT

Declared none.

REFERENCES

[1]   M.U.A. Khan, S.I. Abd Razak, H. Mehboob, M.R. Abdul Kadir, T.J.S. Anand, F. Inam, S.A. Shah, M.E.F. Abdel-Haliem, and R. Amin, "Synthesis and characterization of silver-coated polymeric scaffolds for bone tissue engineering: antibacterial and *in vivo* evaluation of cytotoxicity and biocompatibility", *ACS Omega,* vol. 6, no. 6, pp. 4335-4346, 2021.
[http://dx.doi.org/10.1021/acsomega.0c05596] [PMID: 33623844]

[2]   M.U. Aslam Khan, S.I. Abd Razak, W.S. Al Arjan, S. Nazir, T.J. Sahaya Anand, H. Mehboob, and R. Amin, "Recent Advances in Biopolymeric Composite Materials for Tissue Engineering and Regenerative Medicines: A Review", *Molecules,* vol. 26, no. 3, p. 619, 2021.
[http://dx.doi.org/10.3390/molecules26030619] [PMID: 33504080]

[3]   A.P. Kishan, and E.M. Cosgriff-Hernandez, "Recent advancements in electrospinning design for tissue engineering applications: A review", *J. Biomed. Mater. Res. A,* vol. 105, no. 10, pp. 2892-2905, 2017.
[http://dx.doi.org/10.1002/jbm.a.36124] [PMID: 28556551]

[4]   M. Umar Aslam Khan, S. Haider, A. Haider, S. Izwan Abd Razak, M. Rafiq Abdul Kadir, S.A. Shah, A. Javed, I. Shakir, and A.A. Al-Zahrani, "Development of porous, antibacterial and biocompatible GO/n-HAp/bacterial cellulose/β-glucan biocomposite scaffold for bone tissue engineering", *Arab. J.*

Chem., vol. 14, no. 2, p. 102924, 2021.
[http://dx.doi.org/10.1016/j.arabjc.2020.102924]

[5]   X. Hu, S. Liu, G. Zhou, Y. Huang, Z. Xie, and X. Jing, "Electrospinning of polymeric nanofibers for drug delivery applications", *J. Control. Release,* vol. 185, pp. 12-21, 2014.
[http://dx.doi.org/10.1016/j.jconrel.2014.04.018] [PMID: 24768792]

[6]   M.U.A. Khan, S. Haider, S.A. Shah, S.I.A. Razak, S.A. Hassan, M.R.A. Kadir, and A. Haider, "Arabinoxylan-co-AA/HAp/TiO$_2$ nanocomposite scaffold a potential material for bone tissue engineering: An *in vivo* study", *Int. J. Biol. Macromol.,* vol. 151, pp. 584-594, 2020.
[http://dx.doi.org/10.1016/j.ijbiomac.2020.02.142] [PMID: 32081758]

[7]   S. Khorshidi, A. Solouk, H. Mirzadeh, S. Mazinani, J.M. Lagaron, S. Sharifi, and S. Ramakrishna, "A review of key challenges of electrospun scaffolds for tissue-engineering applications", *J. Tissue Eng. Regen. Med.,* vol. 10, no. 9, pp. 715-738, 2016.
[http://dx.doi.org/10.1002/term.1978] [PMID: 25619820]

[8]   Y. Liu, R. Deng, M. Hao, H. Yan, and W. Yang, "Orthogonal design study on factors effecting on fibers diameter of melt electrospinning", *Polym. Eng. Sci.,* vol. 50, no. 10, pp. 2074-2078, 2010.
[http://dx.doi.org/10.1002/pen.21753]

[9]   D. Kai, S.S. Liow, and X.J. Loh, "Biodegradable polymers for electrospinning: Towards biomedical applications", *Mater. Sci. Eng. C,* vol. 45, pp. 659-670, 2014.
[http://dx.doi.org/10.1016/j.msec.2014.04.051] [PMID: 25491875]

[10]   A.M. Loordhuswamy, V.R. Krishnaswamy, P.S. Korrapati, S. Thinakaran, and G.D.V. Rengaswami, "Fabrication of highly aligned fibrous scaffolds for tissue regeneration by centrifugal spinning technology", *Mater. Sci. Eng. C,* vol. 42, pp. 799-807, 2014.
[http://dx.doi.org/10.1016/j.msec.2014.06.011] [PMID: 25063182]

[11]   D. Sundaramurthi, U.M. Krishnan, and S. Sethuraman, "Electrospun nanofibers as scaffolds for skin tissue engineering", *Polym. Rev. (Phila. Pa.),* vol. 54, no. 2, pp. 348-376, 2014.
[http://dx.doi.org/10.1080/15583724.2014.881374]

[12]   R. Vasireddi, J. Kruse, M. Vakili, S. Kulkarni, T.F. Keller, D.C.F. Monteiro, and M. Trebbin, "Solution blow spinning of polymer/nanocomposite micro-/nanofibers with tunable diameters and morphologies using a gas dynamic virtual nozzle", *Sci. Rep.,* vol. 9, no. 1, p. 14297, 2019.
[http://dx.doi.org/10.1038/s41598-019-50477-6] [PMID: 31586141]

[13]   G. Kim, and W. Kim, "H ighly porous 3D nanofiber scaffold using an electrospinning technique", *J. Biomed. Mater. Res. B Appl. Biomater.,* vol. 81B, no. 1, pp. 104-110, 2007.
[http://dx.doi.org/10.1002/jbm.b.30642] [PMID: 16924612]

[14]   C. Ribeiro, V. Sencadas, J.L.G. Ribelles, and S. Lanceros-Méndez, "Influence of processing conditions on polymorphism and nanofiber morphology of electroactive poly (vinylidene fluoride) electrospun membranes", *Soft Mater.,* vol. 8, no. 3, pp. 274-287, 2010.
[http://dx.doi.org/10.1080/1539445X.2010.495630]

[15]   F. Zamani, M. Amani-Tehran, A. Zaminy, and M.A. Shokrgozar, "Conductive 3D structure nanofibrous scaffolds for spinal cord regeneration", *Fibers Polym.,* vol. 18, no. 10, pp. 1874-1881, 2017.
[http://dx.doi.org/10.1007/s12221-017-7349-7]

[16]   M.D. Calisir, and A. Kilic, "A comparative study on SiO$_2$ nanofiber production *via* two novel non-electrospinning methods: Centrifugal spinning *vs* solution blowing", *Mater. Lett.,* vol. 258, p. 126751, 2020.
[http://dx.doi.org/10.1016/j.matlet.2019.126751]

[17]   W.E. Teo, and S. Ramakrishna, "Electrospun nanofibers as a platform for multifunctional, hierarchically organized nanocomposite", *Compos. Sci. Technol.,* vol. 69, no. 11-12, pp. 1804-1817, 2009.
[http://dx.doi.org/10.1016/j.compscitech.2009.04.015]

[18] D. Thassu, M. Deleers, and Y.V. Pathak, "Incorporation of drugs in an amorphous state into electrospun nanofibers composed of a water-insoluble, nonbiodegradable polymer, Journal of controlled release", In: *Nanoparticulate drug delivery systems.* CRC Press, 2007.
[http://dx.doi.org/10.1201/9781420008449]

[19] G. Verreck, I. Chun, J. Rosenblatt, J. Peeters, A.V. Dijck, J. Mensch, M. Noppe, and M.E. Brewster, "Incorporation of drugs in an amorphous state into electrospun nanofibers composed of a water-insoluble, nonbiodegradable polymer", *J. Control. Release,* vol. 92, no. 3, pp. 349-360, 2003.
[http://dx.doi.org/10.1016/S0168-3659(03)00342-0] [PMID: 14568415]

[20] S. Thenmozhi, N. Dharmaraj, K. Kadirvelu, and H.Y. Kim, "Electrospun nanofibers: New generation materials for advanced applications", *Mater. Sci. Eng. B,* vol. 217, pp. 36-48, 2017.
[http://dx.doi.org/10.1016/j.mseb.2017.01.001]

[21] S. Zhu, and L. Nie, "Progress in fabrication of one-dimensional catalytic materials by electrospinning technology", *J. Ind. Eng. Chem.,* 2020.

[22] K.M. Zadeh, A.S. Luyt, L. Zarif, R. Augustine, A. Hasan, M. Messori, M.K. Hassan, and H.C. Yalcin, "Electrospun polylactic acid/date palm polyphenol extract nanofibres for tissue engineering applications", *Emergent Materials,* vol. 2, no. 2, pp. 141-151, 2019.
[http://dx.doi.org/10.1007/s42247-019-00042-8]

[23] M.M. Hohman, M. Shin, G. Rutledge, and M.P. Brenner, "Electrospinning and electrically forced jets. II. Applications", *Phys. Fluids,* vol. 13, no. 8, pp. 2221-2236, 2001.
[http://dx.doi.org/10.1063/1.1384013]

[24] H.S. SalehHudin, E.N. Mohamad, W.N.L. Mahadi, and A. Muhammad Afifi, "Multiple-jet electrospinning methods for nanofiber processing: A review", *Mater. Manuf. Process.,* vol. 33, no. 5, pp. 479-498, 2018.
[http://dx.doi.org/10.1080/10426914.2017.1388523]

[25] R. Augustine, S.R.U. Rehman, R. Ahmed, A.A. Zahid, M. Sharifi, M. Falahati, and A. Hasan, "Electrospun chitosan membranes containing bioactive and therapeutic agents for enhanced wound healing", *Int. J. Biol. Macromol.,* vol. 156, pp. 153-170, 2020.
[http://dx.doi.org/10.1016/j.ijbiomac.2020.03.207] [PMID: 32229203]

[26] N. Bhardwaj, and S.C. Kundu, "Electrospinning: A fascinating fiber fabrication technique", *Biotechnol. Adv.,* vol. 28, no. 3, pp. 325-347, 2010.
[http://dx.doi.org/10.1016/j.biotechadv.2010.01.004] [PMID: 20100560]

[27] S. Xie, and Y. Zeng, "Effects of electric field on multineedle electrospinning: experiment and simulation study", *Ind. Eng. Chem. Res.,* vol. 51, no. 14, pp. 5336-5345, 2012.
[http://dx.doi.org/10.1021/ie2020763]

[28] R. Nayak, I.L. Kyratzis, Y.B. Truong, R. Padhye, L. Arnold, G. Peeters, L. Nichols, and M. O'Shea, "Fabrication and characterisation of nanofibres by meltblowing and melt electrospinning", *Adv. Mat. Res.,* vol. 472-475, pp. 1294-1299, 2012. [Trans Tech Publ.].
[http://dx.doi.org/10.4028/www.scientific.net/AMR.472-475.1294]

[29] A. Balogh, B. Farkas, K. Faragó, A. Farkas, I. Wagner, I. Van assche, G. Verreck, Z.K. Nagy, and G. Marosi, "Melt-blown and electrospun drug-loaded polymer fiber mats for dissolution enhancement: a comparative study", *J. Pharm. Sci.,* vol. 104, no. 5, pp. 1767-1776, 2015.
[http://dx.doi.org/10.1002/jps.24399] [PMID: 25761776]

[30] I.M. Hutten, *Handbook of nonwoven filter media.* Elsevier, 2007.

[31] R. Hufenus, Y. Yan, M. Dauner, D. Yao, and T. Kikutani, "Bicomponent fibers", In: *Handbook of Fibrous Materials*, 2020, pp. 281-313.
[http://dx.doi.org/10.1002/9783527342587.ch11]

[32] S.S. Rajgarhia, R.E. Benavides, and S.C. Jana, "Morphology control of bi-component polymer nanofibers produced by gas jet process", *Polymer (Guildf.),* vol. 93, pp. 142-151, 2016.

[http://dx.doi.org/10.1016/j.polymer.2016.04.018]

[33]   S.A. Tuin, B. Pourdeyhimi, and E.G. Loboa, "Interconnected, microporous hollow fibers for tissue engineering: Commercially relevant, industry standard scale-up manufacturing", *J. Biomed. Mater. Res. A,* vol. 102, no. 9, pp. 3311-3323, 2014.
        [http://dx.doi.org/10.1002/jbm.a.35002] [PMID: 24142629]

[34]   C. Thellen, S. Cheney, and J.A. Ratto, "Melt processing and characterization of polyvinyl alcohol and polyhydroxyalkanoate multilayer films", *J. Appl. Polym. Sci.,* vol. 127, no. 3, pp. 2314-2324, 2013.
        [http://dx.doi.org/10.1002/app.37850]

[35]   Z. Ustunol, "Edible films and coatings for meat and poultry", In: *Edible films and coatings for food applications.* Springer, 2009, pp. 245-268.
        [http://dx.doi.org/10.1007/978-0-387-92824-1_8]

[36]   M. Shang, H. Matsuyama, M. Teramoto, D.R. Lloyd, and N. Kubota, "Preparation and membrane performance of poly(ethylene-co-vinyl alcohol) hollow fiber membrane *via* thermally induced phase separation", *Polymer (Guildf.),* vol. 44, no. 24, pp. 7441-7447, 2003.
        [http://dx.doi.org/10.1016/j.polymer.2003.08.033]

[37]   H.A. Tsai, C.Y. Kuo, J.H. Lin, D.M. Wang, A. Deratani, C. Pochat-Bohatier, K.R. Lee, and J.Y. Lai, "Morphology control of polysulfone hollow fiber membranes *via* water vapor induced phase separation", *J. Membr. Sci.,* vol. 278, no. 1-2, pp. 390-400, 2006.
        [http://dx.doi.org/10.1016/j.memsci.2005.11.029]

[38]   X. Fu, H. Matsuyama, M. Teramoto, and H. Nagai, "Preparation of polymer blend hollow fiber membrane *via* thermally induced phase separation", *Separ. Purif. Tech.,* vol. 52, no. 2, pp. 363-371, 2006.
        [http://dx.doi.org/10.1016/j.seppur.2006.05.018]

[39]   I. Alghoraibi, and S. Alomari, "Different methods for nanofiber design and fabrication", In: *Handbook of nanofibers*, 2018, pp. 1-46.
        [http://dx.doi.org/10.1007/978-3-319-42789-8_11-2]

[40]   C. Liu, *Microporous carbon nanofibers prepared by combining electrospinning and phase separation methods for supercapacitor*, 2016.
        [http://dx.doi.org/10.1016/j.jechem.2016.03.017]

[41]   H.W. Liang, Q.F. Guan, L.F. Chen, Z. Zhu, W.J. Zhang, and S.H. Yu, "Macroscopic-scale template synthesis of robust carbonaceous nanofiber hydrogels and aerogels and their applications", *Angew. Chem. Int. Ed.,* vol. 51, no. 21, pp. 5101-5105, 2012.
        [http://dx.doi.org/10.1002/anie.201200710] [PMID: 22505338]

[42]   Y. Wang, M. Zheng, H. Lu, S. Feng, G. Ji, and J. Cao, "Template synthesis of carbon nanofibers containing linear mesocage arrays", *Nanoscale Res. Lett.,* vol. 5, no. 6, pp. 913-916, 2010.
        [http://dx.doi.org/10.1007/s11671-010-9562-9] [PMID: 20671793]

[43]   Kenry, and C.T. Lim, "Nanofiber technology: current status and emerging developments", *Prog. Polym. Sci.,* vol. 70, pp. 1-17, 2017.
        [http://dx.doi.org/10.1016/j.progpolymsci.2017.03.002]

[44]   S. Liu, H. Shan, S. Xia, J. Yan, J. Yu, and B. Ding, "Polymer Template Synthesis of Flexible SiO_2 Nanofibers to Upgrade Composite Electrolytes", *ACS Appl. Mater. Interfaces,* vol. 12, no. 28, pp. 31439-31447, 2020.
        [http://dx.doi.org/10.1021/acsami.0c06922] [PMID: 32589014]

[45]   R. Jin, Y. Yang, Y. Xing, L. Chen, S. Song, and R. Jin, "Facile synthesis and properties of hierarchical double-walled copper silicate hollow nanofibers assembled by nanotubes", *ACS Nano,* vol. 8, no. 4, pp. 3664-3670, 2014.
        [http://dx.doi.org/10.1021/nn500275d] [PMID: 24617673]

[46]   G. Che, B.B. Lakshmi, C.R. Martin, E.R. Fisher, and R.S. Ruoff, "Chemical vapor deposition based

synthesis of carbon nanotubes and nanofibers using a template method", *Chem. Mater.,* vol. 10, no. 1, pp. 260-267, 1998.
[http://dx.doi.org/10.1021/cm970412f]

[47]   M. Aghayan, I. Hussainova, K. Kirakosyan, and M.A. Rodríguez, "The template-assisted wet-combustion synthesis of copper oxide nanoparticles on mesoporous network of alumina nanofibers", *Mater. Chem. Phys.,* vol. 192, pp. 138-146, 2017.
[http://dx.doi.org/10.1016/j.matchemphys.2017.01.068]

[48]   J.D. Hartgerink, E. Beniash, and S.I. Stupp, "Peptide-amphiphile nanofibers: A versatile scaffold for the preparation of self-assembling materials", *Proc. Natl. Acad. Sci. USA,* vol. 99, no. 8, pp. 5133-5138, 2002.
[http://dx.doi.org/10.1073/pnas.072699999] [PMID: 11929981]

[49]   Y. Nagai, L.D. Unsworth, S. Koutsopoulos, and S. Zhang, "Slow release of molecules in self-assembling peptide nanofiber scaffold", *J. Control. Release,* vol. 115, no. 1, pp. 18-25, 2006.
[http://dx.doi.org/10.1016/j.jconrel.2006.06.031] [PMID: 16962196]

[50]   Y. Gao, Y. Kuang, Z.F. Guo, Z. Guo, I.J. Krauss, and B. Xu, "Enzyme-instructed molecular self-assembly confers nanofibers and a supramolecular hydrogel of taxol derivative", *J. Am. Chem. Soc.,* vol. 131, no. 38, pp. 13576-13577, 2009.
[http://dx.doi.org/10.1021/ja904411z] [PMID: 19731909]

[51]   Z. Zheng, P. Chen, M. Xie, C. Wu, Y. Luo, W. Wang, J. Jiang, and G. Liang, "Cell environment-differentiated self-assembly of nanofibers", *J. Am. Chem. Soc.,* vol. 138, no. 35, pp. 11128-11131, 2016.
[http://dx.doi.org/10.1021/jacs.6b06903] [PMID: 27532322]

[52]   X. Zhang, and Y. Lu, "Centrifugal spinning: an alternative approach to fabricate nanofibers at high speed and low cost", *Polym. Rev. (Phila. Pa.),* vol. 54, no. 4, pp. 677-701, 2014.
[http://dx.doi.org/10.1080/15583724.2014.935858]

[53]   G.T.V. Prabu, and B. Dhurai, "A novel profiled multi-pin electrospinning system for nanofiber production and encapsulation of nanoparticles into nanofibers", *Sci. Rep.,* vol. 10, no. 1, p. 4302, 2020.
[http://dx.doi.org/10.1038/s41598-020-60752-6] [PMID: 32152364]

[54]   C.J. Luo, S.D. Stoyanov, E. Stride, E. Pelan, and M. Edirisinghe, "Electrospinning *versus* fibre production methods: from specifics to technological convergence", *Chem. Soc. Rev.,* vol. 41, no. 13, pp. 4708-4735, 2012.
[http://dx.doi.org/10.1039/c2cs35083a] [PMID: 22618026]

[55]   A.L. Szentivanyi, H. Zernetsch, H. Menzel, and B. Glasmacher, "A review of developments in electrospinning technology: new opportunities for the design of artificial tissue structures", *Int. J. Artif. Organs,* vol. 34, no. 10, pp. 986-997, 2011.
[http://dx.doi.org/10.5301/ijao.5000062] [PMID: 22161282]

[56]   R.J. Stoddard, A.L. Steger, A.K. Blakney, and K.A. Woodrow, "In pursuit of functional electrospun materials for clinical applications in humans", *Ther. Deliv.,* vol. 7, no. 6, pp. 387-409, 2016.
[http://dx.doi.org/10.4155/tde-2016-0017] [PMID: 27250537]

[57]   I.S. Chronakis, "Novel nanocomposites and nanoceramics based on polymer nanofibers using electrospinning process—A review", *J. Mater. Process. Technol.,* vol. 167, no. 2-3, pp. 283-293, 2005.
[http://dx.doi.org/10.1016/j.jmatprotec.2005.06.053]

[58]   X. Zhang, "Fundamentals of fiber science", *DEStech Publications, Inc.,* 2014.

[59]   X. Zhang, J. Chen, and Y. Zeng, "Morphology development of helical structure in bicomponent fibers during spinning process", *Polymer (Guildf.),* vol. 201, p. 122609, 2020.
[http://dx.doi.org/10.1016/j.polymer.2020.122609]

[60]   Y. Li, N. Grishkewich, L. Liu, C. Wang, K.C. Tam, S. Liu, Z. Mao, and X. Sui, "Construction of functional cellulose aerogels *via* atmospheric drying chemically cross-linked and solvent exchanged

cellulose nanofibrils", *Chem. Eng. J.,* vol. 366, pp. 531-538, 2019.
[http://dx.doi.org/10.1016/j.cej.2019.02.111]

[61]   J. Ding, K. Zhong, S. Liu, X. Wu, X. Shen, S. Cui, and X. Chen, "Flexible and super hydrophobic polymethylsilsesquioxane based silica aerogel for organic solvent adsorption *via* ambient pressure drying technique", *Powder Technol.,* vol. 373, pp. 716-726, 2020.
[http://dx.doi.org/10.1016/j.powtec.2020.07.024]

[62]   X. Hu, M. Liao, H. Gong, L. Zhang, H. Cox, T.A. Waigh, and J.R. Lu, "Recent advances in short peptide self-assembly: from rational design to novel applications", *Curr. Opin. Colloid Interface Sci.,* vol. 45, pp. 1-13, 2020.
[http://dx.doi.org/10.1016/j.cocis.2019.08.003]

[63]   D. Andrade, L.B.A. Oliveira, and G. Colherinhas, "Design and analysis of polypeptide nanofiber using full atomistic Molecular Dynamic", *J. Mol. Liq.,* vol. 302, p. 112610, 2020.
[http://dx.doi.org/10.1016/j.molliq.2020.112610]

[64]   B. Sachin Kumar, S.K. Kalpathy, and S. Anandhan, "Synergism of fictitious forces on nickel cobaltite nanofibers: electrospinning forces revisited", *Phys. Chem. Chem. Phys.,* vol. 20, no. 7, pp. 5295-5304, 2018.
[http://dx.doi.org/10.1039/C7CP07435B] [PMID: 29405210]

[65]   S. Noroozi, H. Alamdari, W. Arne, R.G. Larson, and S.M. Taghavi, "Regularized string model for nanofibre formation in centrifugal spinning methods", *J. Fluid Mech.,* vol. 822, pp. 202-234, 2017.
[http://dx.doi.org/10.1017/jfm.2017.279]

[66]   R. Gorkin, J. Park, J. Siegrist, M. Amasia, B.S. Lee, J.M. Park, J. Kim, H. Kim, M. Madou, and Y.K. Cho, "Centrifugal microfluidics for biomedical applications", *Lab Chip,* vol. 10, no. 14, pp. 1758-1773, 2010.
[http://dx.doi.org/10.1039/b924109d] [PMID: 20512178]

[67]   R. Grayson, "Fine Gold Recovery–Alternatives to", *WORLD,* vol. 7, pp. 66-161, 2007.

[68]   M. Ardanuy, J. Claramunt, and R.D. Toledo Filho, "Cellulosic fiber reinforced cement-based composites: A review of recent research", *Constr. Build. Mater.,* vol. 79, pp. 115-128, 2015.
[http://dx.doi.org/10.1016/j.conbuildmat.2015.01.035]

[69]   M.L. Siriwardane, K. DeRosa, G. Collins, and B.J. Pfister, "Controlled formation of cross-linked collagen fibers for neural tissue engineering applications", *Biofabrication,* vol. 6, no. 1, p. 015012, 2014.
[http://dx.doi.org/10.1088/1758-5082/6/1/015012] [PMID: 24589999]

[70]   A. Barhoum, K. Pal, H. Rahier, H. Uludag, I.S. Kim, and M. Bechelany, "Nanofibers as new-generation materials: From spinning and nano-spinning fabrication techniques to emerging applications", *Appl. Mater. Today,* vol. 17, pp. 1-35, 2019.
[http://dx.doi.org/10.1016/j.apmt.2019.06.015]

[71]   D.M. dos Santos, D.S. Correa, E.S. Medeiros, J.E. Oliveira, and L.H.C. Mattoso, "Advances in Functional Polymer Nanofibers: From Spinning Fabrication Techniques to Recent Biomedical Applications", *ACS Appl. Mater. Interfaces,* vol. 12, no. 41, pp. 45673-45701, 2020.
[http://dx.doi.org/10.1021/acsami.0c12410] [PMID: 32937068]

[72]   R. Al-Attabi, L.F. Dumée, J.A. Schütz, and Y. Morsi, "Pore engineering towards highly efficient electrospun nanofibrous membranes for aerosol particle removal", *Sci. Total Environ.,* vol. 625, pp. 706-715, 2018.
[http://dx.doi.org/10.1016/j.scitotenv.2017.12.342] [PMID: 29306158]

[73]   P. Ravishankar, A. Khang, M. Laredo, and K. Balachandran, "Using dimensionless numbers to predict centrifugal jet-spun nanofiber morphology", *Journal of Nanomaterials,* 2019.
[http://dx.doi.org/10.1155/2019/4639658]

[74]   E.S. Medeiros, G.M. Glenn, A.P. Klamczynski, W.J. Orts, and L.H.C. Mattoso, "Solution blow

spinning: A new method to produce micro- and nanofibers from polymer solutions", *J. Appl. Polym. Sci.,* vol. 113, no. 4, pp. 2322-2330, 2009.
[http://dx.doi.org/10.1002/app.30275]

[75]   S. Padron, A. Fuentes, D. Caruntu, and K. Lozano, "Experimental study of nanofiber production through forcespinning", *J. Appl. Phys.,* vol. 113, no. 2, p. 024318, 2013.
[http://dx.doi.org/10.1063/1.4769886]

[76]   Y. Polat, E.S. Pampal, E. Stojanovska, R. Simsek, A. Hassanin, A. Kilic, A. Demir, and S. Yilmaz, "Solution blowing of thermoplastic polyurethane nanofibers: A facile method to produce flexible porous materials", *J. Appl. Polym. Sci.,* vol. 133, no. 9, 2016.
[http://dx.doi.org/10.1002/app.43025]

[77]   M.R. Badrossamay, K. Balachandran, A.K. Capulli, H.M. Golecki, A. Agarwal, J.A. Goss, H. Kim, K. Shin, and K.K. Parker, "Engineering hybrid polymer-protein super-aligned nanofibers *via* rotary jet spinning", *Biomaterials,* vol. 35, no. 10, pp. 3188-3197, 2014.
[http://dx.doi.org/10.1016/j.biomaterials.2013.12.072] [PMID: 24456606]

[78]   K. Jia, X. Zhuang, B. Cheng, S. Shi, Z. Shi, and B. Zhang, "Solution blown aligned carbon nanofiber yarn as supercapacitor electrode", *J. Mater. Sci. Mater. Electron.,* vol. 24, no. 12, pp. 4769-4773, 2013.
[http://dx.doi.org/10.1007/s10854-013-1472-z]

[79]   T. O'Haire, S.J. Russell, and C.M. Carr, "Centrifugal melt spinning of polyvinylpyrrolidone (PVP)/triacontene copolymer fibres", *J. Mater. Sci.,* vol. 51, no. 16, pp. 7512-7522, 2016.
[http://dx.doi.org/10.1007/s10853-016-0030-5]

[80]   M.A. Hassan, B.Y. Yeom, A. Wilkie, B. Pourdeyhimi, and S.A. Khan, "Fabrication of nanofiber meltblown membranes and their filtration properties", *J. Membr. Sci.,* vol. 427, pp. 336-344, 2013.
[http://dx.doi.org/10.1016/j.memsci.2012.09.050]

[81]   H.Y. Liu, Y. Chen, G-S. Liu, S-G. Pei, J-Q. Liu, H. ji, and R-D. Wang, "Preparation of high-quality zirconia fibers by super-high rotational centrifugal spinning of inorganic sol", *Mater. Manuf. Process.,* vol. 28, no. 2, pp. 133-138, 2013.
[http://dx.doi.org/10.1080/10426914.2012.746786]

[82]   H. Liu, Y. Chen, S. Pei, G. Liu, and J. Liu, "Preparation of nanocrystalline titanium dioxide fibers using sol–gel method and centrifugal spinning", *J. Sol-Gel Sci. Technol.,* vol. 65, no. 3, pp. 443-451, 2013.
[http://dx.doi.org/10.1007/s10971-012-2956-7]

[83]   L. Li, W. Kang, Y. Zhao, Y. Li, J. Shi, and B. Cheng, "Preparation of flexible ultra-fine Al_2O_3 fiber mats *via* the solution blowing method", *Ceram. Int.,* vol. 41, no. 1, pp. 409-415, 2015.
[http://dx.doi.org/10.1016/j.ceramint.2014.08.085]

[84]   A. Oya, and T. Sando, "Method of producing carbon nanomaterials and centrifugal melt spinning apparatus", *Google Patents,* 2010.

[85]   M.W. Lee, S.S. Yoon, and A.L. Yarin, "Solution-blown core–shell self-healing nano-and microfibers", *ACS Appl. Mater. Interfaces,* vol. 8, no. 7, pp. 4955-4962, 2016.
[http://dx.doi.org/10.1021/acsami.5b12358] [PMID: 26836581]

[86]   M.R. Badrossamay, H.A. McIlwee, J.A. Goss, and K.K. Parker, "Nanofiber assembly by rotary jet-spinning", *Nano Lett.,* vol. 10, no. 6, pp. 2257-2261, 2010.
[http://dx.doi.org/10.1021/nl101355x] [PMID: 20491499]

[87]   E.L.G. Medeiros, A.L. Braz, I.J. Porto, A. Menner, A. Bismarck, A.R. Boccaccini, W.C. Lepry, S.N. Nazhat, E.S. Medeiros, and J.J. Blaker, "Porous bioactive nanofibers *via* cryogenic solution blow spinning and their formation into 3D macroporous scaffolds", *ACS Biomater. Sci. Eng.,* vol. 2, no. 9, pp. 1442-1449, 2016.
[http://dx.doi.org/10.1021/acsbiomaterials.6b00072] [PMID: 33440582]

[88]   H. Peng, Y. Liu, and S. Ramakrishna, "Recent development of centrifugal electrospinning", *J. Appl. Polym. Sci.,* vol. 134, no. 10, 2017.
[http://dx.doi.org/10.1002/app.44578]

[89]   F. Dabirian, S.A. Hosseini Ravandi, A.R. Pishevar, and R.A. Abuzade, "A comparative study of jet formation and nanofiber alignment in electrospinning and electrocentrifugal spinning systems", *J. Electrost.,* vol. 69, no. 6, pp. 540-546, 2011.
[http://dx.doi.org/10.1016/j.elstat.2011.07.006]

[90]   Q. Zhang, J. Welch, H. Park, C.Y. Wu, W. Sigmund, and J.C.M. Marijnissen, "Improvement in nanofiber filtration by multiple thin layers of nanofiber mats", *J. Aerosol Sci.,* vol. 41, no. 2, pp. 230-236, 2010.
[http://dx.doi.org/10.1016/j.jaerosci.2009.10.001]

[91]   K. Molnar, and Z.K. Nagy, "Corona-electrospinning: Needleless method for high-throughput continuous nanofiber production", *Eur. Polym. J.,* vol. 74, pp. 279-286, 2016.
[http://dx.doi.org/10.1016/j.eurpolymj.2015.11.028]

[92]   N. Bu, Y. Huang, X. Wang, and Z. Yin, "Continuously tunable and oriented nanofiber direct-written by mechano-electrospinning", *Mater. Manuf. Process.,* vol. 27, no. 12, pp. 1318-1323, 2012.
[http://dx.doi.org/10.1080/10426914.2012.700145]

[93]   F. Zhu, Q. Xin, Q. Feng, Y. Zhou, and R. Liu, "Novel poly(vinylidene fluoride)/thermoplastic polyester elastomer composite membrane prepared by the electrospinning of nanofibers onto a dense membrane substrate for protective textiles", *J. Appl. Polym. Sci.,* vol. 132, no. 26, p. n/a, 2015.
[http://dx.doi.org/10.1002/app.42170]

[94]   M. Obaid, M.A. Abdelkareem, S. Kook, H-Y. Kim, N. Hilal, N. Ghaffour, and I.S. Kim, "Breakthroughs in the fabrication of electrospun-nanofiber-supported thin film composite/nanocomposite membranes for the forward osmosis process: A review", *Crit. Rev. Environ. Sci. Technol.,* vol. 50, no. 17, pp. 1727-1795, 2020.
[http://dx.doi.org/10.1080/10643389.2019.1672510]

[95]   D. Feldman, "Polyblend Nanocomposites", *J. Macromol. Sci. Part A Pure Appl. Chem.,* vol. 52, no. 8, pp. 648-658, 2015.
[http://dx.doi.org/10.1080/10601325.2015.1050638]

[96]   S. Tan, X. Huang, and B. Wu, "Some fascinating phenomena in electrospinning processes and applications of electrospun nanofibers", *Polym. Int.,* vol. 56, no. 11, pp. 1330-1339, 2007.
[http://dx.doi.org/10.1002/pi.2354]

[97]   X. Lu, C. Wang, and Y. Wei, "One-dimensional composite nanomaterials: synthesis by electrospinning and their applications", *Small,* vol. 5, no. 21, pp. 2349-2370, 2009.
[http://dx.doi.org/10.1002/smll.200900445] [PMID: 19771565]

[98]   Y. Lu, Y. Li, S. Zhang, G. Xu, K. Fu, H. Lee, and X. Zhang, "Parameter study and characterization for polyacrylonitrile nanofibers fabricated *via* centrifugal spinning process", *Eur. Polym. J.,* vol. 49, no. 12, pp. 3834-3845, 2013.
[http://dx.doi.org/10.1016/j.eurpolymj.2013.09.017]

[99]   R. Leidy, and Q-C. Maria Ximena, "Use of electrospinning technique to produce nanofibres for food industries: A perspective from regulations to characterisations", *Trends Food Sci. Technol.,* vol. 85, pp. 92-106, 2019.
[http://dx.doi.org/10.1016/j.tifs.2019.01.006]

[100]   D. Lv, M. Zhu, Z. Jiang, S. Jiang, Q. Zhang, R. Xiong, and C. Huang, "Green electrospun nanofibers and their application in air filtration", *Macromol. Mater. Eng.,* vol. 303, no. 12, p. 1800336, 2018.
[http://dx.doi.org/10.1002/mame.201800336]

[101]   Y. Lu, K. Fu, S. Zhang, Y. Li, C. Chen, J. Zhu, M. Yanilmaz, M. Dirican, and X. Zhang, "Centrifugal spinning: A novel approach to fabricate porous carbon fibers as binder-free electrodes for electric

double-layer capacitors", *J. Power Sources,* vol. 273, pp. 502-510, 2015.
[http://dx.doi.org/10.1016/j.jpowsour.2014.09.130]

[102]   M. Tang, F. Liu, X. He, K. Sun, T. Wang, K. Liu, Z. Huang, and M. Fan, "Effective carbon dioxide stabilization of nanofibers electrospun from raw coal tar and polyacrylonitrile", *J. Clean. Prod.,* vol. 276, p. 123229, 2020.
[http://dx.doi.org/10.1016/j.jclepro.2020.123229]

[103]   K. Mondal, M.A. Ali, C. Singh, G. Sumana, B.D. Malhotra, and A. Sharma, "Highly sensitive porous carbon and metal/carbon conducting nanofiber based enzymatic biosensors for triglyceride detection", *Sens. Actuators B Chem.,* vol. 246, pp. 202-214, 2017.
[http://dx.doi.org/10.1016/j.snb.2017.02.050]

[104]   C. Guo, M. Hu, Z. Li, F. Duan, L. He, Z. Zhang, F. Marchetti, and M. Du, "Structural hybridization of bimetallic zeolitic imidazolate framework (ZIF) nanosheets and carbon nanofibers for efficiently sensing α-synuclein oligomers", *Sens. Actuators B Chem.,* vol. 309, p. 127821, 2020.
[http://dx.doi.org/10.1016/j.snb.2020.127821]

[105]   W. Zhang, R. Zhu, L. Ke, X. Liu, B. Liu, and S. Ramakrishna, "Anatase mesoporous TiO_2 nanofibers with high surface area for solid-state dye-sensitized solar cells", *Small* vol. 6. , 2010, no. 19, pp. 2176-2182.
[http://dx.doi.org/10.1002/smll.201000759]

[106]   H. Zheng, K. Liu, H. Cao, and X. Zhang, "L-Lysine-assisted synthesis of ZrO_2 nanocrystals and their application in photocatalysis", *J. Phys. Chem. C,* vol. 113, no. 42, pp. 18259-18263, 2009.
[http://dx.doi.org/10.1021/jp9057324]

[107]   C. Chen, M. Dirican, and X. Zhang, "Centrifugal spinning—high rate production of nanofibers", In: *in Electrospinning: Nanofabrication and Applications* Elsevier, 2019, pp. 321-338.
[http://dx.doi.org/10.1016/B978-0-323-51270-1.00010-8]

[108]   M.J. Divvela, A.C. Ruo, Y. Zhmayev, and Y.L. Joo, "Discretized modeling for centrifugal spinning of viscoelastic liquids", *J. Non-Newt. Fluid Mech.,* vol. 247, pp. 62-77, 2017.
[http://dx.doi.org/10.1016/j.jnnfm.2017.06.005]

[109]   L. Ren, "Centrifugal jet spinning of polymer nanofiber assembly: process characterization and engineering applications", *Rensselaer Polytechnic Institute.,* 2014.

[110]   S. Qi, and D. Craig, "Recent developments in micro- and nanofabrication techniques for the preparation of amorphous pharmaceutical dosage forms", *Adv. Drug Deliv. Rev.,* vol. 100, pp. 67-84, 2016.
[http://dx.doi.org/10.1016/j.addr.2016.01.003] [PMID: 26776230]

[111]   S.M. Taghavi, and R.G. Larson, "Regularized thin-fiber model for nanofiber formation by centrifugal spinning", *Phys. Rev. E Stat. Nonlin. Soft Matter Phys.,* vol. 89, no. 2, p. 023011, 2014.
[http://dx.doi.org/10.1103/PhysRevE.89.023011] [PMID: 25353575]

[112]   A.A. Almetwally, M. El-Sakhawy, M. Elshakankery, and M. Kasem, "Technology of nano-fibers: Production techniques and properties-Critical review", *J. Text. Assoc,* vol. 78, no. 1, pp. 5-14, 2017.

[113]   L. Ren, R. Ozisik, S.P. Kotha, and P.T. Underhill, "Highly efficient fabrication of polymer nanofiber assembly by centrifugal jet spinning: process and characterization", *Macromolecules,* vol. 48, no. 8, pp. 2593-2602, 2015.
[http://dx.doi.org/10.1021/acs.macromol.5b00292]

[114]   J. Merchiers, W. Meurs, W. Deferme, R. Peeters, M. Buntinx, and N.K. Reddy, "Influence of Polymer Concentration and Nozzle Material on Centrifugal Fiber Spinning", *Polymers (Basel),* vol. 12, no. 3, p. 575, 2020.
[http://dx.doi.org/10.3390/polym12030575] [PMID: 32150836]

[115]   R.T. Weitz, L. Harnau, S. Rauschenbach, M. Burghard, and K. Kern, "Polymer nanofibers *via* nozzle-free centrifugal spinning", *Nano Lett.,* vol. 8, no. 4, pp. 1187-1191, 2008.

[http://dx.doi.org/10.1021/nl080124q] [PMID: 18307320]

[116] Y. Fang, "High performance nonwoven fiber production *via* UV-reactive and melt state centrifugal spinning", 2016,

[117] B. Ghorani, A. Alehosseini, and N. Tucker, "Nanocapsule formation by electrospinning", In: *Nanoencapsulation technologies for the food and nutraceutical industries.* Elsevier, 2017, pp. 264-319.
[http://dx.doi.org/10.1016/B978-0-12-809436-5.00008-2]

[118] L. Wang, J. Shi, L. Liu, E. Secret, and Y. Chen, "Fabrication of polymer fiber scaffolds by centrifugal spinning for cell culture studies", *Microelectron. Eng.,* vol. 88, no. 8, pp. 1718-1721, 2011.
[http://dx.doi.org/10.1016/j.mee.2010.12.054]

[119] M. Khamforoush, and T. Asgari, "A modified electro-centrifugal spinning method to enhance the production rate of highly aligned nanofiber", *Nano,* vol. 10, no. 2, p. 1550016, 2015.
[http://dx.doi.org/10.1142/S1793292015500162]

[120] N. Aliheidari, N. Aliahmad, M. Agarwal, and H. Dalir, "Electrospun nanofibers for label-free sensor applications", *Sensors (Basel),* vol. 19, no. 16, p. 3587, 2019.
[http://dx.doi.org/10.3390/s19163587] [PMID: 31426538]

[121] B. Zhang, J. Sun, U. Salahuddin, and P.X. Gao, "Hierarchical and scalable integration of nanostructures for energy and environmental applications: a review of processing, devices, and economic analyses", *Nano Futures,* vol. 4, no. 1, p. 012002, 2020.
[http://dx.doi.org/10.1088/2399-1984/ab75ad]

[122] Y. Gao, J. Zhang, Y. Su, H. Wang, X.X. Wang, L.P. Huang, M. Yu, S. Ramakrishna, and Y.Z. Long, "Recent progress and challenges in solution blow spinning", *Mater. Horiz.,* vol. 8, no. 2, pp. 426-446, 2021.
[http://dx.doi.org/10.1039/D0MH01096K] [PMID: 34821263]

[123] S. Ramakrishna, R. Jose, P.S. Archana, A.S. Nair, R. Balamurugan, J. Venugopal, and W.E. Teo, "Science and engineering of electrospun nanofibers for advances in clean energy, water filtration, and regenerative medicine", *J. Mater. Sci.,* vol. 45, no. 23, pp. 6283-6312, 2010.
[http://dx.doi.org/10.1007/s10853-010-4509-1]

[124] W. Sigmund, J. Yuh, H. Park, V. Maneeratana, G. Pyrgiotakis, A. Daga, J. Taylor, and J.C. Nino, "Processing and structure relationships in electrospinning of ceramic fiber systems", *J. Am. Ceram. Soc.,* vol. 89, no. 2, pp. 395-407, 2006.
[http://dx.doi.org/10.1111/j.1551-2916.2005.00807.x]

[125] Y. Gao, H-F. Xiang, X-X. Wang, K. Yan, Q. Liu, X. Li, R-Q. Liu, M. Yu, and Y-Z. Long, "A portable solution blow spinning device for minimally invasive surgery hemostasis", *Chem. Eng. J.,* vol. 387, p. 124052, 2020.
[http://dx.doi.org/10.1016/j.cej.2020.124052]

[126] E.N. Bolbasov, K.S. Stankevich, E.A. Sudarev, V.M. Bouznik, V.L. Kudryavtseva, L.V. Antonova, V.G. Matveeva, Y.G. Anissimov, and S.I. Tverdokhlebov, "The investigation of the production method influence on the structure and properties of the ferroelectric nonwoven materials based on vinylidene fluoride – tetrafluoroethylene copolymer", *Mater. Chem. Phys.,* vol. 182, pp. 338-346, 2016.
[http://dx.doi.org/10.1016/j.matchemphys.2016.07.041]

[127] R. Ramaseshan, *Decontamination of chemical warfare simulants using electrospun media.,* 2011.

[128] Q.S. Li, H.W. He, Z.Z. Fan, R.H. Zhao, F.X. Chen, R. Zhou, and X. Ning, "Preparation and Performance of Ultra-Fine Polypropylene Antibacterial Fibers *via* Melt Electrospinning", *Polymers (Basel),* vol. 12, no. 3, p. 606, 2020.
[http://dx.doi.org/10.3390/polym12030606] [PMID: 32155928]

[129] J.P. Vacanti, and R. Langer, "Tissue engineering: the design and fabrication of living replacement

devices for surgical reconstruction and transplantation", *Lancet,* vol. 354, suppl. Suppl. 1, pp. S32-S34, 1999.
[http://dx.doi.org/10.1016/S0140-6736(99)90247-7] [PMID: 10437854]

[130] C. Xu, R. Inai, M. Kotaki, and S. Ramakrishna, "Electrospun nanofiber fabrication as synthetic extracellular matrix and its potential for vascular tissue engineering", *Tissue Eng.,* vol. 10, no. 7-8, pp. 1160-1168, 2004.
[http://dx.doi.org/10.1089/ten.2004.10.1160] [PMID: 15363172]

[131] D.J. Behonick, and Z. Werb, "A bit of give and take: the relationship between the extracellular matrix and the developing chondrocyte", *Mech. Dev.,* vol. 120, no. 11, pp. 1327-1336, 2003.
[http://dx.doi.org/10.1016/j.mod.2003.05.002] [PMID: 14623441]

[132] R.T. Franceschi, and B.S. Iyer, "Relationship between collagen synthesis and expression of the osteoblast phenotype in MC3T3-E1 cells", *J. Bone Miner. Res.,* vol. 7, no. 2, pp. 235-246, 1992.
[http://dx.doi.org/10.1002/jbmr.5650070216] [PMID: 1373931]

[133] C.W. Lan, F.F. Wang, and Y.J. Wang, "Osteogenic enrichment of bone-marrow stromal cells with the use of flow chamber and type I collagen-coated surface", *J. Biomed. Mater. Res.,* vol. 66A, no. 1, pp. 38-46, 2003.
[http://dx.doi.org/10.1002/jbm.a.10507] [PMID: 12833429]

[134] X.M. Mo, C.Y. Xu, M. Kotaki, and S. Ramakrishna, "Electrospun P(LLA-CL) nanofiber: a biomimetic extracellular matrix for smooth muscle cell and endothelial cell proliferation", *Biomaterials,* vol. 25, no. 10, pp. 1883-1890, 2004.
[http://dx.doi.org/10.1016/j.biomaterials.2003.08.042] [PMID: 14738852]

[135] L.A. Smith, and P.X. Ma, "Nano-fibrous scaffolds for tissue engineering", *Colloids Surf. B Biointerfaces,* vol. 39, no. 3, pp. 125-131, 2004.
[http://dx.doi.org/10.1016/j.colsurfb.2003.12.004] [PMID: 15556341]

[136] B. Sharma, and J.H. Elisseeff, "Engineering structurally organized cartilage and bone tissues", *Ann. Biomed. Eng.,* vol. 32, no. 1, pp. 148-159, 2004.
[http://dx.doi.org/10.1023/B:ABME.0000007799.60142.78] [PMID: 14964730]

[137] D.W. Hutmacher, "Scaffolds in tissue engineering bone and cartilage", *Biomaterials,* vol. 21, no. 24, pp. 2529-2543, 2000.
[http://dx.doi.org/10.1016/S0142-9612(00)00121-6] [PMID: 11071603]

[138] F. Rosso, G. Marino, A. Giordano, M. Barbarisi, D. Parmeggiani, and A. Barbarisi, "Smart materials as scaffolds for tissue engineering", *J. Cell. Physiol.,* vol. 203, no. 3, pp. 465-470, 2005.
[http://dx.doi.org/10.1002/jcp.20270] [PMID: 15744740]

[139] R.G. Flemming, C.J. Murphy, G.A. Abrams, S.L. Goodman, and P.F. Nealey, "Effects of synthetic micro- and nano-structured surfaces on cell behavior", *Biomaterials,* vol. 20, no. 6, pp. 573-588, 1999.
[http://dx.doi.org/10.1016/S0142-9612(98)00209-9] [PMID: 10213360]

[140] R.L. Price, L.G. Gutwein, L. Kaledin, F. Tepper, and T.J. Webster, "Osteoblast function on nanophase alumina materials: Influence of chemistry, phase, and topography", *J. Biomed. Mater. Res. A,* vol. 67A, no. 4, pp. 1284-1293, 2003.
[http://dx.doi.org/10.1002/jbm.a.20011] [PMID: 14624515]

[141] Z. Ma, M. Kotaki, R. Inai, and S. Ramakrishna, "Potential of nanofiber matrix as tissue-engineering scaffolds", *Tissue Eng.,* vol. 11, no. 1-2, pp. 101-109, 2005.
[http://dx.doi.org/10.1089/ten.2005.11.101] [PMID: 15738665]

[142] A.I. Teixeira, G.A. Abrams, P.J. Bertics, C.J. Murphy, and P.F. Nealey, "Epithelial contact guidance on well-defined micro- and nanostructured substrates", *J. Cell Sci.,* vol. 116, no. 10, pp. 1881-1892, 2003.
[http://dx.doi.org/10.1242/jcs.00383] [PMID: 12692189]

[143] F. Yang, R. Murugan, S. Ramakrishna, X. Wang, Y.X. Ma, and S. Wang, "Fabrication of nano-

structured porous PLLA scaffold intended for nerve tissue engineering", *Biomaterials,* vol. 25, no. 10, pp. 1891-1900, 2004.
[http://dx.doi.org/10.1016/j.biomaterials.2003.08.062] [PMID: 14738853]

[144] K.L. Elias, R.L. Price, and T.J. Webster, "Enhanced functions of osteoblasts on nanometer diameter carbon fibers", *Biomaterials,* vol. 23, no. 15, pp. 3279-3287, 2002.
[http://dx.doi.org/10.1016/S0142-9612(02)00087-X] [PMID: 12102199]

CHAPTER 5

Electrospun Nanofibers Scaffolds: Fabrication, Characterization and Biomedical Applications

Murtada A. Oshi[1], Abdul Muhaymin[2], Ammara Safdar[2], Meshal Gul[3], Kainat Tufail[3], Fazli Khuda[3], Sultan Ullah[4], Fakhar-ud-Din[5], Fazli Subhan[6] and Muhammad Naeem[6,*]

[1] *Department of Pharmaceutics, Faculty of Pharmacy, Omdurman Islamic University, Omdurman, Sudan*

[2] *Preston Institute of Nanoscience and Technology, Preston University Kohat, Islamabad Campus, Islamabad, Pakistan*

[3] *Department of Pharmacy, University of Peshawar, Khyber Pakhtoonkhwa, Pakistan*

[4] *Department of Molecular Medicine, The Scripps Research Institute, Florida, USA*

[5] *Department of Pharmacy, Quaid-i- Azam University, Islamabad, Pakistan*

[6] *Department of Biological Sciences, National University of Medical Sciences, Rawalpindi, Pakistan*

Abstract: The electrospinning (ES) technique in the fabrication of biomaterials-based electrospun nanofibers (ESNFs) has risen to prominence because of its accessibility, cost-effectiveness, high production rate and diverse biomedical applications. The ESNFs have unique characteristics, such as stability and mechanical performance, high permeability, porosity, high surface area to volume ratio, and ease of functionalization. The characteristics of ESNFs can be controlled by varying either process variables or biomaterial solution properties. The active pharmaceutical agents can be introduced into ESNFs by blending, surface modification, or emulsion formation. In this chapter, in the first part, we briefly discuss the fundamental aspects of the fabrication, commonly used materials, process parameters, and characterization of ESNFs. In the second part, we discuss in detail the biomedical applications of ESNFs in drug delivery, tissue engineering, and wound healings, cancer therapy, dentistry, medical filtration, biosensing and imaging of disease.

Keywords: Biomedical Applications, Electrospinning, Electrospun Nanofibers.

* **Corresponding author Muhammad Naeem:** Department of Biological Sciences, National University of Medical Sciences, Rawalpindi, Pakistan; E-mail: m.naeem@numspak.edu.pk

Adnan Haider & Sajjad Haider (Eds.)

INTRODUCTION

Nanotechnology deals with the fabrication of materials ranging from 1 nm to about 1000 nm (nanomaterials) and represents one of the newest approaches in medicine and science. Due to their unique physicochemical properties and biocompatibility, nanomaterials have increasingly been used in a variety of biological applications, including drug delivery, wound healing, and tissue engineering [1]. Electrospun nanofibers (ESNFs) are an example of nanomaterials that are mostly fabricated *via* the electrospinning technique [2]. Rayleigh introduced electrospinning in 1897. This is a flexible process in which ESNFs are produced from polymer solutions using an electric field [3]. It is worth mentioning that the ESNFs can be produced using either natural polymers, such as chitosan, alginates, collagen, and gelatin or synthetic polymers, such as poly (lactic-co-glycolic acid) (PLGA), poly(ethylene-co-vinyl acetate) (PEVA), poly(lactic acid) (PLA) polyvinyl alcohol (PVA) and polycaprolactone (PCL) [4, 5].

ESNFs are used in various fields, such as air and water filtration [6], semiconductors and sensors [7], sound absorptions [8], chemical resistance [9, 10], and clean energy [11]. However, the most pivotal applications of ESNFs lie in the biomedical fields, which include cancer therapy, drug delivery, dentistry, wound dressing, tissue engineering and diagnosis of disease [12]. The versatile biomedical application of ESNFs is attributed to their unique properties, such as large surface area and variable porosity. Moreover, the unique chemical composition and physicochemical characteristics of ESNFs usually facilitate the incorporation of hydrophilic and hydrophobic drugs [4]. The usefulness of ESNFs using polymers (biocompatible and biodegradable/non biodegradable) and other compounds can be predicted from the fact that research and review articles are published regularly. This book chapter highlights the aforementioned promising biomedical advances of ESNFs reported in literature.

Fundamental Aspects of Electrospinning

Electrospinning is a simple and versatile nanofiber fabrication process that uses a strong electric field to transform a viscoelastic fluid (*e.g.*, a polymer solution) into continuous nanosized fibers. The polymer solution is pushed from a syringe towards the tip of a metallic needle. The fiber jets are generated from the Taylor cone (formed at the tip of the metallic needle) when high electrostatic forces overcome the cohesive forces [13].

The instrument used in electrospinning consists of a syringe pump and a syringe with a metallic needle, a high voltage power supply as a power source, and a collector plate (grounded metal plate) (Fig. **1**). To operate the instrument, the syringe is filled with the polymer solution and the orifice of the needle is connected to one terminal of the high voltage power supply, and the other terminal of the power supply is connected to the collector [2, 10]. The main function of the syringe is to pump the polymer solution at a constant flow rate (mL/h) to produce continuous ESNFs. The electrostatic forces overcome the surface tension and form fibrous jets (the Taylor cone formed at the tip of the needle), which are collected at the collector. The electric voltage range is from about 10 to 50 kV approximately [10].

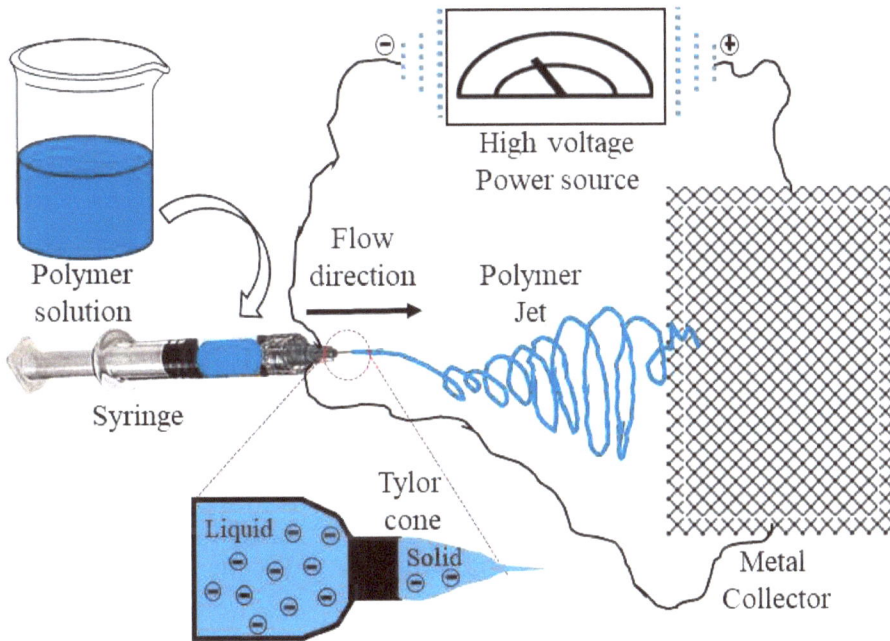

Fig. (1). Schematic diagram representing the fabrication of ESNFs.

Electrospinning Techniques for ESNFs Fabrication

Blending Electrospinning

In the blending approach, the drug is dissolved or distributed in a polymeric solution which is then subjected to the process of electrospinning. The relationship between the mechanical and physicochemical properties of the obtained ESNFs can be enhanced mainly by using the polymer blend. The polymeric blend is an effective means to control the release rate of the drug from the ESNFs [14, 15].

Coaxial Electrospinning

The coaxial electrospinning produces ESNFs with a core-shell structure, which encapsulates drugs in the core regions, enhancing biomolecule functionality, bioactivities and drug release [10]. In the coaxial electrospinning technique, a coaxial spinneret with a small capillary fitted inside a larger outer capillary is used. To prepare core-shell structured ESNFs, the inner capillary of the spinneret is filled with the core-forming polymer and the outer shell with the shell-forming polymer. The spinneret orifice discharges the polymer solutions into an electric field; an optimized electric field applied to the polymer solution induces the Taylor cone to generate core/shell ESNFs [14, 16]. The properties of core-shell ESNFs depend on the rheological, chemical and physical differences between the core-forming polymer and shell-forming polymer. A uniform core-shell ESNFs can be easily obtained if a stable Taylor cone is maintained [14, 16].

Emulsion Electrospinning

In this ESNFs preparation technique, the oil phase of the emulsion is generated by using the drug or aqueous protein solution in the polymer solution, followed by electrospinning [10]. Compared to blending electrospinning, this technique is characterized by the ability to eliminate the necessity for a common solvent. As a result, a variety of hydrophilic and hydrophobic drugs and polymeric solutions are used [17]. Hu *et al.* used emulsion electrospinning to load metformin hydrochloride or metoprolol in Polycaprolactone (PCL and poly (3-hydroxybutyric acid-co-3-hydroxyvaleric acid) (PHBV). By regulating the water and oil phase of the emulsions, the obtained ESNFs demonstrated a good drug release. However, among the tested polymers, PCL had shown better drug delivery contrary to PHBV [18].

Melt Electrospinning

The melt electrospinning technique is used for the polymer, which tends to block the metallic needle, thus hampering the flow of the polymeric solution, which in turn hinders the electrospinning process [19]. In this technique prior to electrospinning, the polymer is melted. Compared to solutions electrospinning, in melt electrospinning, thicker diameter nanofibers are obtained, which in turn impact the morphology of the ESNFs [20].

Materials used for Fabrication of ESNFs

Both natural and synthetic polymers are usually employed to make ESNFs. Polysaccharides and proteins are the most commonly employed natural polymers in electrospinning for the fabrication of ESNFs. Among the polysaccharides, cellulose, alginate derivatives, and chitosan have the potential to fabricate ESNFs. Proteins are the second most frequent natural polymer used in the production of ESNFs. The most common proteins used for the preparation of ESNFs are collagen, elastin, and silk proteins. Besides polysaccharides and proteins, other various types of natural polymers have also been used for the fabrication of ESNFs, such as poly (hydroxybutyrate) (PHB) and PCL. PHB is a synthetic polyester produced by a variety of micro-organisms and belongs to the class of polyhydroxyalkanoates [21]. It is a biocompatible and biodegradable polymer combined with other polymers to enhance its qualities [22]. PCL is a biodegradable polyester with a glass transition temperature of -60°C and a low melting point of 60°C. PCL is most commonly used in the manufacturing of polyurethane composites for various biomedical purposes [23]. ESNFs have been produced by combining PHB and PCL solutions for a variety of biomedical applications, including the integration of Spirulina sp [24].

Synthetic polymers predominate in the manufacture of ESNFs, including biological products. Polyethylene oxide (PEO), polyvinyl alcohol (PVA), PCL and its co-polymers, polyvinylpyrrolidone (PVP), and PLA are the most popular synthetic polymers used to produce ESNFs. The US Food and Drug Administration approved all of the above-mentioned polymers in different biomedical applications such as drug delivery and tissue engineering [25]. Such polymers can be employed alone and with other synthetic or natural polymers.

Table **1** shows additional types of materials used in the fabrication of ESNFs, with their typical biomedical applications.

Table 1. Additional examples of polymers used in ESNFs fabrication with their applications.

Drug/Cell Loaded	Polymer(s) Type	Application	Refs.
Curcumin	Poly L-lactide	Drug delivery	[26]
Tranexamic acid	Chitosan	Drug delivery	[27]
Metformin hydrochloride	PCL + poly(3-hydroxybutyric acid-co-3-hydroxyvaleric acid)	Drug delivery	[18]
Insulin	PVA/sodium alginate	Drug delivery	[28]
Ciprofloxacin	Gelatin + PLC	Drug delivery	[29]
	Chitosan + PVA	Wound healing	[30]
	gelatin/polyurethane	Wound healing	[31]

(Table 1) cont.....

Drug/Cell Loaded	Polymer(s) Type	Application	Refs.
Curcumin	PLGA	Wound healing	[32]
	PVA + HA + L-arginine	Wound healing	[33]
	PLGA	Tissue engineering	[34]
	Gellan/PVA	Tissue engineering	[35]
	PEG + PCL	Tissue engineering	[36]
	PLGA + Poly(acrylic acid) (PAA)	Tissue engineering	[37]
Adipose-derived stem cells	PLA	Tissue engineering	[38]

Parameters Affect Electrospinning

Solution Parameters

The conversion of polymer solutions into ESNFs is influenced by solution parameters, such as the solvent used, solution viscosity, polymer concentration, conductivity, and surface tension (Fig. **2**). In general, these parameters directly affect the preparation of smooth, bead-free ESNFs. Therefore, it is necessary to carefully control these parameters to produce the desired ESNFs. In the following sections, each of these parameters is briefly explained.

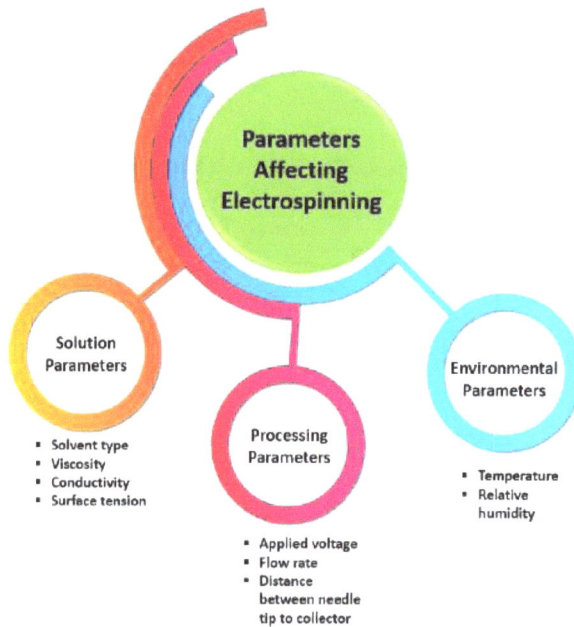

Fig. (2). Depicitng Parameters affecting electrospinning processes such as types of a solution, processing, and environmental parameters.

i. Solvent: The production of smooth ESNFs depends on the solvent. The solvents used in electrospinning must be fully miscible with the chosen polymers. In addition, the solvents should have a low to medium boiling point as this directly affects the volatility of the solvent [4]. The use of highly volatile solvents with a low boiling point would cause the polymer solution to dry out at the tip of the metal needle. As a result, the solution would clog the tip of the needle, leading to an interruption of the electrospinning process. On the other hand, if low-volatility solvents with high boiling points are used, the solvent evaporates slowly, resulting in beaded ESNFs [4, 39].

ii. Solution viscosity and concentration: In the production of ESNFs, the selection of an appropriate solution viscosity is a crucial parameter for a successful electrospinning process. At a low solution viscosity, the charged jets lose their intermolecular attraction and beads are formed instead of smooth, uniform ESNFs, while at high viscosity, the diameter of the produced nanofibers is larger [40, 41]. The reported optimal viscosity range for successful electrospinning of polymer solution is about 1 to 200 poise.

iii. Solution conductivity: The conductivity of a solution depends mainly on the type of polymer and solvent used. It is reported that the conductivity of the solution has a direct effect on the formation of the Taylor cone. When a solution with low conductivity is used in electrospinning, the charges on the surface of the pendant droplet hinder the formation of the Taylor cone [42]. On the other hand, the use of a solution with a conductivity above the threshold value leads to an increased accumulation of surface charges, which in turn leads to the formation of the Taylor cone [4, 42].

Processing Parameters

The conversion of polymer solutions into ESNFs is influenced by electrospinning processing parameters such as needle diameter, applied voltage, the distance between needle tip and collector, and flow rate (Fig. **2**). In the following sections, each of these parameters is briefly discussed:

i. Applied voltage: Applying a high voltage electric field to a polymeric solution *via* a metal needle causes the pendent droplet to form a deformed Taylor cone (Fig. **1**). When the electric field voltage reaches a critical value, the droplet forms ultrafine ESNFs [43]. The critical voltage is different for each polymer. Several studies have confirmed that the application of a high electric current leads to the formation of large diameter ESNFs. From the available literature, it is concluded that the electrical voltage has a direct effect on the morphology of ESNFs [44].

ii. Flow rate: In the preparation of ESNFs using electrospinning, the shape and morphology of the resulting ESNFs can be determined by controlling the flow rate of the polymer solution [41]. An optimized flow rate leads to the formation of uniform, beadless ESNFs. If you increase the flow rate beyond the critical value, beaded ESNFs may form (Fig. **1**).

iii. Distance between the needle tip and the collector: The distance between the needle tip and the collector is an important factor that determines the shape and morphology of the ESNFs (Fig. **1**). This is because the distance can easily affect the deposition time and evaporation rate. Therefore, a proper distance between the needle tip and the collector (critical distance) is very important to produce smooth and uniform ESNFs. The study of the distance between the needle tip and the collector has been a hot research topic for many research groups [45]. In most of these studies, the formation of ESNFs with a larger diameter was observed when the distance between the metal needle and the collector was smaller. In contrast, the diameter becomes smaller when the distance between the needle and the collector becomes larger. However, in a few studies, the distance between the collector and the needle was found to have no effect on the morphology of ESNFs [41].

Environmental Parameters

The conversion of polymer solutions into ESNFs by electrospinning is influenced by environmental parameters such as relative humidity and temperature (Fig. **2**). In the following sections, these two parameters are briefly discussed:

i. Temperature: The ambient temperature affects the morphology of ESNFs. At low temperatures, the evaporation rate of the solvent is low, so the solidification of ESNFs takes a long time, which in turn leads to the formation of beads. In addition, the high temperature would lead to a decrease in the viscosity of the polymer solution, resulting in the elongation of the ESNFs [46]. (ii) Humidity: A change in the humidity can affect the morphology of ESNFs. Casper *et al.* showed that pores form on the surface of the nanofibers at high humidity [47, 48]. Therefore, for the fabrication of porous ESNFs, the optimization and control of humidity is of utmost importance [49].

Mechanisms of Drug Loading into ESNFs

Blending Method

Of all the drug-loading mechanisms in ESNFs, blending remains the most popular. In this process, the drug is either dissolved or dispersed in a polymer solution and then encapsulated by a single-phase electrospinning technique (Fig. **3**). To obtain ESNFs with sustained drug release profile, blending and incorporating are carried out using different polymer blends (hydrophilic/hydrophobic polymers). Meng *et al*. prepared PLGA and PLGA/gelatin ESNFs loaded with fenbufen using the ES technique and investigated the effect of gelatin content on the drug release behavior. Drug release was increased in all ESNF composite samples when gelatin content was increased [50]. In another study, Jannesari *et al*. prepared controlled release ESNFs to accelerate wound healing with PVA and poly(vinyl acetate). *In vitro* drug release evaluations showed that the release and release rate of the drug was controlled by using the above polymer and technique to prepare ESNFs [51].

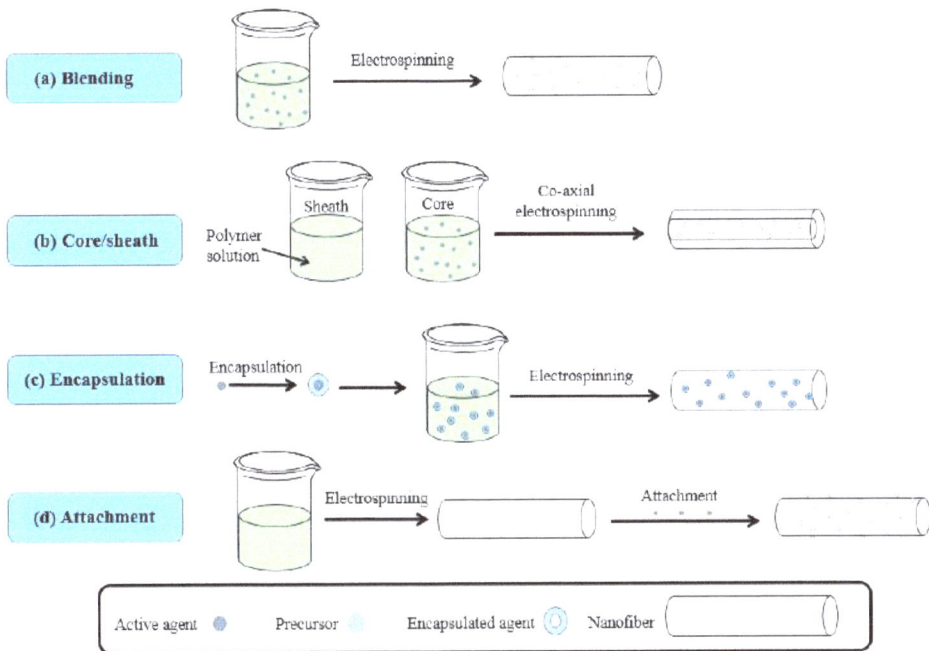

Fig. (3). Mechanisms of drug loading into ESNFs: **(a)** blending, **(b)** core/sheath, **(c)** encapsulation and **(d)** attachment.

Core/Sheath Method

For drug delivery, ESNFs are primarily used to administer the required drug for a specified time, depending on the condition of the disease [52]. A drug-loaded ESNFs can be produced using a drug and a polymer solution. Mostly the fabricated ESNFs show the burst release rate of the loaded drug (initial burst release). In this case, techniques such as cross-linking or chemical modifications of the ESNFs are necessary to overcome/reduce the initial burst release of the loaded drug. However, they can cause toxicity and reduction of biocompatibility [53]. The incorporation of drugs in ESNFs scaffolds remains challenging as, during the process, both the ESNFs and the drug should not lose their scaffolding properties. To solve this problem, many modifications were incorporated into the electrospinning technique to produce ESNFs with improved performance. For example, the preparation of ESNFs uses a core-shell design, in which one polymer fiber is entrapped inside the other polymer fibers, thus, the fabricated scaffolds have properties of both the polymers. Core/sheath design includes the coating of fiber with a shell that could effectively control the release profile of the drugs from the ESNFs (Fig. **3**). In addition, the core-shell design can be used to avoid any damage caused to the incorporated drugs. Recently, many core/sheath ESNFs have been prepared for loading various bioactive molecules of drugs for the slow and sustained release of the drug. The main advantages of core/sheath ESNFs include preventing initial burst drug-release, enabling sustained drug-releasing, and more than one drug can be loaded in the core, which can provide better therapeutic effect and reduced toxicity [52, 54].

Encapsulation Method

In encapsulation methods, the drug solution (aqueous solution) is blended within a polymer solution (oily solution), and after electrospinning, the drug-loaded phase can be dispersed within the ESNFs (Fig. **3**). As compared to the conventional blending method of drug loading into ESNFs, the advantage of the encapsulation method is that the drug and the polymer are dissolved in suitable solvents, thus excluding the need for a common solvent. This also enables the researchers to fabricate numerous hydrophilic drugs and hydrophobic polymer combinations [55]. However, controlling the compatibility between the drug and polymer is still an issue [56].

Attachment Method

Another favorable method for the drug loading into ESNFs is the surface modification of ESNFs with drugs. Using this method, the drugs are attached to the ESNFs surfaces (Fig. **3**) [57]. This method is frequently used to overcome the hurdles of the burst release of drugs from ESNFs. This method is used for biomolecules, such as genes and growth factors, where a sustained and prolonged drug release is required [58]. This technique is also pivotal for the loading of drugs which can get degraded and ultimately results in losing their efficacy [57].

Characterization of ESNFs

Physical Characterization

ESNFs are characterized *via* numerous techniques, however, in this book chapter, we only focused on techniques used for characterizing the morphology and surface area of ESNFs.

Morphology

For determining the morphological features of ESNFs, a scanning electron microscope (SEM) is most commonly used. For SEM, dry samples of ESNFs must be used and coated with gold or platinum to render them electrically conductive. Raghavan *et al.* used SEM to observe the bead morphology of polyacrylonitrile (PAN)-based ESNFs. Also, they demonstrated that the morphology of the ESNFs can be affected by increasing PAN concentration. The formation of beads was observed when the PAN concentration was 8%, while fewer beads formed when the solution concentration was increased by more than 8%. They also investigated the effect of the solution's viscosity on the morphology of ESNFs and concluded that fewer beads formed when the viscosity of the solution was higher [59]. Moreover, SEM is used to investigate the architectural properties and determine whether the ESNFs are smooth, rough, or have a porous structure. Bognitzki *et al.* used PLA, polycarbonate, polyvinyl carbazole polymers and dichloromethane as a solvent for the fabrication of porous ESNFs. When the structure of ESNFs was examined using SEM, pores were observed on the surfaces of the ESNFs. The authors explained the formation of the pores by the rapid evaporation of dichloromethane from the surfaces of ESNFs [60]. Other techniques to analyze the morphology of ESNFs include transmission electron microscope (used to determine fiber diameter up to < 300 nm) and atomic force microscope (used to analyze the surface morphology of ESNFs and the roughness of the fiber [61].

Surface Area

It is well known that ESNFs are characterized by their large surface area. The large surface area of ESNFs can help to determine the adsorption capacity of the surface [4]. In practice, the Brunauer-Emmett-Teller test (BET) is an effective method to quantify the surface area of ESNFs. Hussain *et al.* prepared PAN-based ESNFs to determine their specific surface area; the BET technique was effectively used. They showed that the specific surface area increased when the diameter of ESNFs increased from 150 nm to 1.3 μm [62].

Tool used for Chemical Characterization of ESNFs

To determine the chemical structure of ESNFs, various techniques such as Fourier transform infrared (FTIR) and nuclear magnetic resonance (NMR) can be used. Kiristia *et al.* prepared ESNFs using (PMCh) poly 3,4-ethylenedioxythiophene (PEDOT) and PVA. They examined their samples by FTIR, which confirmed the successful preparation of their samples. The analysis revealed that the formation of hydrogen bonds between the constituents resulted in a stable interaction between the polymer molecules [63]. Other techniques, such as X-ray photoelectron spectroscopy (XPS), are used for qualitative and quantitative analysis of ESNFs. Kayaci *et al.* investigated the surface modification of ESNFs prepared from polyester and cyclodextrin using XPS. They observed an increase in elemental oxygen on the surface of ESNFs, which was attributed to the presence of cyclodextrin [64].

Biomedical Applications of ESNFs

The biomedical applications of ESNFs have increased significantly due to the ease of fabrication of nanofibers in various shapes and structures. Various materials are used to fabricate electrospun nanofibers for biomedical applications, such as polymers, due to their biocompatible and biodegradable properties. Different types of natural and synthetic polymers, such as poly(lactic-co-glycolic acid) (PLGA), polylactic acid (PLA), and polycaprolactone (PCL), which are either hydrophobic or hydrophilic, have been used as polymeric ESNFs for wound healing, tissue engineering, antimicrobial activity, cancer therapy, controlled release of a drug and for biosensing and imaging. The polymers can be further conjugated with ligands to perform various functions [65]. The biomedical applications of ESNFS are immense, and some of them are reviewed in this chapter.

ESNFs as Drug Delivery Carrier

ESNFs can mitigate problems with drug degradation by providing a robust storage network from which a drug can eventually be released. Incorporating the drug into a nanofiber can protect it from heat- and light-induced degradation and keep the drug away from molecules that can lead to instability. The slow degradation of the nanofiber may allow the regulated release of a drug by a polymer-controlled mechanism. For example, titanocene dichloride ($TiCp_2Cl_2$) has been shown to be effective against a number of cancerous tumors, with better results than cisplatin for colon adenomas. Absorption of $TiCp_2Cl_2$ into poly(L-lactic acid) nanofibers may provide a regulated release with a total duration of approximately 7-8 days. Clinical trials are still in progress. However, similar to cisplatin, $TiCp_2Cl_2$ is unstable *in vivo* and has a short half-life in the human body with low water solubility [66]. Due to their novel surface chemistry, nanofibers are used in drug delivery systems as carriers of drugs to specific sites in the human body [67]. They can overcome drug delivery challenges such as poor solubility and stability and site-specific drug delivery. Due to their properties, such as a high surface area to volume ratio and short diffusion passage, ESNFs can transport different types of drugs such as antibiotics [68], anticancer drugs [69], proteins [70], and DNA

[71]. The drugs can either be loaded by mixing with the polymers of ESNFs or by fabrication of core-shell structure through coaxial spinneret [72, 73]. Similarly, the physical and chemical binding of ligand molecules leads to the tailoring of ESNFs for specific applications [74].

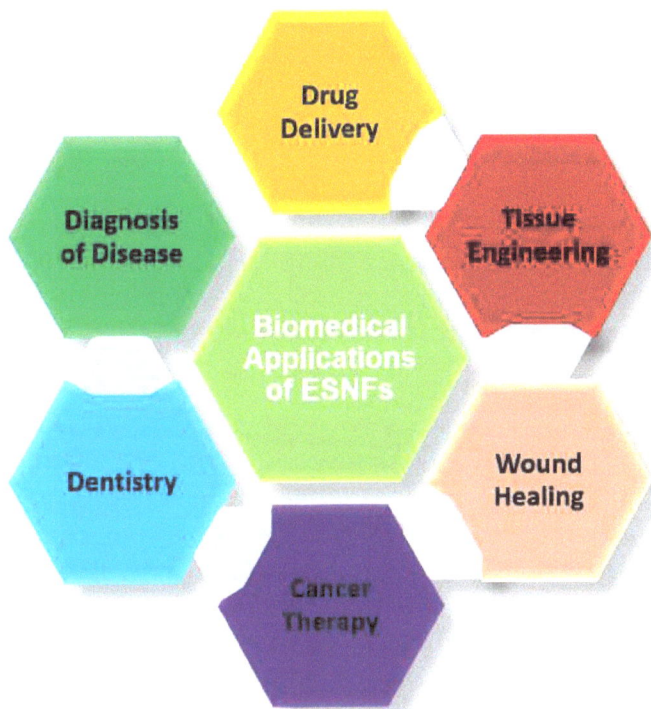

Fig. (4). Biomedical applications of ESNFs.

ESNFs have also been chosen to control the stability and hydrophobicity of drugs in the amorphous state. The arrangement of hydrophobic units depends on the interaction of hydrophilic molecules in solution. The hydrophobic drug and hydrophilic polymeric nanofibers, such as polyvinylpyrrolidone (PVP), are conjugated to form a composite structure. This is usually done by physically mixing the drug and polymeric nanofibers in a suitable solvent. The random movement of the drug and polymer in the solvent allows them to disperse at the molecular level. For example, Yu *et al.* developed ESNFs from PVP K30 containing the active ingredient ibuprofen [75, 76]. A uniform nanofiber solution was prepared from 40 wt% PVP and 7.5 wt% solutions of ibuprofen in ethanol. X-ray diffraction (XRD) and differential scanning calorimetry (DSC) were used to find the drug in an amorphous state in the fiber. Strong hydrogen bonding between ibuprofen and the polymer nanofibers was confirmed by IR spectroscopy, indicating that these two components are compatible with each other. Similarly, Meng *et al.* integrated fenbufen into poly(D, L-lactide-co-glycolide)/chitosan nanofibers in their study [77]. It was reported that the release rate of the drug can be regulated by adjusting the chitosan content of the fibers. Solubility problems often limit the use of antibiotics such as metronidazole. These can be overcome by incorporating them into nanofibers based on poly(ethylene glycol-co-lactide) or poly- DL -lactide co-polymers by electrospinning, resulting in a significant improvement in dissolution properties [78]. Similarly, electrospinning has been used to improve the solubility of active pharmaceutical ingredients (API) used in herbal medicine. Shikonin, for example, is a naturally occurring antibacterial, antioxidant and anti-inflammatory compound found in the roots of Boraginaceae plants. It is not soluble in hydrophilic solvents such as water. Han *et al.* reported the electrospinning of shikonin into poly(-caprolactone) and poly(trimethylene carbonate) fibers. *In vitro* experiments indicated that the nanofibers provided a continuous release of the drug, with over 20% of the incorporated drug distributed within the first 30 minutes [79].

In addition, nanofibers have cell-specific delivery as well as targeting specific organs or tissue. Shen *et al.* reported Eudragit L100–55 nanofibers loaded with diclofenac sodium (DS) [65]. Eudragit L100-55 is a pH-stimulating polymer that degrades only under alkaline conditions. Different compositions of DS with nanofibers were studied in a strongly acidic medium (pH=1) and phosphate-buffered saline (PBS) (pH=6.8). The results showed that the acidic buffer released very little drug, whereas PBS released virtually all of the loaded drug in a period of 60 to 300 minutes, depending on the composition of the nanofibers. Thus, this pH-dependent drug release of Eudragit DS nanofibers could be used in an oral drug release system by ensuring DS release in the intestinal tract (alkaline pH) for systemic circulation while preventing it in the stomach where the pH is more acidic [80]. By controlling parameters such as nanofiber diameter and mesh size, drug release can also be controlled in cancer and autoimmune diseases [81]. In 2002, Kenawy *et al.* studied the release of drugs by using ESNFs. Tetracycline hydrochloride was injected into PLA, poly(ethylene-co-vinyl acetate) (PEVA), and a 50:50 PLA/PEVA blend for the treatment of periodontal disease. The drug was fed into each system at a rate of 5%. The results showed that the drug release profile was influenced by the polymer used and the amount of drug loaded [82].

However, the main concern in drug delivery is the accumulation of drugs at the target site without causing side effects and degradation *in situ*. In recent years, adenoviral vectors (ADV) have been developed for gene delivery to kill tumor cells, but they have the disadvantage that a high accumulation of ADV is required for successful necrosis of the tumor. However, due to the non-specific delivery, the efficacy of ADV treatments is low. Therefore, ADV treatments are combined with other systems such as nanofibers to achieve good efficacy. Electrospinning was used to integrate ADVs into nanofibers made of a poly(ethylene glycol)-chitosan-folic acid co-polymer [83]. The incorporation of poly(ethylene glycol) into the fibers helped to limit nonspecific uptake, while folic acid was reported to enhance ADV transfer into specialized cell types.

ESNFs in Tissue Engineering

Tissue engineering is an interdisciplinary area that combines medical, biological, and engineering expertise to restore or reconstruct healthy tissues and organs. Scaffolds, cells and biomolecules are the three fundamentals of tissue engineering. Scaffolds are essential for tissue engineering. Scaffolds are used to direct the growth of cells, which are either implanted into the porous structure of the scaffold or migrate in from the surrounding tissue. Nanofiber scaffolds have a variety of physicochemical properties, including the ability to maintain and induce differentiation, transport cells, channel tissue formation, promote cell adhesion, and stimulate cell response [84].

The composition (*i.e.*, biomaterials of synthetic or natural origin) and architecture of a tissue-engineered scaffold control cell response. It has long been believed that the ECM should be mimicked to replicate all necessary intercellular responses and facilitate native intracellular responses. The goal is to help the body repair itself *via* a tissue-engineered scaffold. These artificial ECMs or scaffolds are manufactured to meet the desired requirements for cell regeneration. The high performance of ESNF materials in supporting tissue regeneration has been demonstrated [85].

ESNFs have been effectively used as scaffolds for tissue engineering purposes because they mimic the typical fibrous extracellular matrix (ECM) and exhibit a three-dimensional (3D) porous network [86]. Electrospinning is suitable for the fabrication of grafts for small diameter blood vessels. Coaxial electrospinning was used to fabricate tubular grafts with an inner diameter of 4 mm using heparin-loaded PLCL nanofibers [87]. Heparin was released in two stages: a 50 percent explosive release on day 2, followed by a continuous release of up to 72 percent on day 14. Heparin loading was shown to significantly improve the efficacy of small diameter grafts in a canine artery model. Vascular endothelial growth factor (VEGF) is frequently introduced into scaffolds to promote proliferation of endothelial progenitor cells. Emulsion electrospinning was used to incorporate heparin and VEGF into PLCL nanofibers to construct vascular grafts for anticoagulation and rapid endothelialization [88]. Similarly, Haider *et al.* fabricated PLGA/nHA composite scaffolds for drug and growth factor delivery. From the analysis of their results, they concluded that PLGA/nHA composite nanofibers loaded with drugs and growth factors promote the growth of osteoblastic cells (bone-forming cells) [89, 90].

ESNFs in Wound Healing

The treatment of wounds has become a significant public health problem and economic burden in recent decades. Chronic wounds are becoming increasingly common, necessitating the production of successful therapeutic wound treatments. Dry gauze is still the most commonly used wound dressing due to its affordability and availability [91]. However, it has many disadvantages, including high absorbency, which can contribute to wound dehydration and bacterial infections and the risk of reinjuring the newly formed epithelium during or after gauze removal. As a result, more complicated dressings with low adhesion and semipermeable properties have been developed. Hydrocolloids and hydrogels, for example, use their hydrophilic properties to absorb exudate and allow gas exchange while keeping the environment moist and preventing microbial infection. The aforementioned materials possess antimicrobial properties, stimulate local cell migration and proliferation, and increase matrix deposition [92]. However, these materials are unable to accurately mimic the architecture of the skin ECM and provide a viable/native environment for cells [93, 94].

Therefore, the production of alternative and efficient wound dressings is required, as conventional wound dressings do not meet all the requirements of an ideal wound dressing. In this context, ESNFs have been extensively studied due to their demonstrated potential role in wound treatment [95]. The inherent properties of nanofibers, such as their large surface area and inter/intra-fibrous porosity, allow them to provide a native environment for cells [96 - 99]. Charernsriwilaiwat *et al.* developed chitosan-based nanofiber mats filled with lysozyme for wound healing [100]. Kossovich *et al.* proposed the use of chitosan-based ESNF for the treatment of burn wounds. Burn healing is considered one of the most complicated procedures. Grade IIIa and IIIb burns were analysed on the dressings made. The wound aeration, exudate absorption and infection control of these nanofiber dressings were satisfactory [101]. Similar antimicrobial dressings prepared by electrospinning have shown efficacy against both antibiotic-resistant and non-resistant microbes and especially bacteria [102].

ESNFs in Cancer Therapy

To minimize the side effects of numerous chemotherapy drugs, many studies have been conducted to discover treatments that specifically target cancer cells. The use of ESNFs as a local drug delivery system for chemotherapy after resecting a solid tumor is becoming increasingly popular because they offer a large surface-to-volume ratio [103]. Moreover, implantable drug delivery systems based on ESNFs have proven to be one of the most efficient options for localised cancer treatment. They enable on-site cancer diagnosis and therapy while limiting systemic toxicities and adverse effects on healthy cells. ESNFs can be used to limit the recurrence of a given tumor by implanting them into the postoperative tumor cavity, which also helps to prolong the release of the drug at the tumor site. Recently, a PLA fiber was encapsulated with doxorubicin (Dox) using the Dox fiber mat to cover the entire liver with carcinoma to improve the practical efficacy of local chemotherapy against liver carcinoma. The rapid release of Dox from the fiber mat within 24 hours and its localization in the liver tissue demonstrated the efficacy and safety of Dox fiber treatment and the inhibition of tumor growth [104]. Recently, a PLA fiber was encapsulated with doxorubicin (Dox) using the Dox fiber mat to cover the entire liver with carcinoma to improve the practical efficacy of local chemotherapy against liver carcinoma. The rapid release of Dox from the fiber mat within 24 hours and its localization in the liver tissue demonstrated the efficacy and safety of Dox fiber treatment and the inhibition of tumor growth [105]. Ranganath *et al.* introduced a successful electrospun fiber, poly-(D, L-lactide-co-glycolide), for the administration of paclitaxel for 80 days *In vitro* postoperatively in malignant gliomas and proved the suppression of tumor growth [106].

ESNFs also behave like an ECM that mimics the microenvironment of cells. In a solid tumor, cancer cells are closely associated with the ECM, which consists of collagen and fibronectin, elastin, laminin, proteoglycan, and so on. In some cases, the ECM also actively contributes to the regulation of tumor cells and their properties, especially in an abnormal environment that may be caused by various malignancies. In breast cancer, for example, cancer development is significantly affected by the stiffness of the ECM, which in turn also affects tumor migration. Also, in malignancies such as ovarian or small cell lung cancer, the expression of collagen VI correlates with the grade of the tumor, which upregulates cells that are resistant to drugs and makes them chemoresistant due to the adhesion of cancer cells to the ECM. Therefore, a 3D tumor model must be available to recapitulate the cell ECM features in tumor cells for which biomimetic substrates are produced [107].

In a recent study, polyaniline was incorporated into poly(ε-caprolactone) and gelatin (PG) to form nanofibers by electrostatic spinning, which can ablate tumor cells by converting optical energy into thermal energy when exposed to 808nm laser irradiation. In the treatment of hepatoma H22 tumor (*in vivo*), experimental results confirmed effective inhibition of tumor growth by polyaniline PG [108].

The ESNF system for cancer therapy is still in preclinical testing, as there are several challenges faced by healthcare providers or researchers in the application of ESNF in chemotherapy. These include obstacles in evenly distributing drugs in a given ESNF fiber or creating a 3D scaffold with the required porosity is sometimes almost impossible [103].

ESNFs for Dentistry Applications

ESNFs have significant applications in regeneration and bone tissue engineering [109]. For oral and dental tissues, various materials can be electrospun for tissue engineering, such as natural polymers (chitosan, collagen, silk), and synthetic polymers (polydioxanone, polyvinyl alcohol), and nanocomposites (hydroxyapatite blends) [110]. Dental tissues can be damaged by various pathophysiological problems such as caries and trauma. Tooth roots are regenerated by pulp therapies when they have suffered trauma. Various biomaterials such as ferric sulfate, mineral trioxide, and calcium hydroxide aggregates are used as medicine for pulpal and radicular dentin regeneration, but in some cases, complications occur, such as internal resorption of teeth [111]. Electrospun nanofibers scaffolds can induce odontoblast regeneration, which can greatly improve pulp therapy. Kim *et al.* developed electrospun scaffolds of PVA and hydroxyapatite (HA) with potential properties for dentin regeneration [112]. In addition, another study demonstrated the ability of electrospun PCL meshes to stimulate odontogenic differentiation and growth [113].

Endodontic therapy, also known as root canal treatment, also requires the administration of antimicrobial drugs into the pulp chamber and root canal to eradicate bacterial flora [114]. Bottino *et al.* developed electrospun scaffolds of polydioxanone (PDS) in which two antibiotics, *i.e.*, metronidazole and ciprofloxacin in formulations, were used. The results showed significant improvement in antimicrobial drug delivery compared to drugs delivered through pastes [115].

ESNFs in Imaging and Biosensing for Disease Diagnostics/Prognosis

Due to its high sensitivity, selectivity, and ease of miniaturization, the use of ESNFs is being extended to biosensing for various types of diseases [116].

Yoo *et al.* developed biosensors based on ESNFs to monitor temperature and impedance during cell-drug interaction. From their results, they concluded that the fabricated ESFNs can be used as a tool to measure/monitor the real-time response of cells and drugs [117]. Electrospun polyurethane and gelatin fibers have been fabricated to detect/measure glucose during implantation. However, work is in progress to fabricate various types of biosensors based on ESNFs, which may have potential applications in the biomedical field [116].

ESNFs in Membranes for Medical Filtration and Dialysis

The extraction of heavy metals from water has become a serious problem posing a serious threat to human health. Adsorption and filtration are the most commonly used methods to remove these pollutants. Heavy metals can be successfully removed by adsorption using electrospun nanofibers. The large surface-to-volume ratio, large surface area, good gas permeability and inter/intra fiber pore size of these electrospun membranes are crucial for the removal of pollutants from wastewater [118, 119]. Due to the above properties, ESNFs have the potential to be used as a component in dialysis. Tubular nanofiber membranes composed of polysulfone nanofibers were prepared for patients undergoing dialysis to remove urea and creatinine from blood and urine. Initially, these scaffolds were prepared by electrospinning 11.5 wt% polysulfone (PS) using the binary solvent dimethylformamide (DMF)/tetrahydrofuran (THF) in a 70/30 ratio. The removal efficiency of urea and creatinine was analyzed. It was found that the tubular membrane with an inner diameter of 3 mm had the highest efficiency in removing urea and creatinine [120].

A smaller dialyzer with a polymeric nanofiber nonwoven was developed as a portable chip-based hemodialysis system to investigate and improve the filtration efficiency for different particle sizes in a poly(dimethylsiloxane)-based microplatform. These microchip membranes have shown better filtration compared to other membranes where the blood was not affected at the time of filtration or during passage through the chip so that it can be used as a portable hemolysis system [121]. Using the ESNF technique, a zeolite-polymer composite nanofiber fleece was developed to remove uremic toxins from blood, especially for the treatment of patients with renal insufficiency in areas where resources for conventional hemodialysis are very limited. The low-cost ESNFs used in this device were made of blood-compatible poly(ethylene-co-vinyl alcohol) polymer (EVOH) and zeolites, which have the ability to absorb creatinine. Scanning electron microscopy examination revealed that a 7 w/v% EVOH solution produced nonwoven fibers with a continuous and smooth morphology, suggesting that such tools can be used to extract nitrogenous waste products from blood serum [122].

Table 2. Summarizing the biomedical applications of ESNFs [72].

Polymer	Drug	Spray Nozzle	Solvent
Polycaprolactone/ Polylactic acid (PCL/PLA)	Tetracycline	Single	Chloroform: Dimethylformamide
Polycaprolactone (PCL)	Resveratrol and Gentamycin sulphate (Antioxidants)	Coaxial	Ethanol: Chloroform (1:3)
Polyvinyl acetate (PVA)	Ciprofloxacin hydrochloride	Single	Acetic Acid

(Table 2) cont.....

Polymer	Drug	Spray Nozzle	Solvent
Poly lactic-co-glycolic acid (PLGA)	Rifampicin and Fusidic acid	Single	Dimethylformamide/Tetrahydrofuran (THF)
Poly lactic-co-glycolic acid (PLGA)	Cefoxitin	Single	Dimethylformamide (DMF)
Polycaprolactone (PCL)	Metronidazole benzoate	Single	Dichloromethane: Dimethylformamide (DCM: DMF)
Co-polylactic acid/polyethylene glycol	Moxifloxacin hydrochloride	Single	Dichloromethane: Dimethyl sulfoxide (3:1)
Poly l-lactic acid (PLLA)	Mupirocin and Lidocaine	Dual spinneret	Hexa-fluoro-iso-propanol
Polycaprolactone (PCL)	Ornidazole	Single	Dimethylformamide and Chloroform (7:3)
Chitosan/Polyethylene oxides	Potassium 5-nitro-8-quinolinolate	Single	Acetic Acid
Polyurethane (PU)	Ketanserin and Itraconazole	Single	Dimethylformamide, Dimethylacetamide
Polyvinyl acetate (PVA)	Pleurocidin	Single	Distilled water
Polyvinyl acetate/ Polyacrylic acid/Multiwalled Carbon-nanotube	Ketoprofen	Single	Deionized water
PLGA PEG-g-CHN	Ibuprofen	Side-by-side	DMF
PLGA/Gelatin	Fenbufen	Single nozzle	2,2,2-trifluoroethanol
Chitosan NPs/ Poly caprolactone composite	Naproxen/ Rhodamine B	Single (sheath fiber)	Methanol: Chloroform/ Acetic acid (1:3)
Polyvinyl acetate (PVA)	Meloxicam	Single	Distilled Water
PEG-PLA	Doxorubicin Hydrochloride	Emulsion	Chloroform-methano-DMSO
HPCD	Hydroxycamptothecin	Emulsion	DMSO
Hydroxyapatite- Poly lactic-c--glycolic acid/ Chitosan/ Polyethylene oxide	Paclitaxel	Single	Dichloromethane, Dimethylformamide, Distilled water and Acetic acid
Poly lactic acid/ Poly lactic-c--glycolic acid	Cisplatin	Single	Dichloromethane
PLA	Dichloroacetate	Single	Chloroform
PEO and PEG-PLLA	1,3-Bis(2-chloroethyl)-1-nitrosourea	Single nozzle/Emulsion	Chloroform
Cellulose acetate	Curcumin	Single	Acetone/dimethylacetamide (2:1)
Poly caprolactone/ Multi-walled Carbon Nanotubes	Green tea polyphenols (GTP)	Single	Dichloromethane (DCM)
PLLA	Titanocene Dichloride	Single	Dichloromethane (DCM)
PEI-HA	plasmid DNA (pDNA)	Coaxial	N/A
Poly caplactone; Poly e-caprolactone-co-ethyl ethylene phosphate	siRNA	Single	RNase free water, 2,2,2-Trifluoroethanol (TFE);
Poly e-caprolactone-co-ethyl ethylene phosphate	Human glial cell-derived neurotrophic factor	Single	Dichloromethane (DCM)
Poly e-caprolactone-co-ethyl ethylene phosphate	Human β-nerve growth factor	Single	Dichloromethane (DCM)

(Table 2) cont.....

Polymer	Drug	Spray Nozzle	Solvent
Poly e-caprolactone-co-ethyl ethylene phosphate (PCLEEP)	Endothelial growth factor VEGF	Single	N/A
PEO	Bovine Serum Albumin (BSA)	Single	Deionized water
PLA	Lysozyme	Emulsion	Chloroform
Poly(L-lactide-co-caprolactone)	Human-nerve growth factor (NGF)	Emulsion	Chloroform
PLA-PEG and PLGA	DNA	Single nozzle	DMF
Polyurethan (PU)	Growth factors, *i.e.*, PDGF and VEGF	Coaxial	Ethanol: Chloroform (25:75)
Polyvinyl acetate (PVA)/ Polycaprolactone (PCL)	Horseradish Peroxidase	Coaxial	N/A
Collagen-Polyethylene oxide (PEO)	Homeostatic agent	Single	Hydrochloric acid
Poly(ε-caprolactone)	Green fluorescent protein-tagged adenovirus for gene delivery	Coaxial	Ethanol: Chloroform (25:75)
Poly DL-lactide/Poly lactic-c--glycolic acid	Guided tissue regeneration	Single	Dimethylformamide: Chloroform (1:9) or Tetrahydrofuran: Dimethylformamide (1:3)

CONCLUSION

Electrospinning is a simple, innovative, adaptable, and cost-effective technology for producing nonwoven fibers with high and adaptable porosity and large surface area. By manipulating various solutions and processing parameters that affect electrospinning, it is possible to easily produce ESNF scaffolds for the desired purpose. Natural and synthetic polymers are normally used for the fabrication of ESNFS. For this reason, electrospinning is considered important for a variety of biomedical applications. In this book chapter, we have summarized in detail the process of electrospinning and the factors affecting the generation of smooth, bead-free ESNFs. The last part of this book chapter highlights the potential application of ESNFs in the biomedical field.

CONSENT FOR PUBLICATION

Not applicable.

CONFLICT OF INTEREST

The author declares no conflict of interest, financial or otherwise.

ACKNOWLEDGEMENT

Declared none.

REFERENCES

[1] A.P. Ramos, M.A.E. Cruz, C.B. Tovani, and P. Ciancaglini, "Biomedical applications of nanotechnology", *Biophys. Rev.,* vol. 9, no. 2, pp. 79-89, 2017.
[http://dx.doi.org/10.1007/s12551-016-0246-2] [PMID: 28510082]

[2] H. M. Ibrahim, and A. Klingner, "A review on electrospun polymeric nanofibers: Production parameters and potential applications", *Polymer Testing,* vol. 90, p. 106647, 2020.
[http://dx.doi.org/10.1016/j.polymertesting.2020.106647]

[3] S. Sankaran, K. Deshmukh, M. Basheer Ahamed, and S.K. Khadheer Pasha, "Electrospun Polymeric Nanofibers: Fundamental Aspects of Electrospinning Processes, Optimization of Electrospinning Parameters, Properties, and Applications", *Polymer Nanocomposites in Biomedical Engineering.,* Springer International Publishing: Cham, pp. 375-409, 2019.
[http://dx.doi.org/10.1007/978-3-030-04741-2_12]

[4] A. Haider, S. Haider, and I.-K. Kang, "A comprehensive review summarizing the effect of electrospinning parameters and potential applications of nanofibers in biomedical and biotechnology", *Arabian Journal of Chemistry,* vol. 11, no. 8, pp. 1165-1188, 2018.
[http://dx.doi.org/10.1016/j.arabjc.2015.11.015]

[5] T. Roodbar Shojaei, A. Hajalilou, M. Tabatabaei, H. Mobli, and M. Aghbashlo, "Characterization and Evaluation of Nanofiber Materials", *Handbook of Nanofibers.,* Springer International Publishing: Cham, pp. 491-522, 2019.
[http://dx.doi.org/10.1007/978-3-319-53655-2_15]

[6] K. Kosmider, and J. Scott, "Polymeric nanofibres exhibit an enhanced air filtration performance", *Filtration & Separation,* vol. 39, no. 6, pp. 20-22, 2002.
[http://dx.doi.org/10.1016/S0015-1882(02)80187-2]

[7] D. Wang, L. Wang, and G. Shen, "Nanofiber/nanowires-based flexible and stretchable sensors", *Journal of Semiconductors,* vol. 41, no. 4, . 041605.
[http://dx.doi.org/10.1088/1674-4926/41/4/041605]

[8] A. Özkal, and F. Cengiz Çallıoğlu, "Effect of nanofiber spinning duration on the sound absorption capacity of nonwovens produced from recycled polyethylene terephthalate fibers", *Applied Acoustic,* vol. 169, p. 107468, 2020.
[http://dx.doi.org/10.1016/j.apacoust.2020.107468]

[9] M. Najafi, and M.W. Frey, "Electrospun Nanofibers for Chemical Separation", *Nanomaterials (Basel),* vol. 10, no. 5, p. 982, 2020.
[http://dx.doi.org/10.3390/nano10050982] [PMID: 32455530]

[10] Z.-M. Huang, Y. Z. Zhang, M. Kotaki, and S. Ramakrishna, "A review on polymer nanofibers by electrospinning and their applications in nanocomposites", *Composites Science and Technology,* vol. 63, no. 15, pp. 2223-2253, 2003.
[http://dx.doi.org/10.1016/S0266-3538(03)00178-7]

[11] S. Ramakrishna, "Science and engineering of electrospun nanofibers for advances in clean energy, water filtration, and regenerative medicine", *Journal of Materials Science.,* vol. 45, no. 23, pp. 6283-6312, 2010.
[http://dx.doi.org/10.1007/s10853-010-4509-1]

[12] A.M. Al-Enizi, M.M. Zagho, and A.A. Elzatahry, *Polymer-Based Electrospun Nanofibers for Biomedical Applications.,* vol. 8, no. 4, p. 259, 2018.
[http://dx.doi.org/10.3390/nano8040259]

[13] A. G. Kanani, H. J. T. i. b. Bahrami, and A. Organs, *Review on electrospun nanofibers scaffold and biomedical applications.,* vol. 24, 2010.

[14] S. Nagam Hanumantharao, and S. J. F. Rao, *Multi-functional electrospun nanofibers from polymer blends for scaffold tissue engineering.,* vol. 7, no. 7, p. 66, 2019.

[http://dx.doi.org/10.3390/fib7070066]

[15]   I. Alghoraibi, and S.J.H.n. Alomari, *Different methods for nanofiber design and fabrication.*, pp. 1-46, 2018.
[http://dx.doi.org/10.1007/978-3-319-42789-8_11-2]

[16]   H. Zhuo, J. Hu, and S.J.E.P.L. Chen, *Coaxial electrospun polyurethane core-shell nanofibers for shape memory and antibacterial nanomaterials.*, vol. 5, no. 2, pp. 182-187, 2011.
[http://dx.doi.org/10.3144/expresspolymlett.2011.16]

[17]   N. Nikmaram, *Emulsion-based systems for fabrication of electrospun nanofibers: Food, pharmaceutical and biomedical applications.*, vol. 7, no. 46, pp. 28951-28964, 2017.

[18]   J. Hu, M.P. Prabhakaran, L. Tian, X. Ding, and S.J.R.a. Ramakrishna, "Drug-loaded emulsion electrospun nanofibers: characterization, drug release and *in vitro* biocompatibility" vol. 7, no. 46, pp. 28951-28964, 2017.

[19]   M.L. Muerza-Cascante, D. Haylock, D.W. Hutmacher, and P.D. Dalton, "Melt electrospinning and its technologization in tissue engineering", *Tissue Eng. Part B Rev.*, vol. 21, no. 2, pp. 187-202, 2015.
[http://dx.doi.org/10.1089/ten.teb.2014.0347] [PMID: 25341031]

[20]   S. J. Kim, L. Jeong, S. J. Lee, D. Cho, and W. H. Park, "Fabrication and surface modification of melt-electrospun poly(D,L-lactic-co-glycolic acid) microfibers", *Fibers and Polymers.*, vol. 14, no. 9, pp. 1491-1496, 2013.
[http://dx.doi.org/10.1007/s12221-013-1491-7]

[21]   S.G. Kuntzler, A.C.A. Almeida, J.A.V. Costa, and M.G. Morais, "Polyhydroxybutyrate and phenolic compounds microalgae electrospun nanofibers: A novel nanomaterial with antibacterial activity", *Int. J. Biol. Macromol.*, vol. 113, pp. 1008-1014, 2018.
[http://dx.doi.org/10.1016/j.ijbiomac.2018.03.002] [PMID: 29505877]

[22]   A. Bonartsev, "Biosynthesis, biodegradation, and application of poly (3-hydroxybutyrate) and its copolymers-natural polyesters produced by diazotrophic bacteria", vol. 1, pp. 295-307, 2007.

[23]   R. Dwivedi, S. Kumar, R. Pandey, A. Mahajan, D. Nandana, D.S. Katti, and D. Mehrotra, "Polycaprolactone as biomaterial for bone scaffolds: Review of literature", *J. Oral Biol. Craniofac. Res.*, vol. 10, no. 1, pp. 381-388, 2020.
[http://dx.doi.org/10.1016/j.jobcr.2019.10.003] [PMID: 31754598]

[24]   D.A. Schmatz, *Scaffolds containing Spirulina sp. LEB 18 biomass: development, characterization and evaluation of In vitro biodegradation.*, vol. 16, no. 1, pp. 1050-1059, 2016.

[25]   H.M. Mansour, M. Sohn, A. Al-Ghananeem, and P.P.J.I.s. DeLuca, *Materials for pharmaceutical dosage forms: molecular pharmaceutics and controlled release drug delivery aspects.*, vol. 11, no. 9, pp. 3298-3322, 2010.

[26]   E. Thangaraju, N. T. Srinivasan, R. Kumar, P. K. Sehgal, and S. Rajiv, "Fabrication of electrospun Poly L-lactide and Curcumin loaded Poly L-lactide nanofibers for drug delivery", *Fibers and Polymers.*, vol. 13, no. 7, pp. 823-830, 2012.
[http://dx.doi.org/10.1007/s12221-012-0823-3]

[27]   P. Sasmal, and P. J. J. o. D. D. S. Datta, and Technology, *Tranexamic acid-loaded chitosan electrospun nanofibers as drug delivery system for hemorrhage control applications.*, vol. 52, pp. 559-567, 2019.

[28]   A. Sharma, A. Gupta, G. Rath, A. Goyal, R. Mathur, and S.J.J.M.C.B. Dhakate, *Electrospun composite nanofiber-based transmucosal patch for anti-diabetic drug delivery.*, vol. 1, no. 27, pp. 3410-3418, 2013.
[http://dx.doi.org/10.1039/c3tb20487a]

[29]   Q. Sang, G. R. Williams, H. Wu, K. Liu, H. Li, and L.-M. Zhu, "Electrospun gelatin/sodium bicarbonate and poly(lactide-co-ε-caprolactone)/sodium bicarbonate nanofibers as drug delivery systems", *Materials Science and Engineering: C,* vol. 81, pp. 359-365, 2017.

[http://dx.doi.org/10.1016/j.msec.2017.08.007]

[30] M. Wang, A. K. Roy, and T. J. J. F. i. p. Webster, *Development of chitosan/poly (vinyl alcohol) electrospun nanofibers for infection related wound healing.*, vol. 7, p. 683, 2017.
[http://dx.doi.org/10.3389/fphys.2016.00683]

[31] S. E. Kim, "Electrospun gelatin/polyurethane blended nanofibers for wound healing", *Biomedical Materials,* vol. 4, no. 4, p. 044106, 2009.
[http://dx.doi.org/10.1088/1748-6041/4/4/044106]

[32] T. T. T. Nguyen, C. Ghosh, S.-G. Hwang, L. D. Tran, and J. S. Park, "Characteristics of curcumin-loaded poly (lactic acid) nanofibers for wound healing", *Journal of Materials Science,* vol. 48, no. 20, pp. 7125-7133, 2013.
[http://dx.doi.org/10.1007/s10853-013-7527-y]

[33] Y. Hussein, "Electrospun PVA/hyaluronic acid/L-arginine nanofibers for wound healing applications: Nanofibers optimization and *in vitro* bioevaluation", *International journal of biological macromolecules,* vol. 164, pp. 667-676, 2020.
[http://dx.doi.org/10.1016/j.ijbiomac.2020.07.126]

[34] H. Park, K. Y. Lee, S. J. Lee, K. E. Park, and W. H. Park, "Plasma-treated poly(lactic-co-glycolic acid) nanofibers for tissue engineering", *Macromolecular Research.*, vol. 15, no. 3, pp. 238-243, 2007.
[http://dx.doi.org/10.1007/BF03218782]

[35] P. Vashisth, and V. Pruthi, "Synthesis and characterization of crosslinked gellan/PVA nanofibers for tissue engineering application", *Materials Science and Engineering: C.,* vol. 67, pp. 304-312, 2016.
[http://dx.doi.org/10.1016/j.msec.2016.05.049]

[36] V. R. Hokmabad, S. Davaran, M. Aghazadeh, E. Alizadeh, R. Salehi, and A. Ramazani, "Effect of incorporating Elaeagnus angustifolia extract in PCL-PEG-PCL nanofibers for bone tissue engineering", *Frontiers of Chemical Science and Engineering,* vol. 13, no. 1, pp. 108-119, 2019.
[http://dx.doi.org/10.1007/s11705-018-1742-7]

[37] P. Ghaffari-Bohlouli, M. Shahrousvand, P. Zahedi, and M.J.I.m. Shahrousvand, *Performance evaluation of poly (l-lactide-co-d, l-lactide)/poly (acrylic acid) blends and their nanofibers for tissue engineering applications.*, vol. 122, pp. 1008-1016, 2019.
[http://dx.doi.org/10.1016/j.ijbiomac.2018.09.046]

[38] M. Xavier, *PLLA synthesis and nanofibers production: viability by human mesenchymal stem cell from adipose tissue.*, vol. 49, pp. 213-221, 2016.

[39] J. Lannutti, D. Reneker, T. Ma, D. Tomasko, and D. Farson, "Electrospinning for tissue engineering scaffolds", *Materials Science and Engineering,* vol. 27, no. 3, pp. 504-509, 2007.
[http://dx.doi.org/10.1016/j.msec.2006.05.019]

[40] A. Greiner, and J.H.J.A.C.I.E. Wendorff, *Electrospinning: a fascinating method for the preparation of ultrathin fibers.*, vol. 46, no. 30, pp. 5670-5703, 2007.

[41] V. Pillay, "A Review of the Effect of Processing Variables on the Fabrication of Electrospun Nanofibers for Drug Delivery Applications", *Journal of Nanomaterials,* vol. 2013, p. 789289, 2013.
[http://dx.doi.org/10.1155/2013/789289]

[42] B. Sun, *Advances in three-dimensional nanofibrous macrostructures via electrospinning.*, vol. 39, no. 5, pp. 862-890, 2014.
[http://dx.doi.org/10.1016/j.progpolymsci.2013.06.002]

[43] D. Lubasova, and L.J.J.N. Martinova, *Controlled morphology of porous polyvinyl butyral nanofibers.*, vol. 2011, 2011.

[44] Y. Liu, L. Dong, J. Fan, R. Wang, and J.-Y. Yu, "Effect of applied voltage on diameter and morphology of ultrafine fibers in bubble electrospinning", *Journal of Applied Polymer Science,* vol. 120, no. 1, pp. 592-598, 2011.
[http://dx.doi.org/10.1002/app.33203]

[45]   I. Fatimah, T.I. Sari, and D. Anggoro, "Effect of Concentration and Nozzle-Collector Distance on the Morphology of Nanofibers", *Key Eng. Mater.,* vol. 860, pp. 315-319, 2020.
        [http://dx.doi.org/10.4028/www.scientific.net/KEM.860.315]

[46]   S. De Vrieze, T. Van Camp, A. Nelvig, B. Hagström, P. Westbroek, and K. De Clerck, "The effect of temperature and humidity on electrospinning", *J. Mater. Sci.,* vol. 44, no. 5, pp. 1357-1362, 2009.
        [http://dx.doi.org/10.1007/s10853-008-3010-6]

[47]   D. Li, and Y. Xia, "Electrospinning of Nanofibers", *Adv. Mater.,* vol. 16, no. 14, pp. 1151-1170, 2004.
        [http://dx.doi.org/10.1002/adma.200400719]

[48]   J. Pelipenko, J. Kristl, B. Janković, S. Baumgartner, and P.J.I.p. Kocbek, *The impact of relative humidity during electrospinning on the morphology and mechanical properties of nanofibers.,* vol. 456, no. 1, pp. 125-134, 2013.
        [http://dx.doi.org/10.1016/j.ijpharm.2013.07.078]

[49]   H.S. Bae, A. Haider, K.M.K. Selim, D.Y. Kang, E.J. Kim, and I.K. Kang, "Fabrication of highly porous PMMA electrospun fibers and their application in the removal of phenol and iodine", *J. Polym. Res.,* vol. 20, no. 7, p. 158, 2013.
        [http://dx.doi.org/10.1007/s10965-013-0158-9]

[50]   Z.X. Meng, X.X. Xu, W. Zheng, H.M. Zhou, L. Li, Y.F. Zheng, and X. Lou, "Preparation and characterization of electrospun PLGA/gelatin nanofibers as a potential drug delivery system", *Colloids Surf. B Biointerfaces,* vol. 84, no. 1, pp. 97-102, 2011.
        [http://dx.doi.org/10.1016/j.colsurfb.2010.12.022] [PMID: 21227661]

[51]   M. Jannesari, "Composite Poly(vinyl alcohol)/poly(vinyl acetate) electrospun nanofibrous mats asa novel wound dressing matrix for controlled release of drugs", *Internationl journal of nanomedicine.,* vol. 6, no. 1, pp. 993-1003, 2011.

[52]   J. Wang, and M. Windbergs, "Controlled dual drug release by coaxial electrospun fibers – Impact of the core fluid on drug encapsulation and release", *Int. J. Pharm.,* vol. 556, pp. 363-371, 2019.
        [http://dx.doi.org/10.1016/j.ijpharm.2018.12.026] [PMID: 30572080]

[53]   B. Pant, M. Park, and S.J. Park, "Drug Delivery Applications of Core-Sheath Nanofibers Prepared by Coaxial Electrospinning: A Review", *Pharmaceutics,* vol. 11, no. 7, p. 305, 2019.
        [http://dx.doi.org/10.3390/pharmaceutics11070305] [PMID: 31266186]

[54]   M. Naeimirad, A. Zadhoush, R. Kotek, R. Esmaeely Neisiany, S. Nouri Khorasani, and S.J.J.A.P.S. Ramakrishna, "Recent advances in core/shell bicomponent fibers and nanofibers", *RE:view,* vol. 135, no. 21, p. 46265, 2018.

[55]   X. Xu, "Ultrafine medicated fibers electrospun from W/O emulsions", *Journal of controlled release : official journal of the Controlled Release Society,* vol. 108, no. 1, pp. 33-42, 2005.
        [http://dx.doi.org/10.1016/j.jconrel.2005.07.021]

[56]   S. He, T. Xia, H. Wang, L. Wei, X. Luo, and X.J.A.b. Li, *Multiple release of polyplexes of plasmids VEGF and bFGF from electrospun fibrous scaffolds towards regeneration of mature blood vessels,* vol. 8, no. 7, pp. 2659-2669, 2012.
        [http://dx.doi.org/10.1016/j.actbio.2012.03.044]

[57]   Seeram Ramakrishna, M. Zamani, and S. Ramakrishna, "Advances in drug delivery *via* electrospun and electrosprayed nanomaterials"., *Int. J. Nanomedicine,* vol. 8, pp. 2997-3017, 2013.
        [http://dx.doi.org/10.2147/IJN.S43575] [PMID: 23976851]

[58]   F.Z. Volpato, J. Almodóvar, K. Erickson, K.C. Popat, C. Migliaresi, and M.J.J.A.b. Kipper, *Preservation of FGF-2 bioactivity using heparin-based nanoparticles, and their delivery from electrospun chitosan fibers.,* vol. 8, no. 4, pp. 1551-1559, 2012.

[59]   P. Raghavan, *Electrospun polymer nanofibers: The booming cutting edge technology.,* vol. 72, no. 12, pp. 915-930, 2012.

[60]   M. Bognitzki, "Preparation of fibers with nanoscaled morphologies: Electrospinning of polymer blends", *Polymer Engineering & Science,* vol. 41, no. 6, pp. 982-989, 2001.
[http://dx.doi.org/10.1002/pen.10799]

[61]   Kenry, and C.T. Lim, "Nanofiber technology: current status and emerging developments", *Progress in Polymer Sci.,* vol. 70, pp. 1-17, 2017.
[http://dx.doi.org/10.1016/j.progpolymsci.2017.03.002]

[62]   D. Hussain, F. Loyal, A. Greiner, and J. H. Wendorff, "Structure property correlations for electrospun nanofiber nonwovens", *Polymer,* vol. 51, no. 17, pp. 3989-3997, 2010.
[http://dx.doi.org/10.1016/j.polymer.2010.06.036]

[63]   M. Kiristi, A.U. Oksuz, L. Oksuz, and S. Ulusoy, "Electrospun chitosan/PEDOT nanofibers", *Mater. Sci. Eng. C,* vol. 33, no. 7, pp. 3845-3850, 2013.
[http://dx.doi.org/10.1016/j.msec.2013.05.018] [PMID: 23910286]

[64]   F. Kayaci, Z. Aytac, and T. Uyar, "Surface modification of electrospun polyester nanofibers with cyclodextrin polymer for the removal of phenanthrene from aqueous solution", *J Hazardous Mater,* vol. 261, pp. 286-294, 2013.
[http://dx.doi.org/10.1016/j.jhazmat.2013.07.041]

[65]   X. Shen, D. Yu, L. Zhu, C. Branford-White, K. White, and N.P. Chatterton, "Electrospun diclofenac sodium loaded Eudragit® L 100-55 nanofibers for colon-targeted drug delivery"., *Int. J. Pharm.,* vol. 408, no. 1-2, pp. 200-207, 2011.
[http://dx.doi.org/10.1016/j.ijpharm.2011.01.058] [PMID: 21291969]

[66]   P. Chen, Q.S. Wu, Y.P. Ding, M. Chu, Z.M. Huang, and W. Hu, "A controlled release system of titanocene dichloride by electrospun fiber and its antitumor activity *in vitro*"., *Eur. J. Pharm. Biopharm.,* vol. 76, no. 3, pp. 413-420, 2010.
[http://dx.doi.org/10.1016/j.ejpb.2010.09.005] [PMID: 20854905]

[67]   Z.M. Huang, Y.Z. Zhang, M. Kotaki, and S. Ramakrishna, "A review on polymer nanofibers by electrospinning and their applications in nanocomposites", *Compos. Sci. Technol.,* vol. 63, no. 15, pp. 2223-2253, 2003.
[http://dx.doi.org/10.1016/S0266-3538(03)00178-7]

[68]   N. Bölgen, İ. Vargel, P. Korkusuz, Y.Z. Menceloğlu, and E. Pişkin, "*in vivo* performance of antibiotic embedded electrospun PCL membranes for prevention of abdominal adhesions"., *J. Biomed. Mater. Res. B Appl. Biomater.,* vol. 81B, no. 2, pp. 530-543, 2007.
[http://dx.doi.org/10.1002/jbm.b.30694] [PMID: 17041925]

[69]   J. Xie, and C.H. Wang, "Electrospun micro- and nanofibers for sustained delivery of paclitaxel to treat C6 glioma *in vitro*", *Pharm. Res.,* vol. 23, no. 8, pp. 1817-1826, 2006.
[http://dx.doi.org/10.1007/s11095-006-9036-z] [PMID: 16841195]

[70]   T.G. Kim, D.S. Lee, and T.G. Park, "Controlled protein release from electrospun biodegradable fiber mesh composed of poly(ε-caprolactone) and poly(ethylene oxide)", *Int. J. Pharm.,* vol. 338, no. 1-2, pp. 276-283, 2007.
[http://dx.doi.org/10.1016/j.ijpharm.2007.01.040] [PMID: 17321084]

[71]   C. Burger, and B. Chu, "Functional nanofibrous scaffolds for bone reconstruction", *Colloids Surf. B Biointerfaces,* vol. 56, no. 1-2, pp. 134-141, 2007.
[http://dx.doi.org/10.1016/j.colsurfb.2006.09.023] [PMID: 17113762]

[72]   R. Bhattarai, R. Bachu, S. Boddu, and S. Bhaduri, "Biomedical applications of electrospun nanofibers: Drug and nanoparticle delivery", *Pharmaceutics,* vol. 11, no. 1, p. 5, 2018.
[http://dx.doi.org/10.3390/pharmaceutics11010005] [PMID: 30586852]

[73]   M.S. Islam, B.C. Ang, A. Andriyana, and A.M. Afifi, "A review on fabrication of nanofibers *via* electrospinning and their applications"., *SN Applied Sciences,* vol. 1, no. 10, p. 1248, 2019.
[http://dx.doi.org/10.1007/s42452-019-1288-4]

[74]   A. Garcia-Bennett, M. Nees, and B. Fadeel, "In search of the Holy Grail: Folate-targeted nanoparticles for cancer therapy", *Biochem. Pharmacol.,* vol. 81, no. 8, pp. 976-984, 2011.
[http://dx.doi.org/10.1016/j.bcp.2011.01.023] [PMID: 21300030]

[75]   D.G. Yu, X.F. Zhang, X.X. Shen, C. Brandford-White, and L.M. Zhu, "Ultrafine ibuprofen-loaded polyvinylpyrrolidone fiber mats using electrospinning", *Polym. Int.,* vol. 58, no. 9, pp. 1010-1013, 2009.
[http://dx.doi.org/10.1002/pi.2629]

[76]   D.G. Yu, X.X. Shen, C. Branford-White, K. White, L.M. Zhu, and S.W. Annie Bligh, "Oral fast-dissolving drug delivery membranes prepared from electrospun polyvinylpyrrolidone ultrafine fibers", *Nanotechnology,* vol. 20, no. 5, p. 055104, 2009.
[http://dx.doi.org/10.1088/0957-4484/20/5/055104] [PMID: 19417335]

[77]   Z.X. Meng, W. Zheng, L. Li, and Y.F. Zheng, "Fabrication, characterization and *In vitro* drug release behavior of electrospun PLGA/chitosan nanofibrous scaffold"., *Mater. Chem. Phys.,* vol. 125, no. 3, pp. 606-611, 2011.
[http://dx.doi.org/10.1016/j.matchemphys.2010.10.010]

[78]   Y. Yang, X. Zhu, W. Cui, X. Li, and Y. Jin, "Electrospun composite mats of poly[(D, Llactide)-c--glycolide] and collagen with high porosity as potential scaffolds for skin tissue engineering", *Macromol. Mater. Eng.,* vol. 294, no. 9, pp. 611-619, 2009.
[http://dx.doi.org/10.1002/mame.200900052]

[79]   J. Han, T.X. Chen, C.J. Branford-White, and L.M. Zhu, "Electrospun shikonin-loaded PCL/PTMC composite fiber mats with potential biomedical applications", *Int. J. Pharm.,* vol. 382, no. 1-2, pp. 215-221, 2009.
[http://dx.doi.org/10.1016/j.ijpharm.2009.07.027] [PMID: 19660536]

[80]   G.R. Williams, N.P. Chatterton, T. Nazir, and L-m. Zhu, *Tde.12.17,* vol. 3, pp. 515-533, 2012.

[81]   T. Okuda, K. Tominaga, and S. Kidoaki, "Time-programmed dual release formulation by multilayered drug-loaded nanofiber meshes", *J. Control. Release,* vol. 143, no. 2, pp. 258-264, 2010.
[http://dx.doi.org/10.1016/j.jconrel.2009.12.029] [PMID: 20074599]

[82]   E.R. Kenawy, G.L. Bowlin, K. Mansfield, J. Layman, D.G. Simpson, E.H. Sanders, and G.E. Wnek, "Release of tetracycline hydrochloride from electrospun poly(ethylene-co-vinylacetate), poly(lactic acid), and a blend", *J. Control. Release,* vol. 81, no. 1-2, pp. 57-64, 2002.
[http://dx.doi.org/10.1016/S0168-3659(02)00041-X] [PMID: 11992678]

[83]   Y. Park, E. Kang, O.J. Kwon, T. Hwang, H. Park, J.M. Lee, J.H. Kim, and C.O. Yun, "Ionically crosslinked Ad/chitosan nanocomplexes processed by electrospinning for targeted cancer gene therapy", *J. Control. Release,* vol. 148, no. 1, pp. 75-82, 2010.
[http://dx.doi.org/10.1016/j.jconrel.2010.06.027] [PMID: 20637814]

[84]   G. Khang, S.J. Lee, M.S. Kim, and H.B. Lee, "Biomaterials: Tissue Engineering and Scaffolds", *Encyclopedia of Medical Devices and Instrumentation.,* pp. 366-383, 2006.
[http://dx.doi.org/10.1002/0471732877.emd029]

[85]   M. Gorji, M. Karimi, and S. Nasheroahkam, "Electrospun PU/P(AMPS-GO) nanofibrous membrane with dual-mode hydrophobic–hydrophilic properties for protective clothing applications", *J. Ind. Text.,* vol. 47, no. 6, pp. 1166-1184, 2018.
[http://dx.doi.org/10.1177/1528083716682920]

[86]   X. Xie, Y. Chen, X. Wang, X. Xu, Y. Shen, A.R. Khan, A. Aldalbahi, A.E. Fetz, G.L. Bowlin, M. El-Newehy, and X. Mo, "Electrospinning nanofiber scaffolds for soft and hard tissue regeneration", *J. Mater. Sci. Technol.,* vol. 59, pp. 243-261, 2020.
[http://dx.doi.org/10.1016/j.jmst.2020.04.037]

[87]   C. Huang, S. Wang, L. Qiu, Q. Ke, W. Zhai, and X. Mo, "Heparin loading and pre-endothelialization in enhancing the patency rate of electrospun small-diameter vascular grafts in a canine model", *ACS*

Appl. Mater. Interfaces, vol. 5, no. 6, pp. 2220-2226, 2013.
[http://dx.doi.org/10.1021/am400099p] [PMID: 23465348]

[88] X. Chen, J. Wang, Q. An, D. Li, P. Liu, W. Zhu, and X. Mo, "Electrospun poly(l-lactic acid-co-ε-caprolactone) fibers loaded with heparin and vascular endothelial growth factor to improve blood compatibility and endothelial progenitor cell proliferation", *Colloids Surf. B Biointerfaces,* vol. 128, pp. 106-114, 2015.
[http://dx.doi.org/10.1016/j.colsurfb.2015.02.023] [PMID: 25731100]

[89] A. Haider, K.C. Gupta, and I.K. Kang, "PLGA/nHA hybrid nanofiber scaffold as a nanocargo carrier of insulin for accelerating bone tissue regeneration", *Nanoscale Res. Lett.,* vol. 9, no. 1, p. 314, 2014.
[http://dx.doi.org/10.1186/1556-276X-9-314] [PMID: 25024679]

[90] A. Haider, K.C. Gupta, and I-K. Kang, *Morphological effects of HA on the cell compatibility of electrospun HA/PLGA composite nanofiber scaffolds*, 2014.
[http://dx.doi.org/10.1155/2014/308306]

[91] V. Jones, J.E. Grey, and K.G. Harding, "Wound dressings", *BMJ,* vol. 332, no. 7544, pp. 777-780, 2006.
[http://dx.doi.org/10.1136/bmj.332.7544.777] [PMID: 16575081]

[92] S.A. Eming, P. Martin, and M. Tomic-Canic, "Wound repair and regeneration: Mechanisms, signaling, and translation", *Sci. Transl. Med.,* vol. 6, no. 265, p. 265sr6, 2014.
[http://dx.doi.org/10.1126/scitranslmed.3009337] [PMID: 25473038]

[93] D.E. Discher, P. Janmey, and Y. Wang, "Tissue cells feel and respond to the stiffness of their substrate", *Science,* vol. 310, no. 5751, pp. 1139-1143, 2005.
[http://dx.doi.org/10.1126/science.1116995] [PMID: 16293750]

[94] A. Memic, T. Abdullah, H.S. Mohammed, K. Joshi Navare, T. Colombani, and S.A. Bencherif, "Latest Progress in Electrospun Nanofibers for Wound Healing Applications", *ACS Appl. Bio Mater.,* vol. 2, no. 3, pp. 952-969, 2019.
[http://dx.doi.org/10.1021/acsabm.8b00637] [PMID: 35021385]

[95] T.A. Jeckson, Y.P. Neo, S.P. Sisinthy, and B. Gorain, "Delivery of Therapeutics from Layer-by-Layer Electrospun Nanofiber Matrix for Wound Healing: An Update", *J. Pharm. Sci.,* vol. 110, no. 2, pp. 635-653, 2021.
[http://dx.doi.org/10.1016/j.xphs.2020.10.003] [PMID: 33039441]

[96] J.P. Chen, G.Y. Chang, and J.K. Chen, "Electrospun collagen/chitosan nanofibrous membrane as wound dressing", *Colloids Surf. A Physicochem. Eng. Asp.,* vol. 313-314, pp. 183-188, 2008.
[http://dx.doi.org/10.1016/j.colsurfa.2007.04.129]

[97] Y.P. Afsharian, and M. Rahimnejad, "Bioactive electrospun scaffolds for wound healing applications: A comprehensive review", *Polym. Test.,* vol. 93, p. 106952, 2021.
[http://dx.doi.org/10.1016/j.polymertesting.2020.106952]

[98] A. Al-Enizi, M. Zagho, and A. Elzatahry, "Polymer-based electrospun nanofibers for biomedical applications", *Nanomaterials (Basel),* vol. 8, no. 4, p. 259, 2018.
[http://dx.doi.org/10.3390/nano8040259] [PMID: 29677145]

[99] P.I. Morgado, A. Aguiar-Ricardo, and I.J. Correia, "Asymmetric membranes as ideal wound dressings: An overview on production methods, structure, properties and performance relationship", *J. Membr. Sci.,* vol. 490, pp. 139-151, 2015.
[http://dx.doi.org/10.1016/j.memsci.2015.04.064]

[100] N. Charernsriwilaiwat, P. Opanasopit, T. Rojanarata, and T. Ngawhirunpat, "Lysozyme-loaded, electrospun chitosan-based nanofiber mats for wound healing", *Int. J. Pharm.,* vol. 427, no. 2, pp. 379-384, 2012.
[http://dx.doi.org/10.1016/j.ijpharm.2012.02.010] [PMID: 22353400]

[101] L.Y. Kossovich, Y. Salkovskiy, and I.V. Kirillova, "Electrospun chitosan nanofiber materials as burn

dressing", In: *IFMBE Proceedings.* vol. 31. IFMBE, 2010, pp. 1212-1214.
[http://dx.doi.org/10.1007/978-3-642-14515-5_307]

[102] D.J. Leaper, "Silver dressings: their role in wound management", *Int. Wound J.,* vol. 3, no. 4, pp. 282-294, 2006.
[http://dx.doi.org/10.1111/j.1742-481X.2006.00265.x] [PMID: 17199764]

[103] P.P. Mane, R.S. Ambekar, and B. Kandasubramanian, "Electrospun nanofiber-based cancer sensors: A review", *Int. J. Pharm.,* vol. 583, p. 119364, 2020.
[http://dx.doi.org/10.1016/j.ijpharm.2020.119364] [PMID: 32339630]

[104] S. Liu, G. Zhou, D. Liu, Z. Xie, Y. Huang, X. Wang, W. Wu, and X. Jing, "Inhibition of orthotopic secondary hepatic carcinoma in mice by doxorubicin-loaded electrospun polylactide nanofibers", *J. Mater. Chem. B Mater. Biol. Med.,* vol. 1, no. 1, pp. 101-109, 2013.
[http://dx.doi.org/10.1039/C2TB00121G] [PMID: 32260617]

[105] X. Luo, C. Xie, H. Wang, C. Liu, S. Yan, and X. Li, "Antitumor activities of emulsion electrospun fibers with core loading of hydroxycamptothecin *via* intratumoral implantation"., *Int. J. Pharm.,* vol. 425, no. 1-2, pp. 19-28, 2012.
[http://dx.doi.org/10.1016/j.ijpharm.2012.01.012] [PMID: 22265915]

[106] S.H. Ranganath, I. Kee, W.B. Krantz, P.K.H. Chow, and C.H. Wang, "Hydrogel matrix entrapping PLGA-paclitaxel microspheres: drug delivery with near zero-order release and implantability advantages for malignant brain tumour chemotherapy", *Pharm. Res.,* vol. 26, no. 9, pp. 2101-2114, 2009.
[http://dx.doi.org/10.1007/s11095-009-9922-2] [PMID: 19543956]

[107] M. Cavo, F. Serio, N.R. Kale, E. D'Amone, G. Gigli, and L.L. del Mercato, "Electrospun nanofibers in cancer research: from engineering of *in vitro* 3D cancer models to therapy"., *Biomater. Sci.,* vol. 8, no. 18, pp. 4887-4905, 2020.
[http://dx.doi.org/10.1039/D0BM00390E] [PMID: 32830832]

[108] Y. Chen, C. Li, Z. Hou, S. Huang, B. Liu, F. He, L. Luo, and J. Lin, "Polyaniline electrospinning composite fibers for orthotopic photothermal treatment of tumors *in vivo*"., *New J. Chem.,* vol. 39, no. 6, pp. 4987-4993, 2015.
[http://dx.doi.org/10.1039/C5NJ00327J]

[109] X. Zhang, M.R. Reagan, and D.L. Kaplan, "Electrospun silk biomaterial scaffolds for regenerative medicine", *Adv. Drug Deliv. Rev.,* vol. 61, no. 12, pp. 988-1006, 2009.
[http://dx.doi.org/10.1016/j.addr.2009.07.005] [PMID: 19643154]

[110] J. Lannutti, D. Reneker, T. Ma, D. Tomasko, and D. Farson, "Electrospinning for tissue engineering scaffolds", *Mater. Sci. Eng. C,* vol. 27, no. 3, pp. 504-509, 2007.
[http://dx.doi.org/10.1016/j.msec.2006.05.019]

[111] S. Damle, H. Bhattal, and A. Loomba, "Apexification of anterior teeth: a comparative evaluation of mineral trioxide aggregate and calcium hydroxide paste", *J. Clin. Pediatr. Dent.,* vol. 36, no. 3, pp. 263-268, 2012.
[http://dx.doi.org/10.17796/jcpd.36.3.02354g044271t152] [PMID: 22838228]

[112] G.M. Kim, P. Simon, J-S. Kim, P. Simon, and J.S. Kim, "Electrospun PVA/HAp nanocomposite nanofibers: biomimetics of mineralized hard tissues at a lower level of complexity", *Bioinspir. Biomim.,* vol. 3, no. 4, p. 046003, 2008.
[http://dx.doi.org/10.1088/1748-3182/3/4/046003] [PMID: 18812653]

[113] J.J. Kim, W.J. Bae, J.M. Kim, J.J. Kim, E.J. Lee, H.W. Kim, and E.C. Kim, "Mineralized polycaprolactone nanofibrous matrix for odontogenesis of human dental pulp cells", *J. Biomater. Appl.,* vol. 28, no. 7, pp. 1069-1078, 2014.
[http://dx.doi.org/10.1177/0885328213495903] [PMID: 23839784]

[114] M.S. Zafar, "Assessment of Antimicrobial Efficacy of MTAD, Sodium Hypochlorite, EDTA and Chlorhexidine for Endodontic Applications: An *In vitro* Study"., *Middle East J. Sci. Res.,* vol. 21, pp.

353-357, 2014.
[http://dx.doi.org/10.5829/idosi.mejsr.2014.21.02.524]

[115] M.C. Bottino, K. Kamocki, G.H. Yassen, J.A. Platt, M.M. Vail, Y. Ehrlich, K.J. Spolnik, and R.L. Gregory, "Bioactive nanofibrous scaffolds for regenerative endodontics", *J. Dent. Res.,* vol. 92, no. 11, pp. 963-969, 2013.
[http://dx.doi.org/10.1177/0022034513505770] [PMID: 24056225]

[116] Y. Liu, M. Hao, Z. Chen, L. Liu, Y. Liu, W. Yang, and S. Ramakrishna, "A review on recent advances in application of electrospun nanofiber materials as biosensors", *Curr. Opin. Biomed. Eng.,* vol. 13, pp. 174-189, 2020.
[http://dx.doi.org/10.1016/j.cobme.2020.02.001]

[117] T. Yoo, K. Lim, M.T. Sultan, J.S. Lee, J. Park, H.W. Ju, C. Park, and M. Jang, "The real-time monitoring of drug reaction in HeLa cancer cell using temperature/impedance integrated biosensors", *Sens. Actuators B Chem.,* vol. 291, pp. 17-24, 2019.
[http://dx.doi.org/10.1016/j.snb.2019.03.145]

[118] S. Haider, and S.Y. Park, "Preparation of the electrospun chitosan nanofibers and their applications to the adsorption of Cu(II) and Pb(II) ions from an aqueous solution", *J. Membr. Sci.,* vol. 328, no. 1-2, pp. 90-96, 2009.
[http://dx.doi.org/10.1016/j.memsci.2008.11.046]

[119] V. Thavasi, G. Singh, and S. Ramakrishna, "Electrospun nanofibers in energy and environmental applications", *Energy Environ. Sci.,* vol. 1, no. 2, pp. 205-221, 2008.
[http://dx.doi.org/10.1039/b809074m]

[120] F. Mohammadi, A. Valipouri, D. Semnani, and F. Alsahebfosoul, "Nanofibrous Tubular Membrane for Blood Hemodialysis", *Appl. Biochem. Biotechnol.,* vol. 186, no. 2, pp. 443-458, 2018.
[http://dx.doi.org/10.1007/s12010-018-2744-0] [PMID: 29644596]

[121] K.H. Lee, D.J. Kim, B.G. Min, and S.H. Lee, "Polymeric nanofiber web-based artificial renal microfluidic chip", *Biomed. Microdevices,* vol. 9, no. 4, pp. 435-442, 2007.
[http://dx.doi.org/10.1007/s10544-007-9047-5] [PMID: 17265147]

[122] K. Namekawa, M. Tokoro Schreiber, T. Aoyagi, and M. Ebara, "Fabrication of zeolite–polymer composite nanofibers for removal of uremic toxins from kidney failure patients", *Biomater. Sci.,* vol. 2, no. 5, pp. 674-679, 2014.
[http://dx.doi.org/10.1039/c3bm60263j] [PMID: 32481844]

CHAPTER 6

3D Printed Biomaterials and their Scaffolds for Biomedical Engineering

Rabail Zehra Raza[1], Arun Kumar Jaiswal[2,3], Muhammad Faheem[1], Sandeep Tiwari[2], Raees Khan[1], Siomar de Castro Soares[3], Asmat Ullah Khan[4], Vasco Azevedo[2] and Syed Babar Jamal[1,*]

[1] *Department of Biological Sciences, National University of Medical Sciences, Rawalpindi, Pakistan*

[2] *PG Program in Bioinformatics, Institute of Biological Sciences, Federal University of Minas Gerais, Belo Horizonte, MG, Brazil*

[3] *Department of Immunology, Microbiology and Parasitology, Institute of Biological Sciences and Natural Sciences, Federal University of Triângulo Mineiro (UFTM), Uberaba, MG, Brazil*

[4] *Department of Zoology, Shaheed Benazir Bhutto University, Sheringal, Dir Upper, KPK, Pakistan*

Abstract: Over the past decade, three-dimensional printing (3DP) has gained popularity among the public and the scientific community in a variety of disciplines, including engineering, medicine, manufacturing arts, and, more recently, education. The advantage of this technology is that it is capable of designing and printing almost any object shape using various materials such as ceramics, polymers, metals and bio-inks. This has further favored the use of this technology for biomedical applications in both clinical and research settings. In biomedicine, there has been a remarkable development of a variety of biomaterials, which in turn has accelerated the significant role of this technology as synthetic scaffolds in various forms such as scaffolds, constructs or matrices. In this chapter, we would like to review the trailblazing literature on the application of 3DP technology in biomedical engineering. This chapter focuses on various 3DP techniques and biomaterials for tissue engineering applications (TE). 3DP technology has a variety of applications in biomedicine and TE (B- TE). Customized structures for B- TE applications using 3DP have several advantages, *e.g.*, they are easy to fabricate and are inexpensive. On the other hand, conventional technologies, which are costly, time-consuming, and labor intensive, are generally not compatible with 3DP. Therefore, the capabilities of 3DP, which is a novel fabrication technology, need to be explored for many other potential applications. Here, we provide a comprehensive overview of the different types of 3DP technologies and how they can potentially be used.

---

* **Corresponding author Syed Babar Jamal:** Department of Biological Sciences, National University of Medical Sciences, Rawalpindi, Pakistan; E-mail: babar.jamal@numspak.edu.pk

Adnan Haider & Sajjad Haider (Eds.)

Keywords: Three-Dimensional Printing (3DP), Scaffolds, Biomedical Engineering, Tissue Engineering.

INTRODUCTION

Tissue engineering (TE) has greatly changed the need to design complicated 3D biomedical devices. Reconstruction of 3D anatomical defects, scaffolds for stem cell differentiation, and reconstruction of complicated organs with sophisticated 3D microarchitecture (*e.g.*, lymphoid, liver organs) are some of the applications for 3D biomedical devices. For example, anatomical defects in the craniomaxillo facial complex as a result of cancer, trauma, or congenital defects require functional restoration of important elements of our body systems, such as nerves, vessels, muscles, ligaments, cartilage, bones, and lymph nodes, to name a few.

In recent years, several new approaches have been explored that rely on TE principles to restore and reanimate functional tissues that are highly important in maxillofacial tissue regeneration. In the field of TE, scaffolds are important for a variety of functions, including providing structural support for cell infiltration and proliferation, providing space for extracellular matrix regeneration and remodeling, controlling cell behavior by extending biochemical cues, and reinforcing physical connections for destroyed tissue. Scaffold fabrication requires design at the macro, micro and nano levels of architecture, which in turn are important for cell structural integrity, nutrient transfer and cell-matrix interactions [1, 2, 3]. The macroarchitecture dictates the overall structure of the device, which can be complex considering the various anatomical features as well as patient specificity and organ specificity. The architecture of the tissue with features such as pore size, porosity, shape, spatial distribution and interconnectivity, is replicated at the micro-architectural level. Finally, the nanoarchitecture reflects changes at the surface level, such as the attachment of a biomolecule to ensure cell adhesion, proliferation, and differentiation. Traditional manufacturing uses formative (molding) and subtractive (machine) techniques. These techniques are a multi-step process and require an inefficient infrastructure that makes it impossible to make changes to the final product in a timely manner [4]. Moreover, these conventional techniques limit the scope for fabrication of highly complicated patterns and geometries which are more commonly required in biomedical engineering applications [1].

Over the last four decades, 3D printing or additive manufacturing (AM) has emerged as a robust tool to reconstruct geometrically complicated objects in a short time and in an economical manner [4, 5, 6]. 3D printing, developed in the 1980s, uses a computer-aided model to deposit material layer by layer in a 3D space [7]. This breakthrough paved the way for the adoption and reproduction of complex 3D structures that would have been impossible to achieve using traditional manufacturing methods. Various industries have adopted this technology due to the creation of complex designs and the far-reaching impact of 3D printing technology on healthcare [4]. Due to its direct application in drug delivery [8, 9, 10], surgical planning [11], implant design [12], and tissue engineering [13, 14, 15], 3D printing's function in healthcare is increasingly becoming critical.

Another rapidly expanding application of additive manufacturing is bioprinting, which allows cells to be seeded in a 3D space while taking into account spatial organization [16]. Bioprinting enables the fabrication of replicates *in vitro* for drug screening, disease modeling, and biofabrication of implantable tissues such as skin [17], bone [18] or cartilage [19]. In this review, we aim to highlight AM fabrication methods, printing materials used in biomedicine and their use in health-related applications. The main focus of this review is on the advanced 3D printing technologies currently used to build scaffolds, with emphasis on their ability to align cells and a wide range of materials along intricate 3D gradients. Most of these technologies have been used to date as surgical templates for formulating patient-specific models, preoperative planning, and prosthesis fabrication. Some of the aforementioned technologies have also received FDA approval for implantable device fabrication. In this chapter, we will mainly highlight the work done in the last five years to show the recent progress the field has made [20].

Three-Dimensional Printing (3DP) Technologies

3DP technology and its applications have made several advances, focusing on the suitability of material processing. Different states such as solid, liquid and powder form the basis for different classes of 3DP technology. The materials used for printing are primarily differentiated by the specific technology used in 3DP. However, all 3DP techniques have one thing in common: the combination of a device with 3D modeling software. The processes involved are [21]:

• CAD sketch is obtained, and interpretation is made by the 3DP device of the data retrieved from the CAD file.
• A layer upon layer structure is built *via* plastic, paper sheet, liquid or powder filaments, all of which make up the printing materials.

Widely used 3DP technologies such as material jetting, photopolymerization, binder jetting, powder bed fusion and material extrusion are shown in Fig. (**1a**) [22]. Photo-polymerization uses ultraviolet (UV) light to stiffen each layer of liquid photocurable natural and synthetic resins (Fig. **1a**). Steriolithography is the most commonly used photo-polymerization technique (SLA). According to Juskova *et al.* (2018), advances in technology such as lower cost light sources, SLA techniques, and advanced mirror-lens systems have led to significant improvements in speed and resolution. To deposit similarly shaped layers of photopolymer resin, material jetting uses a print head identical to inkjet printers [23]. The substrates surrounding each layer are treated with UV light (Fig. **1b**). This system method uses durable resin artifacts to support complex acrylic specifics with high resolution and precision without the use of column scanners/complex lasers [24]. The jetting process appears to be limited by the viscosity of the material, which decreases as the print head and nozzle plate are heated. The binder jetting machine used to deposit a powder layer on a build platform is shown in Fig. (**1c**). Binder jetting machines use inkjet print heads to apply liquid binders to the build platform and bond the particles together [25]. Any powder content can be used for this technique. The process is fast, simple and cost effective as it requires only the bonding of powder particles together to stack and create the 3D structure. Binder jetting machines use colored binders to create and print full color parts (inks). This technique binds the powder particles together, resulting in weak parts with inadequate mechanical properties. Therefore, binder jetting parts are usually used as a starting point for building parts, with subsequent processes such as casting, sintering, or infiltration improving the mechanical properties [26]. FDM and direct ink writing are the most popular methods of material extrusion (Fig. **1d**). This approach is mainly used in the fabrication of 3D scaffolds and devices for TE applications. SLS (Fig. **1e**) is a powder-based 3D printing technique in which different types of powders such as thermoplastic polymers and ceramics are sintered using a CO_2 laser. Sheet metal lamination is one of the 3DP methods, as well as ultrasonic additive manufacturing (UAM) and laminated object fabrication (Fig. **1f**). The materials processed in UAM are metal strips or sheets that are ultrasonically welded together. To remove the metal that is not bonded during welding, this method often requires computer-controlled cutting. Driven energy deposition (DED) is a method that uses high energy power sources such as a laser, electron beam, or plasma welding torch to moderate powder or metal wire to create a layered material (Fig. **1g**). Finally, Fig. (**1h**) shows the melt electrospinning process (MEW), a fiber-based fabrication method used to produce scaffolds for various applications TE.

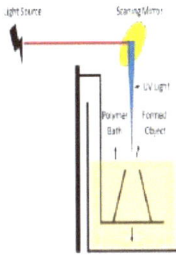

a: Photo-polymerization.

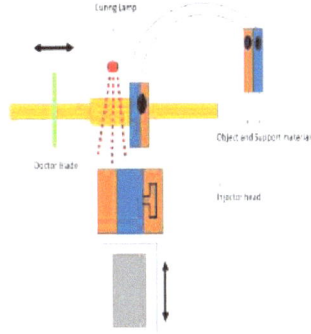

b.   Material Jetting.

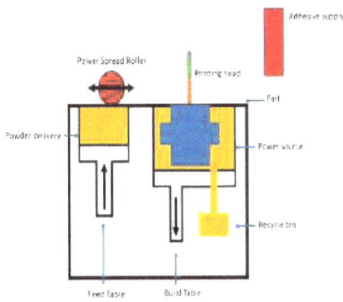

c: Binder Jetting.

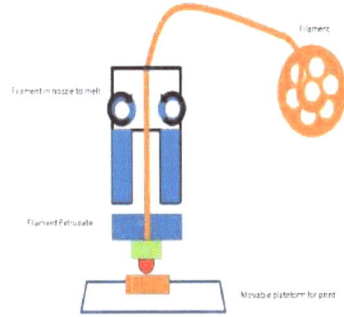

d: Material Extrusion.

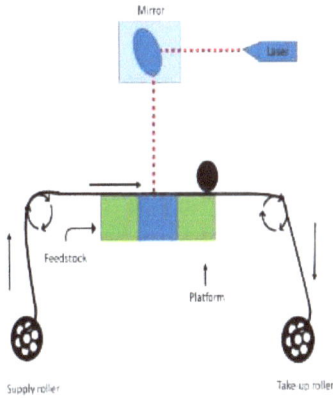

e: Powder Bed Fusion.   f: Sheet Lamination.

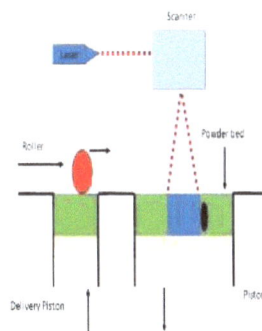

(Fig. 1) contd.....

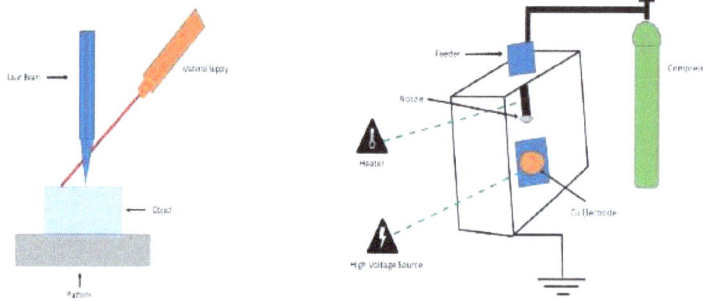

g: Directed energy deposition.   **h: Melt Electrospinning writing.**

Fig. (1). Diagrammatic elucidation of different 3D printing processes. a: Photo-polymerization. b. Material Jetting. c: Binder Jetting. d: Material Extrusion. e: Powder Bed Fusion. f: Sheet Lamination. g: Directed energy deposition. h: Melt Electrospinning writing.

3D Printing Types in Biomedical Applications

Since the 1980s, many methods have been developed for layer-by-layer fabrication/printing of various materials. To date, seven methods are established and in use, defined in the standard ISO/ASTM 52900 [27]. In the following sections, we describe each process briefly and succinctly.

Powder-Based Printing

Powder-based printing involves two techniques, 1) binder jetting and 2) powder bed fusion. The base material for fabricating constructs is a powder bed in both these techniques. However, the powder is modified in different ways to obtain a consistent structure.

Binder Jetting: In this technique, binding material droplets are jetted over a powder bed. After spraying over the powder bed, a fresh powder layer is spread on the top, and the next binder layer is smeared [27, 28].

Powder Bed Fusion: In this technique, lasers are employed to melt the powder. Once the powder is changed into its molten form, a new powder layer is deposited on top, and the process is reiterated [27, 29].

Material Deposition: In these techniques, nozzles are employed to spread the material to be printed in layers.

Direct Energy Deposition: In this method, a constant stream of metal is kept onto a print surface through a nozzle, and a laser is utilized to soften it consistently [27, 30].

Material Extrusion: In this procedure, thermoplastics are softened to a semi-fluid state and are spread on the printing bed constantly. Plastic layers bind due to their semi-fluid consistency [27, 31]. In this method for co-printing with living cells, pastes and hydrogels are used.

Material Jetting (MJ): In this method, jets are utilized for spraying the liquid resin layer while likewise relieving the layer with UV light before the following layer is printed [27, 31].

Bio-Printing: This technique uses various AM techniques to 3D print living cells, and is not itself an AM technique. The cells suspended in the bio-ink are deposited by material extrusion or jetting and also by nozzle-less laser-assisted techniques. In laser-assisted bioprinting, the cells suspended in the bioprinting ink are transferred to focus a laser on a membrane coated with the bioprinting ink and containing the cells on the side of the membrane facing the printing surface [32]. This technique has the advantage of lower shear forces on the cells since there is no aperture, and it achieves resolution at the microscopic level [33]. Another technique is inkjet printing or drop-on-demand (DOD), in which droplets containing a picoliter volume of cell-containing bioink are precisely dispersed to fuse into fibers. These fibers are then cross-linked before further layers are deposited to create a 3D architecture [34, 35]. The very small volume of the droplets ensures a resolution of less than 100 μm [36].

Liquid Reservoir

Stereolithography (SLA): In this technique, the print bed is lowered into a container of liquid photopolymer resin. The resin bed is irradiated with either UV or visible light to solidify it [7]. At each exposure, the bed with the polymerized layer is lifted out of the resin and submerged again to repeat the process [7, 27].

Sheets of Material

Sheet Lamination: It is a process in which sheets of material, such as plastic, paper, or metal are cut with a knife or laser to represent part of the computer-aided design model (CAD). After cutting all the sheets, they are fixed with a binder and glued. The cut parts are separated to reveal the three-dimensional interior design [27, 37].

Nanofabrication

Nanofabrication is a technique that can be used to construct structures less than 100 nm in size, making it extremely useful for medicine and electronics. Nanofabrication is not considered a commonly used AM technique, but it is based on many similar principles. The fabrication process for nanofabrication can be divided into two approaches [38]. One is a "top-down" technique in which a larger material is deconstructed to form the corresponding nanostructure. The principle consists of complicated and costly steps and offers little room for change. Due to uncontrollable variability in the fabrication process, nanostructures made using the top-down approach have limited reproducibility. The other approach to nanofabrication is the bottom-up method, in which the building blocks self-align to create a nanostructure [38]. This approach ensures structures with atomic resolution. Nanofabrication has several applications in the field of tissue design and biomedicine, including (but not limited to): saving the immunogenicity of mixtures in antibodies; limiting transfer dismissal through immuno-confinement; forming biomaterials with exceptional organic and mechanical properties; drug sequestration and conveyance; and flowing poison and waste folios This approach, like most other AM techniques, is also additive in nature. Nanofabrication has numerous potential applications in the biomedical and TE fields. These include preserving the immunogenicity of compounds in vaccines, minimizing transplant rejection through immunoisolation, fabricating biomaterials with exclusive biomechanical properties, drug sequestration and delivery, and toxin circulation [39].

Biocompatible 3D Printing Materials

Biofabrication involves direct printing of cell-populated material called bioink. In this method, cell-free scaffolds of biomaterial ink are first printed, and then cells are seeded [40, 41]. The choice of ink depends on the role performed by a part and the printing methodology used. Biomaterial inks are used for durable and slowly degrading structural stabilization. However, bio-inks are used to produce a milder scaffold that is replaced by the additional deposition of a new extracellular material (ECM) by the embedded cell population [41, 42].

Bioinks

Hydrogels have been used to develop bioinks suitable as materials for 3D cell culture [43, 44, 45]. Hydrogels are biocompatible and highly adaptable in their biophysical and biomechanical properties [45, 46, 47]. The non-Newtonian, shear-thinning behavior of hydrogels makes them ideal for extrusion bioprinting. If they remain unchanged, they have deficiencies in terms of printing properties, which was demonstrated by Malda *et al.* [42]. Hydrogels exhibit reduced thickness before cross-linking, resulting in an inferior shape after extrusion-based printing. This limits the ability of hydrogels to form larger structures without distorting the properties [42, 48, 49]. This has paved the way for the introduction of novel bioinks and TE methods for biofabrication of cell-populated materials

[50]. Biopolymer hydrogels used in bioprinting include agarose, alginate, agarose, cellulose, fibrin, collagen, gelatin and gellan gum. Alginate is a polysaccharide derived from brown algae, carries a negative charge and is the most commonly used hydrogel in bioprinting and TE [51, 52]. Alginate cross-links when divalent cations are added, and it can be functionalized by adding arginine-glycin--aspartate moieties that activate cell binding to the extracellular matrix [53], which in turn facilitates its interaction with the scaffold. To modify both its printing and biological properties, alginate is often combined with other biopolymers. In a recent study, it was shown that the incorporation of hydroxyapatite (HAp) into alginate results in the production of a bioink that produces a calcified cartilage matrix [54]. Incorporation of HAp into alginate decreased the secretion of glucosaminoglycan by chondrocytes, increased the production of Col II and resulted in a marked increase in the calcified cartilage markers alizarin red, Col-X and alkaline phosphatase. The addition of sodium citrate resulted in a more homogeneous distribution of Hap, which improved the pressure characteristics by ensuring that the needle was not clogged [54].

Gelatin, a natural biopolymer, is commonly used for culturing cells. In unmodified form, it undergoes a thermoreversible sol-gel transition. However, the gelation rate is so low that shape fidelity is not guaranteed. To avoid this, many research groups have used gelatin methacrylate (GelMA) to ensure UV cross-linking [55]. This indicates better biocompatibility, cost-effectiveness and degradability of bio-ink [56, 57]. In a study by Byambaa *et al.*, GelMA was operationalized with vascular endothelial growth factors to enable bioprinting of spatially characterized vascular structures in a bone-like construct [58, 59]. To prepare a bio-resin for digital light processing (DLP) lithography, GelMA is combined with methacrylated polyvinyl alcohol (PVA-MA) and a detectable light photoinitiator. This method allows the fabrication of a free-form without the constraints of a grid at 25-50 m resolution. Prior to printing, human bone marrow-derived MSCs were seeded into the bio-resin, which exhibited 85% viability 24 hours after UV polymerization. GelMA has been shown to be a crucial addition to PVA-MA for long-term (04 weeks) cell survival [60]. Components of ECM (gelatin, fibrin, and collagen) have been used for many years to mimic ECM for cell culture scaffolds, but lack most of the biological and chemical signals of total ECM. Tissue-derived decellularized ECM can be a very useful bioink. However, due to the decellularization process, it remains mechanically soft and therefore needs to be mixed with either synthetic or natural materials to improve its integrity after printing [61]. In one study, the preparation of decellularized dentin was demonstrated by removing the pulp and enamel from molar teeth and then grinding and decalcifying them. Sodium alginate and SCAPs were then added to produce a printable bio-ink. By increasing the concentration of dentin matrix molecules, significant upregulation of dentin markers such as ALP and RUNX2 was observed [60].

Biomaterial Inks

Biomaterial inks often have to be processed under cytotoxic conditions, such as temperature limits or the use of solvents. Nevertheless, they can be stacked with helpful therapeutic molecules that can withstand the aforementioned conditions [9]. Biomaterial inks, *e.g.*, from ceramics, thermoplastics, metals and composites, have been produced for use in biomedicine [9, 14, 15].

Synthetic Hydrogels

To be printed by extrusion, it is often not appropriate to use synthetic hydrogels for the direct seeding of cells. Pluronic is a commonly used synthetic ink with non-Newtonian properties similar to biopolymer hydrogels. It is thermally operationalized for UV cross-linking *via* chemical groups. It is used as a sacrificial ink to fabricate hollow structures and also serves as a support material for structures with overhangs [62, 63, 64, 65]. Elastomers have mechanical properties that mimic the viscoelasticity of local tissue and can therefore be used for bioprinting. Embedded techniques, such as Freeform Reversible Embedding (FRE) and printing in microgels, have been used to print polydimethylsiloxane (PDMS) and silicon, respectively, to produce branched, hollow structures that mimic airways and larger vessels [66, 67]. These structures can subsequently be made in part to model flow in larger vessels [67].

Thermoplastics and Resins

Thermoplastics are used by hobbyists and are common 3D printing materials in many industries. In bioprinting, thermoplastics can be prepared and subjected to various thermal cycles to include components and form fibers for extrusion, melt resins for photolithography, or polymers for electrospinning. We have recently shown that commercially available materials can be used for tissue design applications [14, 15]. Polycaprolactone (PCL), polylactic acid (PLA), and polyvinyl alcohol (PVA) have been printed as supports for cell-seeded hydrogels that require mechanical reinforcement and also for direct implantation *in vivo* [63, 68]. They are printed by extrusion from filaments to create structures with high resolution and shape fidelity, also taking into account porosity that can affect the mechanical properties of the scaffold [69]. PORO-LAY is a family of PVA-polyurethane elastomer blends that can be printed by extrusion from filaments before being rinsed with water to remove the PVA component and expose the underlying nanoporous elastomer. Over the course of seven days, doxorubicin and zoledronate (antiresorptive agents) could be entrapped in 3D-printed scaffolds [9, 10]. The complete release of doxorubicin was also enhanced by using different varieties of PORO-LAY with increased porosity, which further decreased metabolic activity in a prostate cancer cell line [9]. Cancer medications are often administered intravenously to specific tissues, but work still needs to be done to prevent the venous spread of excess amounts of drug to surrounding healthy cells/tissues. Continuous fluid interface processing was used to print a thermoplastic mesh-like filter and then coated it with a polystyrene sulfonate absorber. A printed cylindrical filter system was inserted into the iliac vein of a pig. When doxorubicin was injected upstream of the needle, the level of doxorubicin in the blood was significantly lower after the blood passed through the coated filter than after the blood passed through the uncoated control filter [70]. This could be interpreted as a method of removing excess chemotherapeutic agents by implanting a system into large veins derived from diseased tissue.

Ceramics

Ceramic products consist of salts (inorganic) such as calcium and phosphate. Due to their osteoconductivity, they are often used for bone and dental applications. In biofabrication, ceramics are mixed with a polymeric binder for 3D powder printing or extrusion bioprinting to compensate for the fragility of ceramics [71, 55]. Commonly printed ceramics and polymers include hydroxyapatite (HAp), tricalcium phosphate (TCP), polymethyl methacrylate (PMMA), bioglass, and biphasic calcium phosphate (BCP). Tetracalcium phosphate (TTCP) has been shown to be more absorbable at low pH, making it a promising choice for bone replacement. In a study, Mandal *et al.* demonstrated the usefulness of TTCP with a phytic acid binder. It was found that TTCP was the most abundant ceramic phase after printing and post-curing with an additional binder solution, but a liquid calcium phosphate phase, indicated a reaction between the binder and the ceramic powder [72]. Ceramics were used to exploit PCL by combining polymer and powdered ceramic to form a liquid melt, which was then extruded into filaments to produce a layered scaffold [73]. In this analysis, natural ceramics mixed with PCL outperformed synthetic ceramics in terms of osteoinductive ability [73].

Metal Implants

Stainless steel, cobalt-chromium-molybdenum, and titanium alloys have historically been used to cast, weld, and machine metal implants for orthopedic, dental, and craniofacial applications. Recent advances in AM technology have made it possible to design patient-specific implants based on reconstructed 3D image data. Expanding the versatility of these implants is one of the most important topics of recent research. In a recent study, researchers demonstrated how antibiotic-containing cement can be placed in a central cavity within metal implants that cannot be fabricated using conventional methods [74]. Selective laser melting (SLM) of metal powders can achieve high resolution, allowing the development of intricate lattices [12, 75]. This is also being investigated to overcome the stress shielding problems in hip implants. To evaluate the theoretical mechanical efficiency of porous implants, finite element analysis (FEA) and theoretical bone loss reductions were first performed. Then they were fabricated using SLM, and the FEA was validated by mechanical testing [12].

There are several different research directions for 3DP feedstock materials, and they differ in terms of efficiency, properties, and cost. Obtaining target mechanical properties and various variables, such as longevity and engineered architectures, is one of the most significant properties for the selection of raw material in 3DP of biomedical implants. As a result, these technologies' processing speed and stock material capacities must be thoroughly optimized and recorded both *in vitro* and *in vivo*. To avoid implant failure, the failure mode of porous architectures and their mechanical properties must be well known in implant design. In porous architectures, the length scale ranged from ten to hundreds of microns, which was beyond the resolution capability of the 3DP technology at the time. Finally, in order to achieve the final desired surface, the selection of raw materials is pivotal for surface finish and surface texture requirements. The following are some of the issues that are impeding 3DP's growth: (a) Inability to print rigid components with material properties comparable to those of traditionally manufactured pieces. Between successive printed layers, 3DP materials, in particular, appear to tear or crack more easily. Various approaches to achieving appreciable continuity between these layers have been proposed. (b) The inability to print well-ordered layers at small scales. To control the electrical, mechanical, and optical properties of components, precise control of the location and nanoparticles orientation within a thin film is usually needed. A solution/melt electrohydrodynamic printing device, which has the capacity to achieve higher resolution and oriented nanoparticles within the layer/piece, is one of the approaches. (c) Inadequate control over the 3DP operation can result in substantial-high scrap rates and rework, posing significant challenges to 3DP's long-term viability. (d) Porosity variation, also for the same build parameters, is a major technical challenge for 3DP production efficiency. Relationships between porosity, microstructure properties and process parameters must be defined to solve the problem.

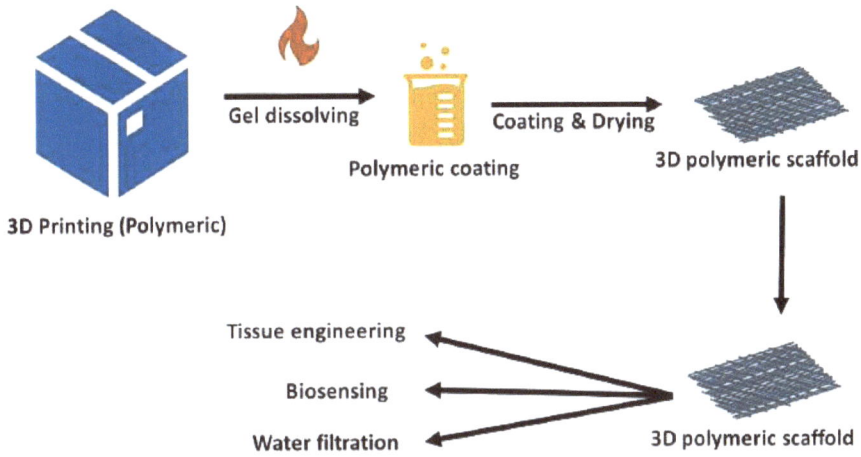

Fig. (2). Depicting biomedical application of the materials fabricated *via* 3D printer.

HEALTHCARE APPLICATIONS

Tissue Engineering

TE is crucial for replacing injured, damaged and non-functioning tissue with biocompatible implants. The field TE is based on the principle of creating implants from biocompatible materials in conjunction with growth factors and living cells, which in turn allow normal tissue growth [76, 77]. AM fits well with TE, as it allows the creation of 3D-printed models that can replicate the connective tissue network at the microscopic level [76]. 3D printing in TE offers a key advantage, namely the provision of geometrically complicated (composite) structures. This is achieved by formulating porous implants that can promote bone regeneration [78], formulating organoids on a complex 3D plane from stem cells, and precisely placing cells or materials in a 3D space [79].

3D Models and Organoids

For clinically applicable tests, TE has laid the groundwork for mimicking physiological models, *e.g.*, Human/Animal TE. The intricate organization of cell types at the 3D level with a complex three-dimensional organization can dictate the formation of organoids and organ system units to study underlying disease mechanisms and treatment responses. It is well known that all cells behave unevenly at the level of gene expression in a conventional 2D environment compared to their behavior in a 3D environment [80, 81]. When treated with the mTOR inhibitor rapamycin, Riedl and colleagues [80] observed inconsistent levels of AKT/mTOR/S6K signaling activity when measuring the response of spheroids compared to 2D cultures of human colon cancer cells. The group detected decreased signaling activity in 3D spheroids, in contrast to increased levels in 2D cultures. The 3D spheroid model proved to be an accurate representation of the *in vivo* response to rapamycin treatment in a xenotransplanted mouse tumor model, demonstrating the higher efficacy of 3D spheroids as more accurate preclinical models for drug testing compared to 2D cell line cultures [80]. Tumor spheroid models can be a cheaper and faster alternative to mimic *in vivo* environments [82], but they also have their drawbacks due to their complexity and size [83]. Due to constraints such as the distribution of nutrients, gasses, and waste materials, spheroids can develop to a certain size before hypoxia and nutrient starvation affect spheroid viability [83]. Moreover, spheroid fabrication methods only consider the creation of simple spheroids without considering the design complexity and are extremely delicate to handle [83]. As an AM method, complex 3D hydrogel models can be fabricated with bioprinting with an exceptional level of control and reproducibility, unlike traditional methods such as spheroid arrangement and molding. Complicated organ-on-a-chip models can be produced with bioprinting, as the creation of functional organs still seems many years away. Biologically printed 3D models with appropriate anatomical, functional, and compositional proximity to host tissues provide a better opportunity to perform interventions in a physiological atmosphere similar to the *in vivo* atmosphere [84]. Replication of the architectural atmosphere and proper mimicry of organ development signaling patterns originally created during the embryonic period by bioprinting is important for shaping developmental and disease processes [85]. Efforts to study stem cell differentiation and self-assembly in organoids have led to the ability to reproduce the functions of many different organs [86, 87]. However, the reliance on stem cell self-assembly leads to limited control over the structure, composition, and final size of the organoid [88, 89]. Controlled spatial deposition of self-assembling stem cells, cell lines, and patient-derived cells is required to achieve structural assembly by bioprinting [79, 85, 88, 89, 90]. Grix *et al.*, used stereolithography to fabricate 4 mm 3D liver lobule models and added GelMA (methacrylated gelatin) cultured with hepatocytes to the liver. Their results showed expanded hepatocyte-specific gene expression within the 3D model compared to 2D monolayer cultures [91]. They also created a vascular network that could be used to perfuse the model by inserting channels into the mode [91]. The high resolution of stereolithography technology allows the printing of details within this model at the microscopic level [91]. In one study, Bulanova *et al.* [92] created a vascularized thyroid model using material extrusion bioprinting technology and a collagen-based bioink. Using bioprinting, they were able to create a model that consisted of thyroid spheroids from embryonic stem cells and epithelial spheroids that were in close proximity to each other, which allowed further invasion and vascularization of the thyroid spheroids by the epithelial cells [93]. After implantation of the bioprinted organoid into hypothyroid mice, thyroid homeostasis was achieved, confirming the neovascularization and functionality of the organoid [92].

In recent years, bioprinting has proven to be a reliable method for the fabrication of laboratory models and highly complicated organoids. To date, this technique is used by countless laboratories to produce biological models from various tissues such as liver [84, 91, 93], mammary epithelium [94], myocardium [95], skeletal muscle [96], kidney [97], skin [17, 90], neurons [98] and malignant tumors [99]. These models are used in laboratories and pharmaceutical companies as efficient tools for drug screening [95, 100, 101, 102]. However, bioprinting also has certain limitations. The current features of bioprinting represent a trade-off between creating a viable environment for cell growth and providing structural support for the printed model. It is believed that flexible biomaterials can be used to create complex models that retain their shape but still limit cell communication and mobility [103]. Relatively viscous biomaterials create a fluid environment for cells, but are less capable of producing larger models without imploding under their own weight [104]. With advances in the field of bioprinting, there is an ongoing need for the introduction/discovery of new biomaterials that can provide architectural/structural support without compromising cell viability and extracellular matrix deposition.

Implants

There is an ever-increasing need for surgical implants and prostheses that incorporate integrative matching of these unfamiliar objects with their surrounding tissues to increase their utility. The development of implants and prostheses requires a multidisciplinary approach that combines the fields of engineering, molecular biology, pharmaceutics, biomechanics, and materials science. This gap between the fields of biology and engineering can be bridged by AM, producing complex biocompatible and bioactive constructs that utilise exceptional material properties such as osteoconductivity and osteoinductivity to ensure tissue recovery and integration of the implant into the surrounding tissue. Computed tomography (CT), a current radiological imaging technology, can be used to create accurate CAD models of a deformity that can be used as a template for 3D printing to ensure an ideal fit to the target tissue [105].

Tissue Regeneration

in vitro and *in vivo* bone regeneration requires implants that can be effectively fabricated by AM [106, 107]. Treatment of bone deformities of a practice nature is both crucial and strenuous, which can often lead to medical complications if neglected [108]. Simply filling a major bone deformity with a bone graft cannot ensure bone cohesion due to the lack of blood supply to the graft [108, 109, 110]. Current surgical alternatives, such as the vascularization of a bone graft, are lengthy and highly specialized. These include performing a Masquelet procedure, which can restore blood supply but requires various surgical procedures and is, therefore, unsafe [108]. AM offers potential implant solutions that could promote both vascularization and bone regeneration [107, 111]. Porosity plays an important role in promoting bone ingrowth. From a variety of materials, including ceramics, biodegradable polymers, and metals, 3D printing offers the technical capability to fabricate porous scaffolds with high resolution [112]. AM-manufactured scaffolds offer many regenerative possibilities in addition to bone. Lee *et al.*, [113] demonstrated the printing and implantation of a doped cellular hydroxyapatite/PCL scaffold with transforming growth factor_3 (TGF_ 3) in rabbits. The articular surface of the proximal humeral joint completely regrew, and the regenerated cartilage exhibited equivalent compressive, histological, and shear properties to native articular cartilage in rabbits [113]. Chang *et al.* reported similar results with hyaline cartilage [114], in which a PCL trachea coated with MSCs and fibrin printed by fused deposition modeling (FDM) showed interactive coordination with native tracheal tissue and production of neocartilage. Tissue regeneration in the periodontium by FDM-printed scaffolds was observed in hydroxyapatite PCL scaffolds [115].

Implant-Tissue Interface

The first interaction between the implant and the surrounding tissue occurs at the surface of the implant. This interaction can jeopardize the success of the implant, *e.g.*, due to different mechanical properties of the implant and the surrounding tissue and implant-induced cell death due to movement at the tissue-implant interface [116]. Therefore, it is extremely important to consider the properties of the implant surface to allow better integration of the implant into the surrounding tissue. For example, in metallic bone implants, surface roughness is an important factor that is inversely related to fixation to the adjacent bone. In their study, MacBarb *et al.* [117] showed that 3D printed titanium spinal implants fabricated by electron beam melting promoted *in vitro* osteoblast proliferation and calcium production on the implant surface better than conventional plasma spraying of titanium onto the implant surface. This was achieved by incorporating a bone-like trabecular network on the surface of the 3D implant model, followed by high-resolution fabrication using AM [117]. Nanofabrication techniques have also been used to produce porous coatings for 3D printed implants. A study by Garcia *et al.* [118] showed how coating titanium implants with fibronectin nanoclusters can improve integrin binding in rats. In another study, Tran and colleagues [119] coated titanium implants with selenium nanoclusters and demonstrated that antitumor activity was enhanced by selenium nanoclusters along with healthy bone production in contrast to uncoated implants [119].

Dentistry

The diverse structural unevenness of patients in the field of dentistry has led to the need for an area that allows for individualized treatment and monitoring of each patient. Recent innovations in imaging technology have led to accurate 3D reconstructions of the oral cavity that can be used as templates for the personalization of oral implants and prostheses [120, 121]. Due to its feasibility and accuracy in the fabrication of prostheses, AM has become clinically established in the field of prosthodontics. Prostheses fabricated using AM are as durable as conventional fabrication methods [122, 123, 124]. Stereolithography is a widely used technique in this field, for which commercial acrylic products are already available [125]. To replicate the final alignment of teeth and print a suitable mold for custom-made silicone prostheses, orthodontics uses CAD models of a patient's oral cavity [126]. AM also plays an important role in oral and maxillofacial surgery, especially in 3D-printed custom osteoinductive and biocompatible implants to compensate for bone deformities and promote bone regeneration, which we discuss in the following section [127].

Orthopedics

Patients with orthopedic complications are helped with implants that restore structural alignment, integrity, and motion. Implants are available in standard sizes that fit the vast majority of patients. However, patients with extreme deformities may require customized implants to ensure the size fits [128]. Similar to dentistry, AM CAD models of the patient's anatomical needs created by radiological imaging can be used to design and fabricate custom implants. To prevent implant failure and provide tissue support, the patient's own bone must also be integrated into the orthopedic implants. There are several factors that control an implant's ability to promote bone regeneration in terms of its structure and composition. One of these factors is porosity, which determines the extent to which an implant penetrates the surrounding tissue and vessel ingrowth [129]. High-resolution AM techniques are used to create implants with high porosity, which allows the meshwork of the bone to connect the implant to the surrounding bone tissue. 3D printing of materials that are osteoconductive and bioabsorbable, such as calcium phosphate cement [130, 131], has been shown to stimulate bone growth more effectively. To avoid systemic exposure to growth factors and drugs, their concentrated and continuous release is ensured by the implant. In addition, 3D printing also requires lower amounts of therapeutics, which increases the overall cost-effectiveness [132]. In addition, the use of 3D printing to fabricate orthoses for patient-specific plaster casts for fractures is a feasible option [133].

Drug Delivery

The field of AM plays a very dynamic role in the pharmaceutical industry [134]. It is important that new techniques are developed to ensure the controlled release of drugs administered through different routes. Drugs are administered through two possible routes: oral tablets and implanted devices, both of which ensure a certain rate of release of the drug in the body. The following sections describe new ways in which AM can play a role in improved drug delivery.

Tablets

To produce tablets in solid form, the powder mixture containing the active pharmaceutical ingredient (API) is pressed into molds [135]. This large-scale industrial method is suitable for producing large quantities of simple tablets, but cannot handle complex structures and dosage variability of tablets. AM can be used to produce specialized tablets for specific patients that can perform complex and unique functions based on the patient's requirements. Due to its cost-effectiveness and rapid production of tablets that can be taken immediately, fused deposition modeling (FDM) is the most commonly used 3D printing technique [136, 137]. The other printing techniques mentioned in this review are mainly used for research purposes [138, 139]. Sadia *et al.*, demonstrated the release mechanics by 3D printing a tablet with holes larger than 0.6 mm. Their study showed that the tablets carried the United States Pharmacopeia (USP) defined "immediate release" designation, as opposed to tablets of the same size. The tablets shown in the study were printed with hydrochlorothiazide-impregnated thermoplastics. These tablets attributed the improved release properties to an expanded surface area to volume ratio achieved through a 3D printed plan [136]. In another study by Khaled *et al.*, a paste based on hydroalcoholic gel and FDM was used to print a three-compartment tablet capable of delivering three different drugs (captopril, nifedipine, glipizide). Interestingly, it was observed that by adjusting the paste, captopril was released in a single tablet with zero-order kinetics, while the other two drugs were released with first-order or different kinetics [140]. In traditional methods of tablet preparation, it is still problematic to provide a wide range of tablet dosages, so there are few traditional dosages of drugs. However, it is worth noting that the physiology of each patient is very different. Considering this wide variability between patients, a drug with a certain dosage may have very different side effects in different patients [141, 142]. Therefore, AM offers a cost-effective solution for producing tablets that can be customized for each patient. In modern studies, printing of requested tablets for emergency situations and tablets with non-durable agents can be achieved by using AM [143].

Transdermal Delivery

Blood flow through the skin allows better diffusion for the controlled release of a variety of drugs, and this particular property is used in transdermal drug delivery [144]. Unlike tablets, injections and insertion methods, transdermal drug delivery is a non-invasive solution [145]. Common methods of transdermal drug delivery include patches and microneedle arrays. Patches provide a continuous supply of a drug to the upper epidermal layers and serve as a drug reservoir that is passively delivered into the vessels of the deeper epidermal layers [145]. Alternatively, microneedle arrays consist of a patch with microscopic needle-shaped projections that can penetrate the upper epidermal layers. This method allows for better penetration of a drug as the integrity of the epidermis is maintained. It also eliminates the risk of infection or other complications that can occur with invasive methods [146]. AM offers an advantage in the production of patches as it allows the formation of patches with a mixture of drugs. AM also enables the fabrication of complex structured patches that allow variable drug delivery from the same patch [145]. SLA, one of the high-resolution AM techniques, has been used to fabricate microneedles from a variety of materials and offers precision and speed [146]. Another AM method is inkjet printing, in which microneedles are coated with a uniform amount of a drug in a controlled manner [147].

Drug-Releasing Implants

In medicine, implants such as joints, screws, stents, or plates play an important role in treating anatomical abnormalities, performing surgery, and also supporting normal body functions. For long-term implants, materials are needed that do not attack the surrounding tissue. For temporary implants, antibacterial coatings are required to prevent infections, such as in catheters for urinary or vascular access [148]. AM provides a high degree of control in the fabrication of complex implants and also incorporates therapeutics into the structure of an implant. By using biocompatible composites in the implant structure, the rate at which a drug leaves these implants can be effectively administered [149]. Studies have shown that antibiotic-infused PLA constructs fabricated using 3D printing are much more effective at preventing bacterial biofilm formation than constructs coated with antibiotics alone [150]. The same drug delivery efficacy compared to direct drug treatment has been demonstrated in *in-vitro* studies using 3D printed scaffolds by extrusion, supporting the possibility of using 3D printed scaffolds for reliable local drug delivery into the tissue.

Surgical Tools

Recent developments in radiologic imaging have greatly facilitated the CAD reconstruction of a patient's anatomy and allow for the customization of patient-specific surgical instruments [151]. Most surgical instruments are designed to fit most patients. However, to accommodate patient-specific anatomy and complicated procedures, customization of instruments is required to ensure a safe, controlled, and simplified surgical procedure and to completely avoid preoperative and postoperative complications [152]. CAD designs can be easily converted into usable instruments with the help of AM, which are helpful for high-risk areas in hospitals, such as operating rooms. The manufactured tools are used for templates, drill and cutting guides [153], and retractors [154] after effective sterilization. In one study, a PLA-based retractor made by 3D printing was shown to effectively mimic its stainless material counterpart while being 10 times cheaper [154]. It is predicted that institutions can save a lot of money if this technology is widely adopted. Interestingly, some studies suggest that 3D-printed surgical instruments are well suited and effective for emergencies during space missions, such as a trip to Mars [155, 156]. It should also be noted that the cost-effectiveness of using 3D-printed surgical instruments or casts could be much more fruitful in the least developed countries and facilities [154].

CONCLUSION

Given the various advances in the field of 3DP, it is likely that the cost of related hardware will decrease in the future. It is also likely that new and different materials will be developed that can contribute to further improvements in tissue and biomedical engineering applications. For example, the use of thermoplastics in low-cost 3D printing manufacturing has emerged as the most commonly used material in various hospitals and research institutes. In addition, a variety of other valuable unique materials are being synthesized and introduced at a rapid pace, which could also be used in hospitals for tissue repair in the near future. Bioprinting has proven to be a very suitable tool for creating 3D models that mimic tissues and whose structure and functions are representative of drug screening and disease modeling. In addition, bioprinting technology has proven successful in the formulation of organoids and simplified miniature organs. The greatest opportunity lies in the ability to print fully functional organs that can be transplanted into a patient. If successful, this has far-reaching public health implications and would greatly reduce long waiting times and nerve-consuming histocompatibility tests, as the printed organ would come from the patient's own stem cells. However, there are still several limitations that bioprinting technology must overcome. These limitations include microvascularization and bioink longevity. However, with the current pace of development in this field, many more innovations can be expected.

CONSENT FOR PUBLICATION

Not applicable.

CONFLICT OF INTEREST

The author declares no conflict of interest, financial or otherwise.

ACKNOWLEDGEMENTS

SBJ acknowledges the contribution of all the co-authors.

REFERENCES

[1]   T.S. Karande, J.L. Ong, and C.M. Agrawal, "Diffusion in musculoskeletal tissue engineering scaffolds: design issues related to porosity, permeability, architecture, and nutrient mixing", *Ann. Biomed. Eng.,* vol. 32, no. 12, pp. 1728-1743, 2004.
        [http://dx.doi.org/10.1007/s10439-004-7825-2] [PMID: 15675684]

[2]   M.M. Stevens, and J.H. George, "Exploring and engineering the cell surface interface", *Science (80-.),* vol. 310, no. 5751, pp. 1135-1138, 2005.
        [http://dx.doi.org/10.1126/science.1106587]

[3]   S.J. Hollister, "Porous scaffold design for tissue engineering", *Nat. Mater.,* vol. 4, no. 7, pp. 518-524, 2005.
        [http://dx.doi.org/10.1038/nmat1421] [PMID: 16003400]

[4]   A.A. Zadpoor, "Meta-biomaterials", *Biomater. Sci.,* vol. 8, no. 1, pp. 18-38, 2020.
        [http://dx.doi.org/10.1039/C9BM01247H] [PMID: 31626248]

[5]   Q. Hu, X.Z. Sun, C.D.J. Parmenter, M.W. Fay, E.F. Smith, G.A. Rance, Y. He, F. Zhang, Y. Liu, D. Irvine, C. Tuck, R. Hague, and R. Wildman, "Additive manufacture of complex 3D Au-containing nanocomposites by simultaneous two-photon polymerisation and photoreduction", *Sci. Rep.,* vol. 7, no. 1, p. 17150, 2017.
        [http://dx.doi.org/10.1038/s41598-017-17391-1] [PMID: 29215026]

[6]   K.J. McHugh, "Fabrication of fillable microparticles and other complex 3D microstructures", *Science (80-.),* vol. 357, no. 6356>, pp. 1138-1142, 2017.
        [http://dx.doi.org/10.1126/science.aaf7447]

[7]   C.W. Hull, "Apparatus for Production of Three-Dmensonal Objects By Stereo Thography", *Patent,* no. 19, p. 16, 1984. https://patents.google.com/patent/US4575330

[8]   J. Norman, R.D. Madurawe, C.M.V. Moore, M.A. Khan, and A. Khairuzzaman, "A new chapter in pharmaceutical manufacturing: 3D-printed drug products", *Adv. Drug Deliv. Rev.,* vol. 108, pp. 39-50, 2017.
        [http://dx.doi.org/10.1016/j.addr.2016.03.001] [PMID: 27001902]

[9]   P. Ahangar, E. Akoury, A. Ramirez Garcia Luna, A. Nour, M. Weber, and D. Rosenzweig, "Nanoporous 3D-printed scaffolds for local doxorubicin delivery in bone metastases secondary to prostate cancer", *Materials (Basel),* vol. 11, no. 9, p. 1485, 2018.
        [http://dx.doi.org/10.3390/ma11091485] [PMID: 30134523]

[10]   E. Akoury, M.H. Weber, and D.H. Rosenzweig, "3D-Printed Nanoporous Scaffolds Impregnated with

Zoledronate for the Treatment of Spinal Bone Metastases", *MRS Adv.,* vol. 4, no. 21, pp. 1245-1251, 2019.
[http://dx.doi.org/10.1557/adv.2019.156]

[11]   G. Coelho, T.M.F. Chaves, A.F. Goes, E.C. Del Massa, O. Moraes, and M. Yoshida, "Multimaterial 3D printing preoperative planning for frontoethmoidal meningoencephalocele surgery", *Childs Nerv. Syst.,* vol. 34, no. 4, pp. 749-756, 2018.
[http://dx.doi.org/10.1007/s00381-017-3616-6] [PMID: 29067504]

[12]   S. Arabnejad, B. Johnston, M. Tanzer, and D. Pasini, "Fully porous 3D printed titanium femoral stem to reduce stress-shielding following total hip arthroplasty", *J. Orthop. Res.,* vol. 35, no. 8, pp. 1774-1783, 2017.
[http://dx.doi.org/10.1002/jor.23445] [PMID: 27664796]

[13]   C. Yang, X. Wang, B. Ma, H. Zhu, Z. Huan, N. Ma, C. Wu, and J. Chang, "3D-Printed Bioactive Ca 3 SiO 5 Bone Cement Scaffolds with Nano Surface Structure for Bone Regeneration", *ACS Appl. Mater. Interfaces,* vol. 9, no. 7, pp. 5757-5767, 2017.
[http://dx.doi.org/10.1021/acsami.6b14297] [PMID: 28117976]

[14]   D. Rosenzweig, E. Carelli, T. Steffen, P. Jarzem, and L. Haglund, "3D-printed ABS and PLA scaffolds for cartilage and nucleus pulposustissue regeneration", *Int. J. Mol. Sci.,* vol. 16, no. 12, pp. 15118-15135, 2015.
[http://dx.doi.org/10.3390/ijms160715118] [PMID: 26151846]

[15]   R. Fairag, D.H. Rosenzweig, J.L. Ramirez-Garcialuna, M.H. Weber, and L. Haglund, "Three-Dimensional Printed Polylactic Acid Scaffolds Promote Bone-like Matrix Deposition *in vitro*"., *ACS Appl. Mater. Interfaces,* vol. 11, no. 17, pp. 15306-15315, 2019.
[http://dx.doi.org/10.1021/acsami.9b02502] [PMID: 30973708]

[16]   A. Vikram Singh, M. Hasan Dad Ansari, S. Wang, P. Laux, A. Luch, A. Kumar, R. Patil, and S. Nussberger, "The adoption of three-dimensional additive manufacturing from biomedical material design to 3D organ printing", *Appl. Sci. (Basel),* vol. 9, no. 4, p. 811, 2019.
[http://dx.doi.org/10.3390/app9040811]

[17]   N. Cubo, M. Garcia, J.F. del Cañizo, D. Velasco, and J.L. Jorcano, "3D bioprinting of functional human skin: production and *in vivo* analysis"., *Biofabrication,* vol. 9, no. 1, p. 015006, 2016.
[http://dx.doi.org/10.1088/1758-5090/9/1/015006] [PMID: 27917823]

[18]   S. Cells, P.E.G. Hydrogel, G. Gao, K. Hubbell, A.F. Schilling, and G. Dai, ""Bioprinting Cartilage Tissue from Mesenchymal," 3D Cell Cult", *Methods Protoc.,* vol. 1612, pp. 391-398, 2017.
[http://dx.doi.org/10.1007/978-1-4939-7021-6]

[19]   X. Zhou, W. Zhu, M. Nowicki, S. Miao, H. Cui, B. Holmes, R.I. Glazer, and L.G. Zhang, "3D Bioprinting a Cell-Laden Bone Matrix for Breast Cancer Metastasis Study", *ACS Appl. Mater. Interfaces,* vol. 8, no. 44, pp. 30017-30026, 2016.
[http://dx.doi.org/10.1021/acsami.6b10673] [PMID: 27766838]

[20]   H.N. Chia, and B.M. Wu, "Recent advances in 3D printing of biomaterials", *J. Biol. Eng.,* vol. 9, no. 1, p. 4, 2015.
[http://dx.doi.org/10.1186/s13036-015-0001-4] [PMID: 25866560]

[21]   C. Guo, M. Zhang, and B. Bhandari, "Model Building and Slicing in Food 3D Printing Processes: A Review", *Compr. Rev. Food Sci. Food Saf.,* vol. 18, no. 4, pp. 1052-1069, 2019.
[http://dx.doi.org/10.1111/1541-4337.12443] [PMID: 33337002]

[22]   J.Y. Lee, J. An, and C.K. Chua, "Fundamentals and applications of 3D printing for novel materials", *Appl. Mater. Today,* vol. 7, pp. 120-133, 2017.
[http://dx.doi.org/10.1016/j.apmt.2017.02.004]

[23]   P. Juskova, A. Ollitrault, M. Serra, J.L. Viovy, and L. Malaquin, "Resolution improvement of 3D stereo-lithography through the direct laser trajectory programming: Application to microfluidic deterministic lateral displacement device", *Anal. Chim. Acta,* vol. 1000, pp. 239-247, 2018.

[http://dx.doi.org/10.1016/j.aca.2017.11.062] [PMID: 29289316]

[24] M. Carve, and D. Wlodkowic, "3D-printed chips: Compatibility of additive manufacturing photopolymeric substrata with biological applications", *Micromachines (Basel),* vol. 9, no. 2, p. 91, 2018.
[http://dx.doi.org/10.3390/mi9020091] [PMID: 30393367]

[25] Y. Bai, and C.B. Williams, "Binder jetting additive manufacturing with a particle-free metal ink as a binder precursor", *Mater. Des.,* vol. 147, no. 2017, pp. 146-156, 2017.
[http://dx.doi.org/10.1016/j.matdes.2018.03.027]

[26] Y. Bai, C. Wall, H. Pham, A. Esker, and C.B. Williams, "Characterizing Binder–Powder Interaction in Binder Jetting Additive Manufacturing *via* Sessile Drop Goniometry"., *J. Manuf. Sci. Eng.,* vol. 141, no. 1, p. 011005, 2019.
[http://dx.doi.org/10.1115/1.4041624]

[27] *ISO/ASTM 52900: Additive manufacturing - General principles and Terminology,* 2015.https://www.iso.org/obp/ui/#iso:std:69669:en%0Ahttps://www.iso.org/standard/69669.html%0A https://www.astm.org/Standards/ISOASTM52900.htm

[28] E. Sachs, M. Cima, and J. Cornie, "Three-Dimensional Printing: Rapid Tooling and Prototypes Directly from a CAD Model", *CIRP Ann.,* vol. 39, no. 1, pp. 201-204, 1990.
[http://dx.doi.org/10.1016/S0007-8506(07)61035-X]

[29] C. Deckard, *SLS Patent,* 1986.https://patents.google.com/patent/US4863538

[30] D.M. Keicher, *Laser Engineered Net Shaping (LENS{trademark}) for additive component processing.*https://www.osti.gov/biblio/231692

[31] P.E. Ruggiero, "4665492 Computer automated manufacturing process and system", *Robot. Comput.- Integr. Manuf.,* vol. 3, no. 4, pp. i-ii, 1987.
[http://dx.doi.org/10.1016/0736-5845(87)90060-3]

[32] P.K. Wu, B.R. Ringeisen, J. Callahan, M. Brooks, D.M. Bubb, H.D. Wu, A. Piqué, B. Spargo, R.A. McGill, and D.B. Chrisey, "The deposition, structure, pattern deposition, and activity of biomaterial thin-films by matrix-assisted pulsed-laser evaporation (MAPLE) and MAPLE direct write", *Thin Solid Films,* vol. 398-399, pp. 607-614, 2001.
[http://dx.doi.org/10.1016/S0040-6090(01)01347-5]

[33] T. Ghidini, *Regenerative medicine and 3D bioprinting for human space exploration and planet colonisation,* 2018.
[http://dx.doi.org/10.21037/jtd.2018.03.19]

[34] X. Cui, and T. Boland, "Human microvasculature fabrication using thermal inkjet printing technology", *Biomaterials,* vol. 30, no. 31, pp. 6221-6227, 2009.
[http://dx.doi.org/10.1016/j.biomaterials.2009.07.056] [PMID: 19695697]

[35] J. Stringer, and B. Derby, "Formation and stability of lines produced by inkjet printing", *Langmuir,* vol. 26, no. 12, pp. 10365-10372, 2010.
[http://dx.doi.org/10.1021/la101296e] [PMID: 20481461]

[36] X. Cui, D. Dean, Z.M. Ruggeri, and T. Boland, "Cell damage evaluation of thermal inkjet printed Chinese hamster ovary cells", *Biotechnol. Bioeng.,* vol. 106, no. 6, pp. 963-969, 2010.
[http://dx.doi.org/10.1002/bit.22762] [PMID: 20589673]

[37] M. Feygin, and S.S. Pak, *Laminated object manufacturing apparatus and method.,* 1999no. 19, p. 52. https://patents.google.com/patent/US5876550A/en

[38] A. Biswas, I.S. Bayer, A.S. Biris, T. Wang, E. Dervishi, and F. Faupel, "Advances in top–down and bottom–up surface nanofabrication: Techniques, applications & future prospects", *Adv. Colloid Interface Sci.,* vol. 170, no. 1-2, pp. 2-27, 2012.
[http://dx.doi.org/10.1016/j.cis.2011.11.001] [PMID: 22154364]

[39] E. Ruiz-Hitzky, P. Aranda, M. Darder, and M. Ogawa, "Hybrid and biohybrid silicate based materials: molecular *vs.* block-assembling bottom–up processes"., *Chem. Soc. Rev.,* vol. 40, no. 2, pp. 801-828, 2011.
[http://dx.doi.org/10.1039/C0CS00052C] [PMID: 21152648]

[40] J. Groll, T. Boland, T. Blunk, J.A. Burdick, D.W. Cho, P.D. Dalton, B. Derby, G. Forgacs, Q. Li, V.A. Mironov, L. Moroni, M. Nakamura, W. Shu, S. Takeuchi, G. Vozzi, T.B.F. Woodfield, T. Xu, J.J. Yoo, and J. Malda, "Biofabrication: reappraising the definition of an evolving field", *Biofabrication,* vol. 8, no. 1, p. 013001, 2016.
[http://dx.doi.org/10.1088/1758-5090/8/1/013001] [PMID: 26744832]

[41] J. Groll, J.A. Burdick, D-W. Cho, B. Derby, M. Gelinsky, S.C. Heilshorn, T. Jüngst, J. Malda, V.A. Mironov, K. Nakayama, A. Ovsianikov, W. Sun, S. Takeuchi, J.J. Yoo, and T.B.F. Woodfield, "A definition of bioinks and their distinction from biomaterial inks", *Biofabrication,* vol. 11, no. 1, p. 013001, 2018.
[http://dx.doi.org/10.1088/1758-5090/aaec52] [PMID: 30468151]

[42] J. Malda, J. Visser, F.P. Melchels, T. Jüngst, W.E. Hennink, W.J.A. Dhert, J. Groll, and D.W. Hutmacher, "25th anniversary article: Engineering hydrogels for biofabrication", *Adv. Mater.,* vol. 25, no. 36, pp. 5011-5028, 2013.
[http://dx.doi.org/10.1002/adma.201302042] [PMID: 24038336]

[43] O. Chaudhuri, L. Gu, D. Klumpers, M. Darnell, S.A. Bencherif, J.C. Weaver, N. Huebsch, H. Lee, E. Lippens, G.N. Duda, and D.J. Mooney, "Hydrogels with tunable stress relaxation regulate stem cell fate and activity", *Nat. Mater.,* vol. 15, no. 3, pp. 326-334, 2016.
[http://dx.doi.org/10.1038/nmat4489] [PMID: 26618884]

[44] N.C. Hunt, and L.M. Grover, "Cell encapsulation using biopolymer gels for regenerative medicine", *Biotechnol. Lett.,* vol. 32, no. 6, pp. 733-742, 2010.
[http://dx.doi.org/10.1007/s10529-010-0221-0] [PMID: 20155383]

[45] J.L. Drury, and D.J. Mooney, "Hydrogels for tissue engineering: scaffold design variables and applications", *Biomaterials,* vol. 24, no. 24, pp. 4337-4351, 2003.
[http://dx.doi.org/10.1016/S0142-9612(03)00340-5] [PMID: 12922147]

[46] J. H. Lee, and H. W. Kim, "Emerging properties of hydrogels in tissue engineering", *J. Tissue Eng,* vol. 9, pp. 0-3, 2018.
[http://dx.doi.org/10.1177/2041731418768285]

[47] J.J. Green, and J.H. Elisseeff, "Mimicking biological functionality with polymers for biomedical applications", *Nature,* vol. 540, no. 7633, pp. 386-394, 2016.
[http://dx.doi.org/10.1038/nature21005] [PMID: 27974772]

[48] A. Ribeiro, "Assessing bioink shape fidelity to aid material development in 3D bioprinting",
[http://dx.doi.org/10.1088/1758-5090/aa90e2]

[49] T. Jungst, W. Smolan, K. Schacht, T. Scheibel, and J. Groll, "Strategies and Molecular Design Criteria for 3D Printable Hydrogels", *Chem. Rev.,* vol. 116, no. 3, pp. 1496-1539, 2016.
[http://dx.doi.org/10.1021/acs.chemrev.5b00303] [PMID: 26492834]

[50] S.R. Moxon, *Suspended manufacture of biological structures.,* 2017.
[http://dx.doi.org/10.1002/adma.201605594]

[51] K.Y. Lee, and D.J. Mooney, "Alginate: Properties and biomedical applications", *Prog. Polym. Sci.,* vol. 37, no. 1, pp. 106-126, 2012.
[http://dx.doi.org/10.1016/j.progpolymsci.2011.06.003] [PMID: 22125349]

[52] M.E. Cooke, M.J. Pearson, R.J.A. Moakes, C.J. Weston, E.T. Davis, S.W. Jones, and L.M. Grover, "Geometric confinement is required for recovery and maintenance of chondrocyte phenotype in alginate", *APL Bioeng.,* vol. 1, no. 1, p. 016104, 2017.
[http://dx.doi.org/10.1063/1.5006752] [PMID: 31069284]

[53]   J.A. Rowley, and D.J. Mooney, "Alginate type and RGD density control myoblast phenotype", *J. Biomed. Mater. Res.,* vol. 60, no. 2, pp. 217-223, 2002.
[http://dx.doi.org/10.1002/jbm.1287] [PMID: 11857427]

[54]   F. You, X. Chen, D.M.L. Cooper, T. Chang, and B.F. Eames, "Homogeneous hydroxyapatite/alginate composite hydrogel promotes calcified cartilage matrix deposition with potential for three-dimensional bioprinting", *Biofabrication,* vol. 11, no. 1, p. 015015, 2018.
[http://dx.doi.org/10.1088/1758-5090/aaf44a] [PMID: 30524110]

[55]   B.J. Klotz, D. Gawlitta, A.J.W.P. Rosenberg, J. Malda, and F.P.W. Melchels, "Gelatin-Methacryloyl Hydrogels: Towards Biofabrication-Based Tissue Repair", *Trends Biotechnol.,* vol. 34, no. 5, pp. 394-407, 2016.
[http://dx.doi.org/10.1016/j.tibtech.2016.01.002] [PMID: 26867787]

[56]   I. Pepelanova, K. Kruppa, T. Scheper, and A. Lavrentieva, "Gelatin-methacryloyl (GelMA) hydrogels with defined degree of functionalization as a versatile toolkit for 3D cell culture and extrusion bioprinting", *Bioengineering (Basel),* vol. 5, no. 3, p. 55, 2018.
[http://dx.doi.org/10.3390/bioengineering5030055] [PMID: 30022000]

[57]   P.A. Levett, F.P.W. Melchels, K. Schrobback, D.W. Hutmacher, J. Malda, and T.J. Klein, "A biomimetic extracellular matrix for cartilage tissue engineering centered on photocurable gelatin, hyaluronic acid and chondroitin sulfate", *Acta Biomater.,* vol. 10, no. 1, pp. 214-223, 2014.
[http://dx.doi.org/10.1016/j.actbio.2013.10.005] [PMID: 24140603]

[58]   B. Byambaa, N. Annabi, K. Yue, G. Trujillo-de Santiago, M.M. Alvarez, W. Jia, M. Kazemzadeh-Narbat, S.R. Shin, A. Tamayol, and A. Khademhosseini, "Bioprinted Osteogenic and Vasculogenic Patterns for Engineering 3D Bone Tissue", *Adv. Healthc. Mater.,* vol. 6, no. 16, p. 1700015, 2017.
[http://dx.doi.org/10.1002/adhm.201700015] [PMID: 28524375]

[59]   Y.L. Park, K. Park, and J.M. Cha, "3D-Bioprinting Strategies Based on *in situ* Bone-Healing Mechanism for Vascularized Bone Tissue Engineering"., *Micromachines (Basel),* vol. 12, no. 3, p. 287, 2021.
[http://dx.doi.org/10.3390/mi12030287] [PMID: 33800485]

[60]   K.S. Lim, R. Levato, P.F. Costa, M.D. Castilho, C.R. Alcala-Orozco, K.M.A. van Dorenmalen, F.P.W. Melchels, D. Gawlitta, G.J. Hooper, J. Malda, and T.B.F. Woodfield, "Bio-resin for high resolution lithography-based biofabrication of complex cell-laden constructs", *Biofabrication,* vol. 10, no. 3, p. 034101, 2018.
[http://dx.doi.org/10.1088/1758-5090/aac00c] [PMID: 29693552]

[61]   D. Choudhury, H.W. Tun, T. Wang, and M.W. Naing, "Organ-Derived Decellularized Extracellular Matrix: A Game Changer for Bioink Manufacturing?", *Trends Biotechnol.,* vol. 36, no. 8, pp. 787-805, 2018.
[http://dx.doi.org/10.1016/j.tibtech.2018.03.003] [PMID: 29678431]

[62]   M. Kesti, C. Eberhardt, G. Pagliccia, D. Kenkel, D. Grande, A. Boss, and M. Zenobi-Wong, "Bioprinting Complex Cartilaginous Structures with Clinically Compliant Biomaterials", *Adv. Funct. Mater.,* vol. 25, no. 48, pp. 7406-7417, 2015.
[http://dx.doi.org/10.1002/adfm.201503423]

[63]   M. de Ruijter, A. Ribeiro, I. Dokter, M. Castilho, and J. Malda, "Simultaneous Micropatterning of Fibrous Meshes and Bioinks for the Fabrication of Living Tissue Constructs", *Adv. Healthc. Mater.,* vol. 8, no. 7, p. 1800418, 2019.
[http://dx.doi.org/10.1002/adhm.201800418] [PMID: 29911317]

[64]   W. Wu, A. DeConinck, and J.A. Lewis, "Omnidirectional printing of 3D microvascular networks", *Adv. Mater.,* vol. 23, no. 24, pp. H178-H183, 2011.
[http://dx.doi.org/10.1002/adma.201004625] [PMID: 21438034]

[65]   D.B. Kolesky, K.A. Homan, M.A. Skylar-Scott, and J.A. Lewis, "Three-dimensional bioprinting of thick vascularized tissues", *Proc. Natl. Acad. Sci. USA,* vol. 113, no. 12, pp. 3179-3184, 2016.

[http://dx.doi.org/10.1073/pnas.1521342113] [PMID: 26951646]

[66] T.J. Hinton, A. Hudson, K. Pusch, A. Lee, and A.W. Feinberg, "3D Printing PDMS Elastomer in a Hydrophilic Support Bath *via* Freeform Reversible Embedding"., *ACS Biomater. Sci. Eng.,* vol. 2, no. 10, pp. 1781-1786, 2016.
[http://dx.doi.org/10.1021/acsbiomaterials.6b00170] [PMID: 27747289]

[67] C.S. O'Bryan, T. Bhattacharjee, S. Hart, C.P. Kabb, K.D. Schulze, I. Chilakala, B.S. Sumerlin, W.G. Sawyer, and T.E. Angelini, "Self-assembled micro-organogels for 3D printing silicone structures", *Sci. Adv.,* vol. 3, no. 5, p. e1602800, 2017.
[http://dx.doi.org/10.1126/sciadv.1602800] [PMID: 28508071]

[68] H.W. Kang, S.J. Lee, I.K. Ko, C. Kengla, J.J. Yoo, and A. Atala, "A 3D bioprinting system to produce human-scale tissue constructs with structural integrity", *Nat. Biotechnol.,* vol. 34, no. 3, pp. 312-319, 2016.
[http://dx.doi.org/10.1038/nbt.3413] [PMID: 26878319]

[69] J. Visser, F.P.W. Melchels, J.E. Jeon, E.M. van Bussel, L.S. Kimpton, H.M. Byrne, W.J.A. Dhert, P.D. Dalton, D.W. Hutmacher, and J. Malda, "Reinforcement of hydrogels using three-dimensionally printed microfibres", *Nat. Commun.,* vol. 6, no. 1, p. 6933, 2015.
[http://dx.doi.org/10.1038/ncomms7933] [PMID: 25917746]

[70] H.J. Oh, M.S. Aboian, M.Y.J. Yi, J.A. Maslyn, W.S. Loo, X. Jiang, D.Y. Parkinson, M.W. Wilson, T. Moore, C.R. Yee, G.R. Robbins, F.M. Barth, J.M. DeSimone, S.W. Hetts, and N.P. Balsara, "3D Printed Absorber for Capturing Chemotherapy Drugs before They Spread through the Body", *ACS Cent. Sci.,* vol. 5, no. 3, pp. 419-427, 2019.
[http://dx.doi.org/10.1021/acscentsci.8b00700] [PMID: 30937369]

[71] L. Moroni, T. Boland, J.A. Burdick, C. De Maria, B. Derby, G. Forgacs, J. Groll, Q. Li, J. Malda, V.A. Mironov, C. Mota, M. Nakamura, W. Shu, S. Takeuchi, T.B.F. Woodfield, T. Xu, J.J. Yoo, and G. Vozzi, "Biofabrication: A Guide to Technology and Terminology", *Trends Biotechnol.,* vol. 36, no. 4, pp. 384-402, 2018.
[http://dx.doi.org/10.1016/j.tibtech.2017.10.015] [PMID: 29137814]

[72] S. Mandal, S. Meininger, U. Gbureck, and B. Basu, "3D powder printed tetracalcium phosphate scaffold with phytic acid binder: fabrication, microstructure and *in situ* X-Ray tomography analysis of compressive failure", *J. Mater. Sci. Mater. Med.,* vol. 29, no. 3, p. 29, 2018.
[http://dx.doi.org/10.1007/s10856-018-6034-8] [PMID: 29520670]

[73] E. Nyberg, A. Rindone, A. Dorafshar, and W.L. Grayson, "Comparison of 3D-Printed Poly-ε-Caprolactone Scaffolds Functionalized with Tricalcium Phosphate, Hydroxyapatite, Bio-Oss, or Decellularized Bone Matrix", *Tissue Eng. Part A,* vol. 23, no. 11-12, pp. 503-514, 2017.
[http://dx.doi.org/10.1089/ten.tea.2016.0418] [PMID: 28027692]

[74] S.C. Cox, P. Jamshidi, N.M. Eisenstein, M.A. Webber, H. Hassanin, M.M. Attallah, D.E.T. Shepherd, O. Addison, and L.M. Grover, "Adding functionality with additive manufacturing: Fabrication of titanium-based antibiotic eluting implants", *Mater. Sci. Eng. C,* vol. 64, pp. 407-415, 2016.
[http://dx.doi.org/10.1016/j.msec.2016.04.006] [PMID: 27127071]

[75] H.E. Burton, "The design of additively manufactured lattices to increase the functionality of medical implants", *Mater. Sci. Eng. C.,* vol. 94, no. September, pp. 901-908, 2019.
[http://dx.doi.org/10.1016/j.msec.2018.10.052]

[76] P. Chocholata, V. Kulda, and V. Babuska, "Fabrication of scaffolds for bone-tissue regeneration", *Materials (Basel),* vol. 12, no. 4, p. 568, 2019.
[http://dx.doi.org/10.3390/ma12040568] [PMID: 30769821]

[77] L.Y. Zhu, L. Li, J.P. Shi, Z.A. Li, and J.Q. Yang, "Mechanical characterization of 3D printed multi-morphology porous Ti_6Al_4V scaffolds based on triply periodic minimal surface architectures", *Am. J. Transl. Res.,* vol. 10, no. 11, pp. 3443-3454, 2018.
[PMID: 30662598]

[78]   Y. Ma, N. Hu, J. Liu, X. Zhai, M. Wu, C. Hu, L. Li, Y. Lai, H. Pan, W.W. Lu, X. Zhang, Y. Luo, and C. Ruan, "Three-Dimensional Printing of Biodegradable Piperazine-Based Polyurethane-Urea Scaffolds with Enhanced Osteogenesis for Bone Regeneration", *ACS Appl. Mater. Interfaces,* vol. 11, no. 9, pp. 9415-9424, 2019.
[http://dx.doi.org/10.1021/acsami.8b20323] [PMID: 30698946]

[79]   H. Zhao, Y. Chen, L. Shao, M. Xie, J. Nie, J. Qiu, P. Zhao, H. Ramezani, J. Fu, H. Ouyang, and Y. He, "Airflow-Assisted 3D Bioprinting of Human Heterogeneous Microspheroidal Organoids with Microfluidic Nozzle", *Small,* vol. 14, no. 39, p. 1802630, 2018.
[http://dx.doi.org/10.1002/smll.201802630] [PMID: 30133151]

[80]   A. Riedl, M. Schlederer, K. Pudelko, M. Stadler, S. Walter, D. Unterleuthner, C. Unger, N. Kramer, M. Hengstschläger, L. Kenner, D. Pfeiffer, G. Krupitza, and H. Dolznig, "Comparison of cancer cells cultured in 2D vs 3D reveals differences in AKT/mTOR/S6-kinase signaling and drug response", *J. Cell Sci.,* vol. 130, no. 1, p. jcs.188102, 2016.
[http://dx.doi.org/10.1242/jcs.188102] [PMID: 27663511]

[81]   L.A. Mathews Griner, X. Zhang, R. Guha, C. McKnight, I.S. Goldlust, M. Lal-Nag, K. Wilson, S. Michael, S. Titus, P. Shinn, C.J. Thomas, and M. Ferrer, "Large-scale pharmacological profiling of 3D tumor models of cancer cells", *Cell Death Dis.,* vol. 7, no. 12, p. e2492, 2016.
[http://dx.doi.org/10.1038/cddis.2016.360] [PMID: 27906188]

[82]   M.A. Theodoraki, C.O. Rezende Jr, O. Chantarasriwong, A.D. Corben, E.A. Theodorakis, and M.L. Alpaugh, "Spontaneously-forming spheroids as an *in vitro* cancer cell model for anticancer drug screening"., *Oncotarget,* vol. 6, no. 25, pp. 21255-21267, 2015.
[http://dx.doi.org/10.18632/oncotarget.4013] [PMID: 26101913]

[83]   S. Nath, and G.R. Devi, "Three-dimensional culture systems in cancer research: Focus on tumor spheroid model", *Pharmacol. Ther.,* vol. 163, pp. 94-108, 2016.
[http://dx.doi.org/10.1016/j.pharmthera.2016.03.013] [PMID: 27063403]

[84]   Y. Kim, K. Kang, S. Yoon, J.S. Kim, S.A. Park, W.D. Kim, S.B. Lee, K.Y. Ryu, J. Jeong, and D. Choi, "Prolongation of liver-specific function for primary hepatocytes maintenance in 3D printed architectures", *Organogenesis,* vol. 14, no. 1, pp. 1-12, 2018.
[http://dx.doi.org/10.1080/15476278.2018.1423931] [PMID: 29359998]

[85]   H. Cui, M. Nowicki, J.P. Fisher, and L.G. Zhang, "3D Bioprinting for Organ Regeneration", *Adv. Healthc. Mater.,* vol. 6, no. 1, p. 1601118, 2017.
[http://dx.doi.org/10.1002/adhm.201601118] [PMID: 27995751]

[86]   G. Schwank, B.K. Koo, V. Sasselli, J.F. Dekkers, I. Heo, T. Demircan, N. Sasaki, S. Boymans, E. Cuppen, C.K. van der Ent, E.E.S. Nieuwenhuis, J.M. Beekman, and H. Clevers, "Functional repair of CFTR by CRISPR/Cas9 in intestinal stem cell organoids of cystic fibrosis patients", *Cell Stem Cell,* vol. 13, no. 6, pp. 653-658, 2013.
[http://dx.doi.org/10.1016/j.stem.2013.11.002] [PMID: 24315439]

[87]   M. Lei, L.J. Schumacher, Y.C. Lai, W.T. Juan, C.Y. Yeh, P. Wu, T.X. Jiang, R.E. Baker, R.B. Widelitz, L. Yang, and C.M. Chuong, "Self-organization process in newborn skin organoid formation inspires strategy to restore hair regeneration of adult cells", *Proc. Natl. Acad. Sci. USA,* vol. 114, no. 34, pp. E7101-E7110, 2017.
[http://dx.doi.org/10.1073/pnas.1700475114] [PMID: 28798065]

[88]   M.A. Lancaster, and J.A. Knoblich, "Organogenesisin a dish: Modeling development and disease using organoid technologies", *Science,* vol. 345, no. 6194, 2014.
[http://dx.doi.org/10.1126/science.1247125]

[89]   X. Yin, B.E. Mead, H. Safaee, R. Langer, J.M. Karp, and O. Levy, "Engineering Stem Cell Organoids", *Cell Stem Cell,* vol. 18, no. 1, pp. 25-38, 2016.
[http://dx.doi.org/10.1016/j.stem.2015.12.005] [PMID: 26748754]

[90]   L.J. Pourchet, A. Thepot, M. Albouy, E.J. Courtial, A. Boher, L.J. Blum, and C.A. Marquette, "Human

Skin 3D Bioprinting Using Scaffold-Free Approach", *Adv. Healthc. Mater.,* vol. 6, no. 4, p. 1601101, 2017.
[http://dx.doi.org/10.1002/adhm.201601101] [PMID: 27976537]

[91]   T. Grix, A. Ruppelt, A. Thomas, A.K. Amler, B. Noichl, R. Lauster, and L. Kloke, "Bioprinting perfusion-enabled liver equivalents for advanced organ-on-a-chip applications", *Genes (Basel),* vol. 9, no. 4, p. 176, 2018.
[http://dx.doi.org/10.3390/genes9040176] [PMID: 29565814]

[92]   E.A. Bulanova, E.V. Koudan, J. Degosserie, C. Heymans, F.D.A.S. Pereira, V.A. Parfenov, Y. Sun, Q. Wang, S.A. Akhmedova, I.K. Sviridova, N.S. Sergeeva, G.A. Frank, Y.D. Khesuani, C.E. Pierreux, and V.A. Mironov, "Bioprinting of a functional vascularized mouse thyroid gland construct", *Biofabrication,* vol. 9, no. 3, p. 034105, 2017.
[http://dx.doi.org/10.1088/1758-5090/aa7fdd] [PMID: 28707625]

[93]   A. Skardal, M. Devarasetty, H.W. Kang, Y.J. Seol, S.D. Forsythe, C. Bishop, T. Shupe, S. Soker, and A. Atala, "Bioprinting cellularized constructs using a tissue-specific hydrogel bioink", *J. Vis. Exp.,* vol. 2016, no. 110, p. e53606, 2016.
[http://dx.doi.org/10.3791/53606] [PMID: 27166839]

[94]   J. A. Reid, P. A. Mollica, R. D. Bruno, and P.C. Sachs, "Erratum: Consistent and reproducible cultures of large-scale 3D mammary epithelial structures using an accessible bioprinting platform", *Breast Cancer Res.,* vol. 20, no. 1, pp. 1-13, 2018.
[http://dx.doi.org/10.1186/s13058-018-1045-4] [http://dx.doi.org/10.1186/s13058-018-1069-9]

[95]   Y.S. Zhang, A. Arneri, S. Bersini, S.R. Shin, K. Zhu, Z. Goli-Malekabadi, J. Aleman, C. Colosi, F. Busignani, V. Dell'Erba, C. Bishop, T. Shupe, D. Demarchi, M. Moretti, M. Rasponi, M.R. Dokmeci, A. Atala, and A. Khademhosseini, "Bioprinting 3D microfibrous scaffolds for engineering endothelialized myocardium and heart-on-a-chip", *Biomaterials,* vol. 110, pp. 45-59, 2016.
[http://dx.doi.org/10.1016/j.biomaterials.2016.09.003] [PMID: 27710832]

[96]   J.H. Kim, Y.J. Seol, I.K. Ko, H.W. Kang, Y.K. Lee, J.J. Yoo, A. Atala, and S.J. Lee, "3D Bioprinted Human Skeletal Muscle Constructs for Muscle Function Restoration", *Sci. Rep.,* vol. 8, no. 1, p. 12307, 2018.
[http://dx.doi.org/10.1038/s41598-018-29968-5] [PMID: 30120282]

[97]   M. Ali, A.K. Pr, J.J. Yoo, F. Zahran, A. Atala, and S.J. Lee, "A Photo-Crosslinkable Kidney ECM-Derived Bioink Accelerates Renal Tissue Formation", *Adv. Healthc. Mater.,* vol. 8, no. 7, p. 1800992, 2019.
[http://dx.doi.org/10.1002/adhm.201800992] [PMID: 30725520]

[98]   W. Lee, J. Pinckney, V. Lee, J.H. Lee, K. Fischer, S. Polio, J.K. Park, and S.S. Yoo, "Three-dimensional bioprinting of rat embryonic neural cells", *Neuroreport,* vol. 20, no. 8, pp. 798-803, 2009.
[http://dx.doi.org/10.1097/WNR.0b013e32832b8be4] [PMID: 19369905]

[99]   X. Dai, L. Liu, J. Ouyang, X. Li, X. Zhang, Q. Lan, and T. Xu, "Coaxial 3D bioprinting of self-assembled multicellular heterogeneous tumor fibers", *Sci. Rep.,* vol. 7, no. 1, p. 1457, 2017.
[http://dx.doi.org/10.1038/s41598-017-01581-y] [PMID: 28469183]

[100]   S. Knowlton, and S. Tasoglu, "A Bioprinted Liver-on-a-Chip for Drug Screening Applications", *Trends Biotechnol.,* vol. 34, no. 9, pp. 681-682, 2016.
[http://dx.doi.org/10.1016/j.tibtech.2016.05.014] [PMID: 27291461]

[101]   H. Kizawa, E. Nagao, M. Shimamura, G. Zhang, and H. Torii, "Scaffold-free 3D bio-printed human liver tissue stably maintains metabolic functions useful for drug discovery", *Biochem. Biophys. Rep.,* vol. 10, pp. 186-191, 2017.
[http://dx.doi.org/10.1016/j.bbrep.2017.04.004] [PMID: 28955746]

[102]   Y. Yang, T. Du, J. Zhang, T. Kang, L. Luo, J. Tao, Z. Gou, S. Chen, Y. Du, J. He, S. Jiang, Q. Mao, and M. Gou, "A 3D-Engineered Conformal Implant Releases DNA Nanocomplexs for Eradicating the Postsurgery Residual Glioblastoma", *Adv. Sci. (Weinh.),* vol. 4, no. 8, p. 1600491, 2017.

[http://dx.doi.org/10.1002/advs.201600491] [PMID: 28852611]

[103] E.O. Osidak, P.A. Karalkin, M.S. Osidak, V.A. Parfenov, D.E. Sivogrivov, F.D.A.S. Pereira, A.A. Gryadunova, E.V. Koudan, Y.D. Khesuani, V.A. Kasyanov, S.I. Belousov, S.V. Krasheninnikov, T.E. Grigoriev, S.N. Chvalun, E.A. Bulanova, V.A. Mironov, and S.P. Domogatsky, "Viscoll collagen solution as a novel bioink for direct 3D bioprinting", *J. Mater. Sci. Mater. Med.,* vol. 30, no. 3, p. 31, 2019.
[http://dx.doi.org/10.1007/s10856-019-6233-y] [PMID: 30830351]

[104] Y. He, F. Yang, H. Zhao, Q. Gao, B. Xia, and J. Fu, "Research on the printability of hydrogels in 3D bioprinting", *Sci. Rep.,* vol. 6, no. 1, p. 29977, 2016.
[http://dx.doi.org/10.1038/srep29977] [PMID: 27436509]

[105] V. U, D. Mehrotra, V. Dichen, V. Anand, and D. Howlader, "Three dimensional reconstruction of late post traumatic orbital wall defects by customized implants using CAD-CAM, 3D stereolithographic models: A case report", *J. Oral Biol. Craniofac. Res.,* vol. 7, no. 3, pp. 212-218, 2017.
[http://dx.doi.org/10.1016/j.jobcr.2017.09.004] [PMID: 29124002]

[106] J.H. Shim, J.Y. Won, J.H. Park, J.H. Bae, G. Ahn, C.H. Kim, D.H. Lim, D.W. Cho, W.S. Yun, E.B. Bae, C.M. Jeong, and J.B. Huh, "Effects of 3D-printed polycaprolactone/β-tricalcium phosphate membranes on guided bone regeneration", *Int. J. Mol. Sci.,* vol. 18, no. 5, p. 899, 2017.
[http://dx.doi.org/10.3390/ijms18050899] [PMID: 28441338]

[107] W. Zhang, C. Feng, G. Yang, G. Li, X. Ding, S. Wang, Y. Dou, Z. Zhang, J. Chang, C. Wu, and X. Jiang, "3D-printed scaffolds with synergistic effect of hollow-pipe structure and bioactive ions for vascularized bone regeneration", *Biomaterials,* vol. 135, pp. 85-95, 2017.
[http://dx.doi.org/10.1016/j.biomaterials.2017.05.005] [PMID: 28499127]

[108] A. Nauth, E. Schemitsch, B. Norris, Z. Nollin, and J.T. Watson, "Critical-Size Bone Defects: Is There a Consensus for Diagnosis and Treatment?", *J. Orthop. Trauma,* vol. 32, no. 3, suppl. Suppl. 1, pp. S7-S11, 2018.
[http://dx.doi.org/10.1097/BOT.0000000000001115] [PMID: 29461395]

[109] A. Kaempfen, A. Todorov, S. Güven, R. Largo, C. Jaquiéry, A. Scherberich, I. Martin, and D. Schaefer, "Engraftment of prevascularized, tissue engineered constructs in a novel rabbit segmental bone defect model", *Int. J. Mol. Sci.,* vol. 16, no. 12, pp. 12616-12630, 2015.
[http://dx.doi.org/10.3390/ijms160612616] [PMID: 26053395]

[110] H. Shao, X. Ke, A. Liu, M. Sun, Y. He, X. Yang, J. Fu, Y. Liu, L. Zhang, G. Yang, S. Xu, and Z. Gou, "Bone regeneration in 3D printing bioactive ceramic scaffolds with improved tissue/material interface pore architecture in thin-wall bone defect", *Biofabrication,* vol. 9, no. 2, p. 025003, 2017.
[http://dx.doi.org/10.1088/1758-5090/aa663c] [PMID: 28287077]

[111] Y. Deng, C. Jiang, C. Li, T. Li, M. Peng, J. Wang, and K. Dai, "3D printed scaffolds of calcium silicate-doped β-TCP synergize with co-cultured endothelial and stromal cells to promote vascularization and bone formation", *Sci. Rep.,* vol. 7, no. 1, p. 5588, 2017.
[http://dx.doi.org/10.1038/s41598-017-05196-1] [PMID: 28717129]

[112] Y. Wen, S. Xun, M. Haoye, S. Baichuan, C. Peng, L. Xuejian, Z. Kaihong, Y. Xuan, P. Jiang, and L. Shibi, "3D printed porous ceramic scaffolds for bone tissue engineering: a review", *Biomater. Sci.,* vol. 5, no. 9, pp. 1690-1698, 2017.
[http://dx.doi.org/10.1039/C7BM00315C] [PMID: 28686244]

[113] C.H. Lee, J.L. Cook, A. Mendelson, E.K. Moioli, H. Yao, and J.J. Mao, "Regeneration of the articular surface of the rabbit synovial joint by cell homing: a proof of concept study", *Lancet,* vol. 376, no. 9739, pp. 440-448, 2010.
[http://dx.doi.org/10.1016/S0140-6736(10)60668-X] [PMID: 20692530]

[114] J.W. Chang, S.A. Park, J.K. Park, J.W. Choi, Y.S. Kim, Y.S. Shin, and C.H. Kim, "Tissue-engineered tracheal reconstruction using three-dimensionally printed artificial tracheal graft: preliminary report", *Artif. Organs,* vol. 38, no. 6, pp. E95-E105, 2014.

[http://dx.doi.org/10.1111/aor.12310] [PMID: 24750044]

[115]   C.H. Lee, J. Hajibandeh, T. Suzuki, A. Fan, P. Shang, and J.J. Mao, "Three-dimensional printed multiphase scaffolds for regeneration of periodontium complex", *Tissue Eng. Part A,* vol. 20, no. 7-8, pp. 1342-1351, 2014.
[http://dx.doi.org/10.1089/ten.tea.2013.0386] [PMID: 24295512]

[116]   Q.H. Zhang, A. Cossey, and J. Tong, "Stress shielding in periprosthetic bone following a total knee replacement: Effects of implant material, design and alignment", *Med. Eng. Phys.,* vol. 38, no. 12, pp. 1481-1488, 2016.
[http://dx.doi.org/10.1016/j.medengphy.2016.09.018] [PMID: 27745873]

[117]   R.F. MacBarb, D.P. Lindsey, C.S. Bahney, S.A. Woods, M.L. Wolfe, and S.A. Yerby, "Fortifying the bone-implant interface part 1: An *in vitro* evaluation of 3D-printed and TPS porous surfaces"., *Int. J. Spine Surg.,* vol. 11, no. 3, p. 15, 2017.
[http://dx.doi.org/10.14444/4015] [PMID: 28765799]

[118]   T.A. Petrie, J.E. Raynor, C.D. Reyes, K.L. Burns, D.M. Collard, and A.J. García, "The effect of integrin-specific bioactive coatings on tissue healing and implant osseointegration", *Biomaterials,* vol. 29, no. 19, pp. 2849-2857, 2008.
[http://dx.doi.org/10.1016/j.biomaterials.2008.03.036] [PMID: 18406458]

[119]   P.A. Tran, L. Sarin, R.H. Hurt, and T.J. Webster, "Titanium surfaces with adherent selenium nanoclusters as a novel anticancer orthopedic material", *J. Biomed. Mater. Res. A,* vol. 93, no. 4, pp. 1417-1428, 2010.
[http://dx.doi.org/10.1002/jbm.a.32631] [PMID: 19918919]

[120]   S.M. Kazzazi, and E.F. Kranioti, "Applicability of 3D-dental reconstruction in cervical odontometrics", *Am. J. Phys. Anthropol.,* vol. 165, no. 2, pp. 370-377, 2018.
[http://dx.doi.org/10.1002/ajpa.23353] [PMID: 29115677]

[121]   Y. Wang, L. Wu, H. Guo, T. Qiu, Y. Huang, B. Lin, and L. Wang, "Computation of tooth axes of existent and missing teeth from 3D CT images", *Biomedical Engineering / Biomedizinische Technik,* vol. 60, no. 6, pp. 623-632, 2015.
[http://dx.doi.org/10.1515/bmt-2014-0111] [PMID: 25941910]

[122]   G. Oberoi, S. Nitsch, M. Edelmayer, K. Janjić, A.S. Müller, and H. Agis, "3D printing-Encompassing the facets of dentistry", *Front. Bioeng. Biotechnol.,* vol. 6, no. NOV, p. 172, 2018.
[http://dx.doi.org/10.3389/fbioe.2018.00172] [PMID: 30525032]

[123]   A. Tahayeri, M. Morgan, A.P. Fugolin, D. Bompolaki, A. Athirasala, C.S. Pfeifer, J.L. Ferracane, and L.E. Bertassoni, "3D printed *versus* conventionally cured provisional crown and bridge dental materials"., *Dent. Mater.,* vol. 34, no. 2, pp. 192-200, 2018.
[http://dx.doi.org/10.1016/j.dental.2017.10.003] [PMID: 29110921]

[124]   N. Alharbi, D. Wismeijer, and R. Osman, "Additive Manufacturing Techniques in Prosthodontics: Where Do We Currently Stand? A Critical Review", *Int. J. Prosthodont.,* vol. 30, no. 5, pp. 474-484, 2017.
[http://dx.doi.org/10.11607/ijp.5079] [PMID: 28750105]

[125]   N. Gan, Y. Ruan, J. Sun, Y. Xiong, and T. Jiao, "Comparison of Adaptation between the Major Connectors Fabricated from Intraoral Digital Impressions and Extraoral Digital Impressions", *Sci. Rep.,* vol. 8, no. 1, p. 529, 2018.
[http://dx.doi.org/10.1038/s41598-017-17839-4] [PMID: 29323129]

[126]   M. Martorelli, S. Gerbino, M. Giudice, and P. Ausiello, "A comparison between customized clear and removable orthodontic appliances manufactured using RP and CNC techniques", *Dent. Mater.,* vol. 29, no. 2, pp. e1-e10, 2013.
[http://dx.doi.org/10.1016/j.dental.2012.10.011] [PMID: 23140842]

[127]   K.R. Hixon, A.M. Melvin, A.Y. Lin, A.F. Hall, and S.A. Sell, "Cryogel scaffolds from patient-specific 3D-printed molds for personalized tissue-engineered bone regeneration in pediatric cleft-craniofacial

defects", *J. Biomater. Appl.,* vol. 32, no. 5, pp. 598-611, 2017.
[http://dx.doi.org/10.1177/0885328217734824] [PMID: 28980856]

[128] J.M. Haglin, A.E.M. Eltorai, J.A. Gil, S.E. Marcaccio, J. Botero-Hincapie, and A.H. Daniels, "Patient-Specific Orthopaedic Implants", *Orthop. Surg.,* vol. 8, no. 4, pp. 417-424, 2016.
[http://dx.doi.org/10.1111/os.12282] [PMID: 28032697]

[129] B. Chang, W. Song, T. Han, J. Yan, F. Li, L. Zhao, H. Kou, and Y. Zhang, "Influence of pore size of porous titanium fabricated by vacuum diffusion bonding of titanium meshes on cell penetration and bone ingrowth", *Acta Biomater.,* vol. 33, pp. 311-321, 2016.
[http://dx.doi.org/10.1016/j.actbio.2016.01.022] [PMID: 26802441]

[130] A. Barba, A. Diez-Escudero, Y. Maazouz, K. Rappe, M. Espanol, E.B. Montufar, M. Bonany, J.M. Sadowska, J. Guillem-Marti, C. Öhman-Mägi, C. Persson, M.C. Manzanares, J. Franch, and M.P. Ginebra, "Osteoinduction by Foamed and 3D-Printed Calcium Phosphate Scaffolds: Effect of Nanostructure and Pore Architecture", *ACS Appl. Mater. Interfaces,* vol. 9, no. 48, pp. 41722-41736, 2017.
[http://dx.doi.org/10.1021/acsami.7b14175] [PMID: 29116737]

[131] R. Trombetta, J.A. Inzana, E.M. Schwarz, S.L. Kates, and H.A. Awad, "3D Printing of Calcium Phosphate Ceramics for Bone Tissue Engineering and Drug Delivery", *Ann. Biomed. Eng.,* vol. 45, no. 1, pp. 23-44, 2017.
[http://dx.doi.org/10.1007/s10439-016-1678-3] [PMID: 27324800]

[132] A. Cipitria, J.C. Reichert, D.R. Epari, S. Saifzadeh, A. Berner, H. Schell, M. Mehta, M.A. Schuetz, G.N. Duda, and D.W. Hutmacher, "Polycaprolactone scaffold and reduced rhBMP-7 dose for the regeneration of critical-sized defects in sheep tibiae", *Biomaterials,* vol. 34, no. 38, pp. 9960-9968, 2013.
[http://dx.doi.org/10.1016/j.biomaterials.2013.09.011] [PMID: 24075478]

[133] A. P Fitzpatrick, "Design of a Patient Specific, 3D printed Arm Cast", *KnE Engineering,* vol. 2, no. 2, p. 135, 2017.
[http://dx.doi.org/10.18502/keg.v2i2.607]

[134] C.I. Gioumouxouzis, C. Karavasili, and D.G. Fatouros, "Recent advances in pharmaceutical dosage forms and devices using additive manufacturing technologies", *Drug Discov. Today,* vol. 24, no. 2, pp. 636-643, 2019.
[http://dx.doi.org/10.1016/j.drudis.2018.11.019] [PMID: 30503803]

[135] P. Anbalagan, P.W.S. Heng, and C.V. Liew, "Tablet compression tooling – Impact of punch face edge modification", *Int. J. Pharm.,* vol. 524, no. 1-2, pp. 373-381, 2017.
[http://dx.doi.org/10.1016/j.ijpharm.2017.04.005] [PMID: 28389365]

[136] M. Sadia, B. Arafat, W. Ahmed, R.T. Forbes, and M.A. Alhnan, "Channelled tablets: An innovative approach to accelerating drug release from 3D printed tablets", *J. Control. Release,* vol. 269, pp. 355-363, 2018.
[http://dx.doi.org/10.1016/j.jconrel.2017.11.022] [PMID: 29146240]

[137] N.G. Solanki, M. Tahsin, A.V. Shah, and A.T.M. Serajuddin, "Formulation of 3D Printed Tablet for Rapid Drug Release by Fused Deposition Modeling: Screening Polymers for Drug Release, Drug-Polymer Miscibility and Printability", *J. Pharm. Sci.,* vol. 107, no. 1, pp. 390-401, 2018.
[http://dx.doi.org/10.1016/j.xphs.2017.10.021] [PMID: 29066279]

[138] P.R. Martinez, A. Goyanes, A.W. Basit, and S. Gaisford, "Influence of Geometry on the Drug Release Profiles of Stereolithographic (SLA) 3D-Printed Tablets", *AAPS PharmSciTech,* vol. 19, no. 8, pp. 3355-3361, 2018.
[http://dx.doi.org/10.1208/s12249-018-1075-3] [PMID: 29948979]

[139] E.A. Clark, M.R. Alexander, D.J. Irvine, C.J. Roberts, M.J. Wallace, S. Sharpe, J. Yoo, R.J.M. Hague, C.J. Tuck, and R.D. Wildman, "3D printing of tablets using inkjet with UV photoinitiation", *Int. J. Pharm.,* vol. 529, no. 1-2, pp. 523-530, 2017.

[http://dx.doi.org/10.1016/j.ijpharm.2017.06.085] [PMID: 28673860]

[140] S.A. Khaled, J.C. Burley, M.R. Alexander, J. Yang, and C.J. Roberts, "3D printing of tablets containing multiple drugs with defined release profiles", *Int. J. Pharm.,* vol. 494, no. 2, pp. 643-650, 2015.
[http://dx.doi.org/10.1016/j.ijpharm.2015.07.067] [PMID: 26235921]

[141] T. Terada, S. Noda, and K. Inui, "Management of dose variability and side effects for individualized cancer pharmacotherapy with tyrosine kinase inhibitors", *Pharmacol. Ther.,* vol. 152, pp. 125-134, 2015.
[http://dx.doi.org/10.1016/j.pharmthera.2015.05.009] [PMID: 25976912]

[142] V. Solhaug, and E. Molden, "Individual variability in clinical effect and tolerability of opioid analgesics – Importance of drug interactions and pharmacogenetics", *Scand. J. Pain,* vol. 17, no. 1, pp. 193-200, 2017.
[http://dx.doi.org/10.1016/j.sjpain.2017.09.009] [PMID: 29054049]

[143] K. Osouli-Bostanabad, and K. Adibkia, "Made-on-demand, complex and personalized 3D-printed drug products", *Bioimpacts,* vol. 8, no. 2, pp. 77-79, 2018.
[http://dx.doi.org/10.15171/bi.2018.09] [PMID: 29977828]

[144] M.N. Pastore, Y.N. Kalia, M. Horstmann, and M.S. Roberts, "Transdermal patches: history, development and pharmacology", *Br. J. Pharmacol.,* vol. 172, no. 9, pp. 2179-2209, 2015.
[http://dx.doi.org/10.1111/bph.13059] [PMID: 25560046]

[145] S.N. Economidou, D.A. Lamprou, and D. Douroumis, "3D printing applications for transdermal drug delivery", *Int. J. Pharm.,* vol. 544, no. 2, pp. 415-424, 2018.
[http://dx.doi.org/10.1016/j.ijpharm.2018.01.031] [PMID: 29355656]

[146] C.P.P. Pere, S.N. Economidou, G. Lall, C. Ziraud, J.S. Boateng, B.D. Alexander, D.A. Lamprou, and D. Douroumis, "3D printed microneedles for insulin skin delivery", *Int. J. Pharm.,* vol. 544, no. 2, pp. 425-432, 2018.
[http://dx.doi.org/10.1016/j.ijpharm.2018.03.031] [PMID: 29555437]

[147] R. Haj-Ahmad, H. Khan, M. Arshad, M. Rasekh, A. Hussain, S. Walsh, X. Li, M.W. Chang, and Z. Ahmad, "Microneedle coating techniques for transdermal drug delivery", *Pharmaceutics,* vol. 7, no. 4, pp. 486-502, 2015.
[http://dx.doi.org/10.3390/pharmaceutics7040486] [PMID: 26556364]

[148] E. Dayyoub, M. Frant, S.R. Pinnapireddy, K. Liefeith, and U. Bakowsky, "Antibacterial and anti-encrustation biodegradable polymer coating for urinary catheter", *Int. J. Pharm.,* vol. 531, no. 1, pp. 205-214, 2017.
[http://dx.doi.org/10.1016/j.ijpharm.2017.08.072] [PMID: 28830785]

[149] J.A. Weisman, D.H. Ballard, U. Jammalamadaka, K. Tappa, J. Sumerel, H.B. D'Agostino, D.K. Mills, and P.K. Woodard, "3D Printed Antibiotic and Chemotherapeutic Eluting Catheters for Potential Use in Interventional Radiology", *Acad. Radiol.,* vol. 26, no. 2, pp. 270-274, 2019.
[http://dx.doi.org/10.1016/j.acra.2018.03.022] [PMID: 29801697]

[150] N. Sandler, I. Salmela, A. Fallarero, A. Rosling, M. Khajeheian, R. Kolakovic, N. Genina, J. Nyman, and P. Vuorela, "Towards fabrication of 3D printed medical devices to prevent biofilm formation", *Int. J. Pharm.,* vol. 459, no. 1-2, pp. 62-64, 2014.
[http://dx.doi.org/10.1016/j.ijpharm.2013.11.001] [PMID: 24239831]

[151] M. George, K.R. Aroom, H.G. Hawes, B.S. Gill, and J. Love, "3D Printed Surgical Instruments: The Design and Fabrication Process", *World J. Surg.,* vol. 41, no. 1, pp. 314-319, 2017.
[http://dx.doi.org/10.1007/s00268-016-3814-5] [PMID: 27822724]

[152] K. Liu, Q. Zhang, X. Li, C. Zhao, X. Quan, R. Zhao, Z. Chen, and Y. Li, "Preliminary application of a multi-level 3D printing drill guide template for pedicle screw placement in severe and rigid scoliosis", *Eur. Spine J.,* vol. 26, no. 6, pp. 1684-1689, 2017.
[http://dx.doi.org/10.1007/s00586-016-4926-1] [PMID: 28028644]

[153] F. Guo, J. Dai, J. Zhang, Y. Ma, G. Zhu, J. Shen, and G. Niu, "Individualized 3D printing navigation template for pedicle screw fixation in upper cervical spine", *PLoS One,* vol. 12, no. 2, p. e0171509, 2017.
[http://dx.doi.org/10.1371/journal.pone.0171509] [PMID: 28152039]

[154] T.M. Rankin, N.A. Giovinco, D.J. Cucher, G. Watts, B. Hurwitz, and D.G. Armstrong, "Three-dimensional printing surgical instruments: are we there yet?", *J. Surg. Res.,* vol. 189, no. 2, pp. 193-197, 2014.
[http://dx.doi.org/10.1016/j.jss.2014.02.020] [PMID: 24721602]

[155] J.Y. Wong, and A.C. Pfahnl, "3D printed surgical instruments evaluated by a simulated crew of a Mars mission", *Aerosp. Med. Hum. Perform.,* vol. 87, no. 9, pp. 806-810, 2016.
[http://dx.doi.org/10.3357/AMHP.4281.2016] [PMID: 27634701]

[156] J.Y. Wong, and A.C. Pfahnl, "3D printing of surgical instruments for long-duration space missions", *Aviat. Space Environ. Med.,* vol. 85, no. 7, pp. 758-763, 2014.
[http://dx.doi.org/10.3357/ASEM.3898.2014] [PMID: 25022166]

CHAPTER 7

Fabrication of Photosensitive Polymers-based Biomaterials through Multiphoton Lithography

Mohammad Sherjeel Javed Khan[1], **Sehrish Manan**[2], **Ronan R. McCarthy**[3] and **Muhammad Wajid Ullah**[2,*]

[1] Department of Chemistry, King Abdulaziz University, Jeddah 21589, Saudi Arabia

[2] Biofuels Institute, School of the Environment and Safety Engineering, Jiangsu University, Zhenjiang 212013, PR China

[3] Division of Biosciences, Department of Life Sciences, College of Health and Life Sciences, Brunel University London, Uxbridge UB8 3PH, UK

Abstract: The use of polymers in the development of biomaterials for various biomedical applications has become increasingly important in recent decades. To match the innate properties of biological tissues, the polymer-based tissue scaffolds must have the desired structural and functional properties. However, the polymer-based hydrogels prepared by conventional methods are often delicate and fragile and require pre-stabilisation. This necessitates the exploration of bio-friendly cross-linkers that promote kinetic or reversible crosslinking in the polymer network of hydrogels and must be nontoxic to cells and tissues. The light initiators with well-organized multiphoton cross sections that are reactive at specific wavelengths could be ideal candidates. This chapter reviews the fabrication of solid or viscoelastic biological scaffolds by multiphoton lithography (MPL) of liquids. It describes the similarities and differences between conventional and MPL photo polymerization of biological scaffolds in terms of synthesis chemistry, properties, and their relevance to biological applications. These photosensitive scaffolds could be useful biomaterials for their biomedical applications.

Keywords: Biomaterials, Biomedical Applications, Cross-Linkers, Hydrogels, Multiphoton Lithography, Photosensitive Polymers.

INTRODUCTION

The emergence of the polymer industry in the early 1950s led to the synthesis of several new products for everyday use [1 - 3]. Currently, the use of various polymers is attracting much attention in the biomedical field, where they are used in the development of drug delivery systems [4 - 6], tissue engineering scaffolds

---

* **Corresponding author Muhammad Wajid Ullah:** Biofuels Institute, School of the Environment and Safety Engineering, Jiangsu University, Zhenjiang 212013, PR China; Email: wajid_kundi@ujs.edu.cn

Adnan Haider & Sajjad Haider (Eds.)

[4, 7 - 9], synthetic organs [10], medical implants [11 - 14], and medical equipment such as biosensors [15 - 17]. The use of various polymers for their biomedical applications requires the development of specialized materials for specific applications by controlling their synthesis process. To this end, advances in photophysics and synthetic chemistry are leading to the synthesis of polymers in a controlled environment, *e.g.*, the initiation and propagation of the polymerization reaction in the presence of light [18, 19].

The use of light as a catalyst during the polymerization reaction allows unique control of the reaction as well as the freedom to perform the experiment at different times and places. Photo polymerization is a reaction carried out in the presence of light that, under suitable conditions, converts the low molecular weight prepolymer solution or monomers into high molecular weight materials. The conventional photo polymerization reaction for material synthesis is usually carried out by the light-induced radical polymerization [20], which requires a suitable light source and at least one precursor solution consisting of a multifunctional monomer and a photo initiator. The light is used to irradiate the precursor solution and produce the photopolymerizable material. The photomask dictates the shape, while the light dose and intensity control the degree and rate of the polymerization reaction [21]. *In vivo* or *in situ* photo polymerization can also be performed by introducing the precursor solution into the body and then initiating the photo polymerization reaction [22]. In this way, a biomaterial corresponding to the desired tissue shape can be rapidly produced. On the other hand, interfacial photo polymerization can be performed by adsorbing or attaching a light initiator to the surface of a polymerizable material that can produce brushes. These photo polymerization approaches are useful for achieving consistent coatings, casting compounds, and *in vivo* implantation of grafts. However, they are limited to planar patterns only and cannot take advantage of the full 3D and spatial resolution offered by light initiation [18].

Over the last couple of decades, photo polymerization has played a crucial role in the establishment, growth, and expansion of several modern industries, such as integrated circuits, coatings and adhesives, and optical devices, due to its unique properties [23, 24]. Even the ancient Egyptians explored photo polymerization by using sunlight to crosslink oily linen to form an environmental barrier during the mummification process [25]. Nowadays, photo polymerization uses monomers and terminal functional polymers to develop functionalized and biocompatible scaffolds and hydrogels [26].

In the field of biomaterials, the photo polymerization process has been used to overcome the limitations of functional design, such as achieving defined shapes, *e.g.*, in bone implants and skin tissues [13, 27 - 29] and sol-gel transitions after application, *e.g.*, in hydrogels developed *in situ* [30 - 32]. The photo polymerized biomaterials are effectively used as cell [33] and drug delivery systems [34], membrane barriers [35, 36], tissue-engineered scaffolds [37, 38], and as coating materials for medicines [26]. These biomedical applications of biomaterials require the development of biocompatible networks or hydrogels, which are related to the crosslinked polymers but differ in their physical state. The former are crosslinked polymers in an undissolved state, while the latter contains a lot of water and are in a swollen state. The high degree of swelling of hydrogels mimics the mechanical properties of biological tissue *in vivo* and facilitates the exchange of nutrients, waste products, and signaling molecules, making them ideal candidates for various biomedical applications [39]. In both cases, the three-dimensional (3D) and sequential control during polymer synthesis enabled by photo polymerization can produce highly structured materials with predetermined shapes and *in situ* polymerization capabilities [40].

With the increasing demand and applications of biomaterials, the old-fashioned monolithic photo polymerization technique cannot meet the desired standards of material production in various disciplines and for various applications. For example, the extracellular matrix (ECM) is a natural environment that supports and controls cellular functions. However, its time-varying structural design at the nanoscale and microscale is very complex [41, 42], and thus cannot be fabricated using conventional techniques. Similarly, many applications require high functional resolution of polymers through 3D objects. Among various material synthesis techniques, photolithography and stereo lithography are widely used for the fabrication of functional biomaterials at micro and nano scales. At the same time, multiphoton lithography (MPL) technology has been applied to photo polymerization to make these necessary tools widely available in the biomedical field [43]. The development of integrated circuits using photolithographic techniques can significantly improve the spatial resolution in the microelectronics industry [44]. The irradiated areas are photo polymerized into non-resolvable blocks, while the non-polymerized areas are eroded after the fabrication process is complete. Then, users create planar structures in the micrometer range and obtain 3D structures by building them layer by layer [45]. The lithographic technique requires high-resolution photo coverage for each shape. It is limited by diffraction and can only produce 3D structures. The photo polymerization technique can also be used for soft lithography [46, 47]. At this time, the main mold is made from the elastomer material, such as polydimethylsiloxane, with a predefined shape. The mold is filled with a precursor solution that photo polymerizes to restore the desired properties. This method has proven successful in the fabrication of pharmaceutical microbial materials [48], tissue engineering scaffolds [49], and microfluidic biosensing [50]. Recent advances in multiphoton technology have led to its application in the fields of materials science, biology, and biomaterials. Therefore, scientists began to investigate the use of multiphoton irradiation to promote nanoscale and microscale developments in photo polymerization and biological environments [51]. Since the focal range of the multiphoton laser is narrow and free from the external environment, the ultrafast multiphoton laser is used for MPL laser scanning lithography (LSL) experiments. This can enhance functional content in 3D down to the nanometer scale, where materials can effectively photo polymerize [52].

This chapter describes the phenomenon of photo polymerization, in which liquids are used as starting materials and converted into solid and viscoelastic materials. It also explains the fundamentals of multiphoton lithography (MPL). The chemical methods and processes used for the photo polymerization of biomaterials and hydrogels by the MPL technique are discussed, and an overview of the similarities between the MPL and photo polymerization techniques is provided. Special emphasis is placed on the use of photopolymers for the development of polymer networks and hydrogels for biological applications.

MULTIPHOTON LITHOGRAPHIC PHOTOPOLYMERIZATION

Maria Goeppert-Mayer was the first to propose the theory of multiphoton excitation. She received the Nobel Prize in Physics in 1963 for her contribution to the description of the nuclear shell model of atomic nuclei [53]. She proposed the principle of multiphoton excitation, which allows the excitation of photoactive molecules by the absorption of many photons (multiphotons) of longer wavelength instead of one photon of shorter wavelength [53]. This well-established and accepted phenomenon of using two photons from a common laser source with energy equal to half of a single excited photon is widely used nowadays [53]. Both photons must be absorbed within femtoseconds as they put the molecule into an excited quantum state. This is usually achieved by a high-energy pulsed femtosecond laser source. Placing a molecule in the excited quantum state increases the possibility of simultaneous absorption of multiple photons in the focal volume [54]. The nonlinear nature of the two-photon process reduces the possibility of excitation as a quadratic function depending on the distance from the focal plane and compresses the axial diffusion of the point spread function [54]. This suggests that the excitation of the photons should be limited to the focal plane only and have a high resolution (<1μm) in the axial or z dimension. If the optical conditions are suitable, this will result in a sub-femto focal length. A schematic representation of a typical MPL septum can be found in Fig. (**1**).

Watt and colleagues first introduced multiphoton technology to biology and biomaterials. This is due to the advancement and spread of multiphoton microscopy [56, 57], which is widely used for imaging complex tissues and living animal models [54]. Multiphoton excitation is also used in fluorescence imaging: for example, for the molecular unwinding of small entities [58], recovery of fluorescence after photobleaching [59], and biophysical studies [60]. Most importantly, multiphoton excitation can be directly integrated with conventional laser scanning microscopy (LSM) or LSL equipment to image or assemble substances at the micro- and nanoscale.

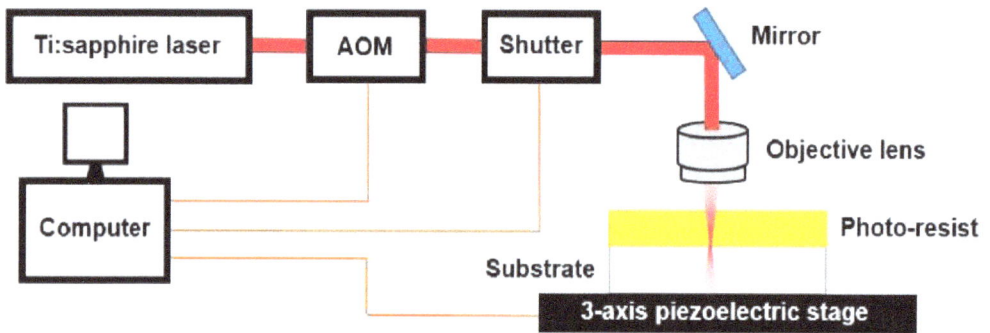

Fig. (1). Schematic illustration of a typical multiphoton lithographic polymerization setup. Fig. reproduced from [55] distributed under the Creative Commons Attribution (CC BY 4.0) license. AOM: Acousto-optic modulator.

With the increasing popularity of multiphoton systems, they are being used in materials science and polymer research for high-resolution 3D photo polymerization [52]. The complex 3D photopolymer structures were successfully generated by MPL photo polymerization. The same principle applies to photo polymerization, in which the simultaneous absorption of two low-energy photons by the same photo initiator leads to the cleavage of the molecule and initiates the formation of free radicals in the focal volume of the sub molecule [52]. Therefore, rapid prototyping with MPL can produce objects with characteristic sizes in the submicron range. Previous studies have developed methods to create open pore structures as biomaterials or photonic crystals [61, 62]. This technology could be applied to all interpretations that can be managed by conventional computer-aided design software (CAD) to fabricate many complex objects, such as cattle, frogs, multicultural carriers, and some others [63 - 65].

MPL has the advantage over LSL in that it reduces the axial resolution of the point spread function, which improves z-resolution, unlike single-photon excitation [54]. In addition, these wavelengths are minimally absorbed and poorly distributed in most media used, such as living tissue, so infrared (IR) or near IR light can be used to increase penetration depth [54]. These advantages enable a true 3D array with a size ratio equivalent to the focal length of multiphoton lasers (< 1 fl or 1 μm3), allowing easy access to the intracellular length scales of biological materials. Although MPL improves the voxel resolution of photopolymerizable materials, especially biocompatible materials and hydrogels, this technology still has some practical limitations. For example, the overall thickness of MPL is limited by the production volume. For millimeter-sized objects with high functionality, the thickness of MPL is even smaller. Therefore, the available photo initiator kits suitable for multiphoton excitation need to be improved. Advances in computational quantum chemistry have led to the development of new photoinitiators that can generate free radicals upon multiphoton excitation and scission [66]. Considerable success has already been achieved in the development of water-soluble initiators with two-photon effective cross sections, allowing them to be used in the biomedical field [64].

MULTIPHOTON POLYMERIZATION TOOLS FOR THE SYNTHESIS OF BIOMATERIALS

Researchers working on the preparation of biomaterials by MPL using the different polymer systems can use commercial LSM or self-built devices. In practice, the main limitations are multiphoton laser sources (such as the femtosecond pulse Ti:sapphire laser), suitable optical paths, sample laser focal volume, and a high-accuracy xyz platform. Commercial multiphoton microscopes meet all these requirements and are primarily designed for imaging biological samples. Such a system is suitable for writing by rasterizing the focal volume with a precursor solution under a specific region of interest (ROI) that describes the shape to be aggregated. The improvements in CAD -supported ROI software have driven the use of commercial microscopes that can easily convert any structure into the 3D photopolymerizable materials. In addition, the MPL instruments require a mechanized platform with independent submicron xyz-dimensional resolution for highly reliable and high-resolution patterning [67].

CHEMISTRY OF MPL PHOTOPOLYMERIZATION

The most important requirement for photo polymerization is the precursor solution, which should contain a photo initiator, an efficient monomer or macromer, and a light source for excitation. The following sections describe the chemical methods and processes used for the photo polymerization of biomaterials, including biocompatible materials and hydrogel systems, using the MPL photo polymerization technique.

Photo Polymerization

photo polymerization is an effective approach for the fabrication of biomaterials by using the MPL. It quickly transforms a liquid solution into solids or viscous biomaterials in the illuminated areas [68]. This technique allows an easy and effective removal of undesirable materials and restores the required microstructure of biological materials.

Illumination is the first step in free radical-mediated photo polymerization. The absorption of a photon by a photo initiator (PI), like in the case of MPL polymerization, the multi-photons simultaneously enter the excited state. The excited photoinitiators, denoted as PI*, are either directly dissociated into one or more first-order radicals (PI type-I) or lead to a bimolecular reaction where PI* reacts with a second molecule and produces a free radical (PI type-II) [69]. The initiation rate Ri of a type-I or PI induced by the two-photon light initiation from a classic pulsed laser is given by the below equation (**1**) [70, 71].

$$Ri = 2 * 1.17 \delta\mu\phi\mu \frac{T}{\tau\rho} \left(\frac{\lambda}{\pi hc\omega 2xy} \right) 2P2avg VF[PI] \qquad (1)$$

δuΦu: cross section of two-photons (express as a function of λ)
T: period of laser pulses
τp: duration of laser pulses
λ: wavelength of laser light
H: Planck's constant
c: light speed
wxy: lateral focal radius of laser
P_{avg}: average laser power
VF: volume factor (0.63 for axial cylinder)
[PI]: local concentration of PI

The factor '2' assumes that each light initiator molecule split off generates two free radicals. The corresponding rate equations can be obtained for other multiphoton excitation processes. Most single-photon photo polymerization is hindered by light decay, which leads to out-of-plane initiation and spatial gradients in light intensity. Due to the limited scattering and absorption of IR light in most biological media, one can assume a uniform light intensity or average power within the spatial range of polymerization. Quantitative evaluation of the initiation rate is important because it affects the polymerization rate, which in turn affects the final properties of the photo polymerized materials. Equation (**1**) shows that the initial polymerization rate can be modulated by adjusting the concentration of PI, the light wavelength, or the laser power, suggesting that a PI with two-photon cross section could be used to produce biomaterials by photo polymerization. Unlike single-photon initiation, the initiation rate of two-photons varies quadratically as a function of laser power. The free radicals generated in the precursor solution immediately disperse to the functional groups of the monomer or macromonomer before recombining in the solvent. A polymerization reaction proceeds *via* a chain or a gradual growth mechanism [69]. The free radicals generated in the precursor solution immediately disperse to the functional groups of the monomer or macromonomer before recombining in the solvent. A polymerization reaction proceeds *via* a chain or a gradual growth mechanism [69]. In traditional gradual growth (or condensation) polymerization reactions initiated by free radicals, the free radicals activate a functional unit (*e.g.*, a thiol group) of the nearby monomer, followed by the addition of another functional unit (*e.g.*, a vinyl group). The monomers and free radicals are transferred to other functional groups in the solution, resulting in the formation of dimers. In the same way, trimers and tetramers are formed when dimers react with the monomer and another dimer, respectively, which then aggregate stepwise one after another. In contrast, when radicals activate the functional groups of the surrounding monomers, such as (meth)acrylates, a conventional radical-triggered chain growth (or additional) polymerization reaction occurs, which is repeated by the addition of monomers, forming a long chain of monomers at each initiation. In both cases, polymerization results in a high molecular weight material that forms the intertwined polymeric solids. Alternatively, if a multifunctional macromonomer is present in the solution, a crosslinked polymer complex is formed. The formation of the free radicals is stopped during this process, resulting in a decrease in the concentration of the free radicals and initiating or continuing the reaction, resulting in a steady rate of the polymerization reaction. Termination of the polymerization reaction occurs by several mechanisms: (1) by the formation of two active chain ends or a bimolecular bond between the active chain end and the initiating group, (2) by disproportionation, in which hydrogen atoms are withdrawn from one active chain by the other atom, (3) by the involvement of inhibitors, such as those that exert oxygen interactions, and (4) by the transfer of chains to non-breeding species [69]. The kinetic rate (Rp) of chain growth polymerization explains both propagation and termination by equation (**2**) below [69]:

$$Rp = kp[M] \left(\frac{Ri}{2kt} \right) 1/2 \qquad\qquad (2)$$

Kp: propagation rate constant

$[M]$: concentration of macromonomer functional group

kt: termination rate constant

In the hydrogels and crosslinked network structures, both kp and kt change with the conversion or polymerization time of the macromonomer due to the changes in the diffusion constants of the reactive species caused by the precursor reaction [72]. During the early transformation, the macromonomer increases the viscosity of the precursor and limits the diffusion of the free radicals, thus limiting the kt value. The decrease in kt mediated by viscosity is directly proportional to the polymerization rate. As the conversion rate of the macromonomer increases and the crosslinked network is formed, the functional groups of the macromonomer are also limited by diffusion, leading to a decrease in kp and an overall decrease in Rp. This phenomenon is referred to as automatic slowdown. Polymerization stops when all available functional groups are consumed.

Photoinitiators

The photoinitiators (PIs) are the organic compounds that convert the absorbed light into an active chemical substance and initiate polymerization [73]. As for the chemical composition and origin of PI, the active substance is essentially a cationic, anionic or free radical, so light is used to control all kinds of chain growth and gradual growth polymerization. Since both the anionic and cationic polymerization reactions are sensitive to moisture, the polymerization reactions in the biological systems are usually carried out by free radical methods. Type I initiators operate mainly autonomously and usually consist of a mixture of aromatic ketones or disilanes. Type II PIs do not self-cleave and are used with co-initiated species (not photosensitive) by hydrogen abstraction to initiate the polymerization reaction. Compared to molecular excitation, high energy is required for bound photolysis. Therefore, type I PIs are mainly limited to ultraviolet light sources with a wavelength less than 400 nm, while type II PIs could be extended to the visible light region with a wavelength less than 600 nm.

Light initiators activated by multiple photoexcitation are commonly used in the fabrication of biomaterials. However, there are some limitations to the fabrication of biomaterials by multiphoton excitation. For example, all materials used in biomaterials fabrication and their associated degradation products or leaching agents must have minimal biocompatibility. This means that all cells, tissues, and organisms, as well as their lysates, must be resistant to photoinitiators. Moreover, when preparing hydrogels for cell encapsulation, the polymerization reaction should be carried out under strict physiological conditions: human body temperature (37°C), pH (7.4), oxygen content (5%), and buffer salts, especially in the presence of viable cells [74]. Moreover, the biological environment requires an additional aqueous environment, which severely limits the choice of available PI. Since most PIs absorb light because they contain conjugated aromatic groups, their large hydrophobicity reduces their water solubility. Generally, light with a wavelength of 365 nm or more is considered suitable for living systems [75]. The energy provided by this wavelength is sufficient to initiate the polymerization reaction. However, it is not sufficient to break the cellular DNA and denature the protein (which is known to occur rapidly at λ=254 nm) [76, 77].

Although photo polymerization occurs in the presence of viable cells under stringent conditions, some useful PIs (Fig. **2**) are capable of polymerization even in the presence of viable cells. For example, Irgacure® 2959 [2-hydroxy-1-(4-(2-ethoxy-hydroxyethyl)phenyl)-2-methylpropan-1-one] is the most commonly used compound that forms two acetone groups and replaces the benzoyl group when exposed to UV light of wavelength 365 nm (Fig. **2i**) [75]. Although these reactive free radicals rapidly initiate chain polymerization in an aqueous medium, the water solubility of the parent compound PI is low (~1 wt%, ~50 mM). In a recent study, 2,4,6-trimethyl-benzoyl-phosphinic acid phenyllithium (LAP) was reported to be water soluble (~8.5 wt%, ~300mM) type-I PI at $300 < \lambda < 400$nm (Fig. **2ii**) [78, 79]. Under UV light, LAP is decomposed into 2,4,6-trimethylbenzoyl and phosphono, both of which effectively triggers the polymerization reactions in the cellular environment. Although LAP is not yet on the market, its two-step preparation method is simple and nearly quantitative [79].

In the presence of viable cells, some II type PIs are used for photo polymerization, such as bengalrose (Fig. **2iii**) and camphorquinone [80]. Upon exposure to light, these compounds are excited and jump to a higher state, *i.e.*, triplet state, which then reacts with a hydrogen donor co-initiator, such as an amine- or thiol-containing molecule, to generate free radicals. Despite their versatility and exclusive potential to work efficiently in the visible region of UV light, the photodynamics of the II type PIs is even less well known than those of the type I PIs [81].

Photopolymer Chemistry

The precursor solution for photo polymerization requires a suitable macromer for the synthesis of biomaterials. The macromer is selected by considering two parameters:(1) determination of backbone chemistry and (2) selection of compound chemistry for network development. In recent decades, the development of biomaterials and hydrogels using single photon light polymerization has become very popular among materials chemists and biologists. If the selected PI and light source provide sufficient free radicals to accelerate polymerization, essentially, the same kit can be used for MPL production.

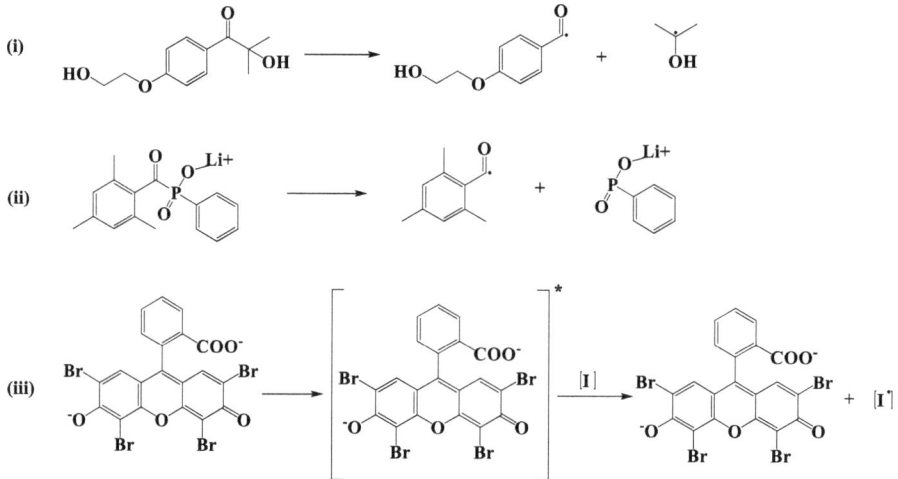

Fig. (2). A selection of type-I and type-II photoinitiators appropriate for MPL polymerization, and can be used in the presence of viable cells. **(i)** Irgacure 2959, **(ii)** Lithium phenyl-2,4,6-trimethylbenzoylphosphinate (LAP), and **(iii)** Eosin Y.

The photopolymerizable biological non-hydrogel materials use solutions such as polymer resins composed of efficient monomers, oligomers or macromonomers that rapidly convert the polymer into a solid material in the presence of a suitable PI. At the same time, efficient oligomers or macromonomers are used for the preparation of biological entities by using non-hydrophilic backbone polymers. This hydrophobic backbone chemistry involves the use of terminally functionalized esters [82] with polymerizable substances such as vinyl and complementary thiol and olefin functional groups [83], carbamates [84], and carbonates [85]. These materials are used for the preparation of microstructured biological materials using MPL [86].

The extensive research conducted in recent decades to develop hydrogels as biocompatible materials has provided a wealth of data on the framework chemistry for their applications. This information is also applicable and useful for the development of MPL hydrogels [87]. Hydrophilic polymers are usually divided into two main categories: synthetic and non-synthetic or natural hydrophilic polymers [88]. Examples of synthetic polymers used in the development of hydrogels are polyethylene glycol (PEG) [89], polyvinyl alcohol

(PVA) [90, 91], polyacrylic acid (PAA) [92], poly caprolactone [93], poly 2-hydroxyethyl methacrylate (PHEMA) [94], and others [95]. Among them, PEG is widely used because of its nonstick properties and its ability to mimic elasticity and natural tissue transport. Other examples of natural polymers used to prepare hydrogels include alginate [96], collagen [97], gelatin [98], hyaluronic acid [99], fibrin [11], chitosan [100, 101], bacterial cellulose [102 - 105], and cellulose derivatives [95]. The addition of various polymerizable groups such as (meth)acrylates, (meth)acrylamides, and thiolenes impart additional functionality to natural and synthetic polymers, while others facilitate the photo polymerization of backbone polymers. The formation of a photo polymerized hydrogel involves the absorption of water by the insoluble polymer network and subsequent swelling [106].

PHOTOCHEMICAL DECOMPOSITION AND POLYMERIZATION

Hundreds of light-mediated reactions are involved in the synthesis and degradation of polymers. This section addresses various subsets of these chemical methods carried out in the biological environment. The additional photoreactions are categorized by PI, photocaged reactive groups, or non-specific use of radicals. Plans for subtractive reduction using the photolabile linkers and strategies for reversible photo functionalization are also described. All of the reactions discussed are shown in Figs. (**3 - 7**).

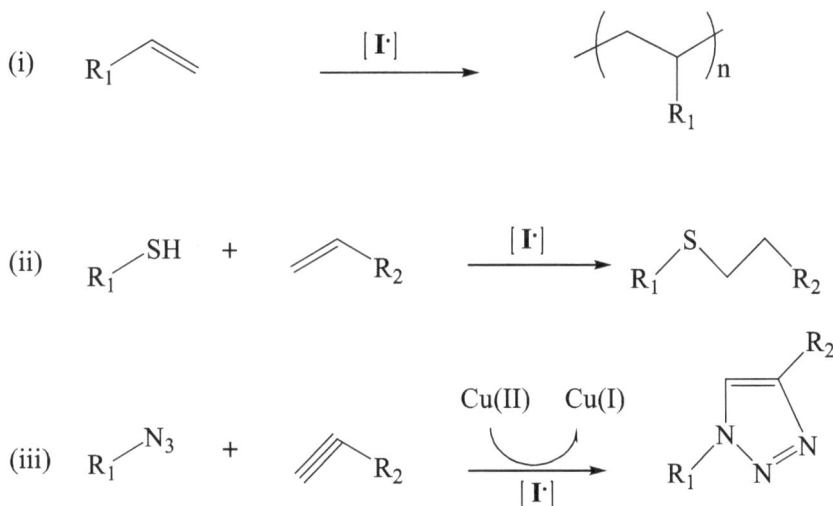

Fig. (3). Photoinitiator-mediated polymerization reactions for the synthesis and modification of biomaterials, such as **(i)** vinyl chain polymerization, **(ii)** thiol–ene reaction, and **(iii)** the photoinducible copper-catalyzed azide–alkyne cycloaddition.

Initiator-Facilitated Photo Polymerization

To date, most photo polymerization reactions used for various applications are supported by initiators. Since PI properties can be carefully and unconventionally chosen for the reaction species, the use of PI for the reaction can improve stability within the range of initial wavelength, solvent composition, and reaction kinetics. Some examples of collective PI -favoured polymerization mechanisms used in the preparation of biomaterials include vinyl chain polymerization, thiol-ene and light-induced copper-catalysed azide-alkyne cycloaddition (PCuAAC) reactions (Fig. **3**).

Photocage-Facilitated Photoconjugation

A photocage is used as a molecular gatekeeper to control chemical reactions. When these compounds are present together, they prevent a polymerization reaction between the two substances by forming a physical cover. During photolysis, the highly reactive functional groups are released and lead to nucleation reactions and photoconjugation. To date, only a few photocage-assisted photoconjugation reactions, such as oxime bond, stress-promoting azide-alkyne cycloadditone (SPAAC), and Michael addition, have been used in the development of biomaterials (Fig. **4**).

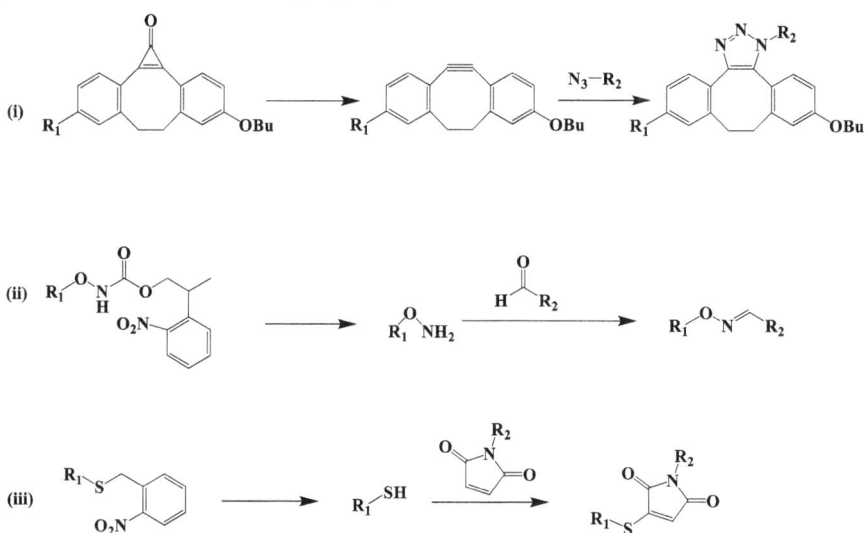

Fig. (4). The photocaged regulation of **(i)** strain-promoted azide–alkyne cycloaddition, **(ii)** oxime ligation, and the **(iii)** Michael-type addition are known to be the most promising conjugation reactions for the synthesis and modification of biopolymer systems.

General Photoconjugation

Activation of various photo reactive groups can be achieved by irradiation with UV light. The random nature of the binding of these compounds to native molecules enables rapid binding with covalent bonds between biopolymers in close proximity in 3D space. Examples of compounds with photo reactive groups are arylazides, diazirines, and benzophenones, which are the well-characterized functions of non-specific photoconjugation (Fig. **5**).

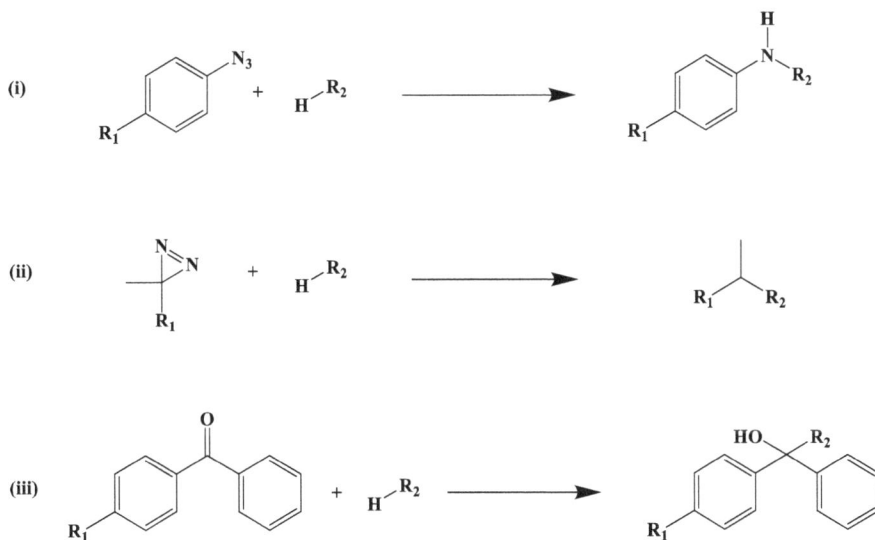

Fig. (5). A range of non-specific chemical conjugation reactions, such as those involving (**i**) aryl azides, (**ii**) diazirines, and (**iii**) benzophenones, are used for light-mediated polymerization for the synthesis of biomaterials.

Light-Induced Decomposition

Although most photochemical reactions result in the formation of covalent bonds, it is increasingly known that photosensitive bonds are cleaved in response to photon irradiation. Once incorporated into the polymer seed main chain, these components provide a tunable sense of network depolymerization. Some examples of compounds containing o-nitrobenzyl esters, coumarin, and disulfide bonds are shown in Fig. (**6**). These compounds are photolytic in nature and can be controlled in biological environments.

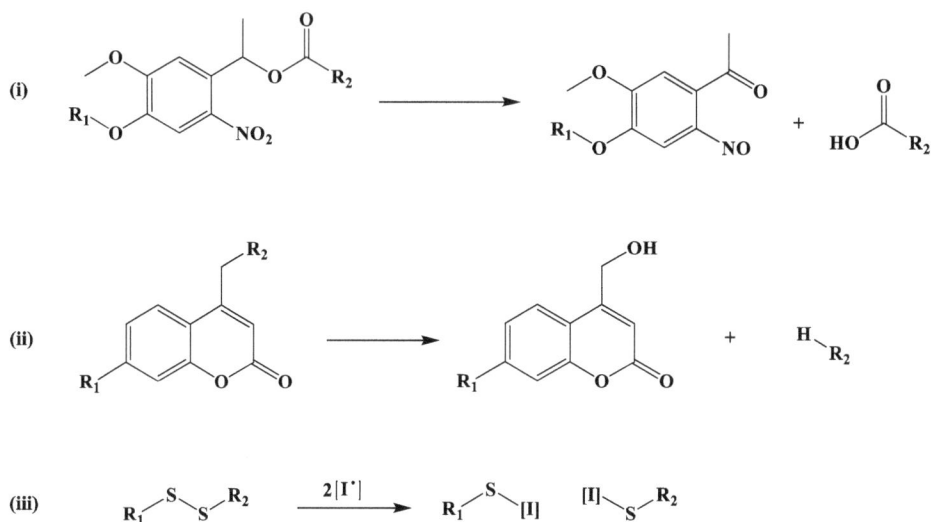

Fig. (6). Different photo labile linkages such as **(i)** o-nitrobenzyl esters, **(ii)** coumarins, and **(iii)** disulfide-containing molecules are photo-chemically cleaved under different physiological conditions of temperature, pH, and ionic strength through MPL, and thus could be useful for the desired depolymerization of a biopolymer.

Revocable Photoconjugation

Most of the chemical methods used focus on either the formation or the breaking of bonds by light. There is considerable interest in developing photochemical methods that are fully reversible. The two most promising biological systems developed to date include reversible dimerization of anthracene and addition-fragmentation chain transfer of allyl sulfide, as shown in Fig. (7).

APPLICATIONS

Biomaterials Synthesis

A previous study addressed the preparation of non-biological materials by MPL polymerization [107]. Various polymerizable resins and negative photoresists are commonly used to prepare MPL materials, using viscous liquids or amorphous solids as starting materials. Such methods can effortlessly fabricate complex structures such as interconnected microchannels [108], cascade chains [109], or Venus de Milo deduction [110]. The field of photonics used to be an entrant operator of the MPL model due to the materials used. MPL photo polymerization is used to synthesize various materials such as functional waveguides [111], interferometers [112], microlenses [113], and microlasers [114]. Due to their ability to produce sub micrometer features, MPL photo polymerization-based methods are used in practice to produce visible-scale photonic crystals.

The viable cells exist in a complex, living, and well-organized ECM. Signals from the components of the ECM control the expression of certain growth factors and thus the metabolic fate of the cells. These signals control various cellular functions such as adhesion, proliferation, growth, migration, and differentiation to specific tissues [115]. However, the exact mechanism of how ECM components control cellular functions has not been fully elucidated. Biomaterial strategy not only provides an attractive platform for analyzing the complex and dynamic cellular ECM communication, but also designs a scaffold to exploit these clues in regenerative medicine applications [41, 116]. To this end, photo polymerization using MPL can accurately reproduce the ECM properties, such as the biophysical and biochemical signals of the ECM components themselves, of 3D biomaterials at the micro- and nanoscale. The use of MPL in the fabrication of photopolymerizable biomaterials requires the investigation of fundamental functions and design constraints of the materials used in cell culture applications. The materials used in the development of MPL-based biomaterials should better support the settlement, translocation, development, and modification of cells and provide the biophysical and biochemical signals appropriate for specific target cell types [117]. Broadly speaking, two types of photopolymerizable MPL materials have been used as biomaterials: the biocompatible bonds in which the cells are seeded after the biomaterials are prepared, and the hydrogels that can directly coat the viable cells.

MPL polymerization produces 3D structures with a high organization that is biocompatible, degradable, and easy to fabricate. These properties allow photo polymerized biomaterials to find important biological applications. MPL polymerization has been used to fabricate cell-compatible microneedles, biodegradable scaffolds, and bone grafts [118 - 120]. In addition, MPL polymerization is used to prepare hydrated polymer networks for cell cultivation in the form of hydrogels for their *in vivo* applications [117]. The structure of hydrogels is similar to natural ECM and thus can serve as an idea material for applications in tissue engineering and regenerative medicine [116]. Some examples of MPL-polymerized biomaterials are pile grids, ultrafine porous structures, and vascular trees [67, 121 - 123]. Other studies have reported the development of natural proteins for biomedical applications [124 - 127].

Modulation of Biological Materials

In addition to synthesizing novel polymers and biomaterials, photochemical reactions could be used to alter the current forum both temporally and spatially according to user-specified parameters. By applying the principles of chemistry, the composition of physical and chemical materials can be altered at a scale smaller than a single mammalian cell. In this way, comprehensive temporal and spatial control of the combination of biomaterials could be developed to achieve the dynamic heterogeneity found in natural tissues [20].

The temporal and spatial definition of the biochemical adaptation of the available materials could be achieved by different strategies based on additive and subtractive light [18]. The most widely used additive chemical method is radical chain polymerization, for example, using acrylic functionalized peptides or proteins. Binding motifs such as the growth hormone arginine glycine aspartate (RGD) and vascular endothelial growth factor (VEGF) are used [128]. Similarly, the interaction between thiols and olefins leads to changes in the function of cells cultured on a polymeric hydrogel scaffold developed by photo patterning thiol-containing biomolecules into the en-functionalized networks [129, 130]. Similarly, co-patterning of certain proteins can be achieved by covalently immobilizing the physically bound partner (*e.g.*, biotin or barnase) of the protein by the photochemical thiol-maleimide reaction [131]. The use of factor XIII for photo activated enzyme crosslinking has also been used in the preparation of photo-patterned peptides for their application in controlling the expansion of mesenchymal stem cells in 3D materials [132].

Like temporal and spatial control, the use of photochemistry also leads to biochemical changes in the material by selectively removing cues in the material. For example, cleavage of o-nitrobenzyl ester (oNB) was used to release RGD and thereby increase cartilage formation of human mesenchymal stem cells [133]. Moreover, oNB lysis has been used to form functional surfaces for studying cooperative and decorated cell movements [134]. Similarly, photolysis of dimers of coumarin byproducts has been used to transport a variety of drugs [135 - 137]. Although the techniques described above have been used to introduce or remove various biochemical signals, a combination of different light-induced reactions can reversibly fix these cues. For example, the fusion of the visible light-promoted thiol-ene addition reaction and UV-mediated oNB cleavage leads to the reversible biochemical production of polymer-based materials [130]. In this method, the binding of cells to the material surface is dynamically monitored by loading the hydrogel in the presence of cells and then removing the binder ligand from the hydrogel.

LIMITATIONS OF MPL TO BIOLOGICAL DESIGN

MPL photo polymerization-assisted synthesis of materials requires an energy balance to start the polymerization reaction. For example, in the synthesis of

PEG-assisted hydrogels, it has been reported that energy higher than 2.3 nJ leads to cavitation of highly water-swollen gels [138]. Moreover, for the synthesis of biomaterials by MPL photo polymerization, a balance between the photo physics of the reaction and the potential for photo damage is also important. For example, sufficient pulse energy has been shown to cleave intracellular actin filaments [60, 138, 139]. Similarly, in the presence of viable cells, a pulsed femtosecond two-photon laser is used, where pulse energy greater than 1.5 nJ causes subcellular optical ablation [139], while an energy greater than 4 nJ can cause cell death [140]. By using pulse energy of less than 1.5 nJ, indeterminate damage to the subcellular scaffold can be avoided and the sustainability of viable cells can be maintained. It should be noted that these recommendations are inflexible in a nonradical system and depend strongly on the nature and concentration of the various free radicals generated during the polymerization process. The complexity of biological systems further limits the functional space of MPL photo polymerization in living cells. In contrast, the advantages of a smaller focal length and low out-of-plane absorption minimize the potential damage to a thin window. There are several other limitations to the use of MPL for biological applications. For example, biomaterials must be efficient and combined with the biological system. They must also contain components that are compatible with the biological system, including all constituents and potential leaching agents. Although most of the chemical reagents discussed in this chapter have already been used for biological applications, the compatibility of new chemical reagents with biological systems should be further evaluated.

CONCLUSION

Combining conventional photo polymerization technology with advanced multiphoton lithography (MPL) can lead to the development of useful photo polymerized biomaterials with defined geometric shapes, structures, and capabilities at the microscale. Due to the increasing application of widely used chemical methods and multiphoton technologies used in the synthesis of biomaterials and hydrogels, biomaterials developed with MPL may have potential applications in tissue engineering and regenerative medicine. The MPL biomaterials could be used to produce customized and functionalized tissue substitutes that could be further adapted for organ development. In addition, the customizable hydrogels with dynamic intracellular biophysical and biochemical patterns could provide new insights into stem cell and therapeutic biology.

CONSENT FOR PUBLICATION

Not applicable.

CONFLICT OF INTEREST

The author declares no conflict of interest, financial or otherwise.

ACKNOWLEDGEMENT

Declared none.

REFERENCES

[1]   P. Irving, and C. Soutis, *Polymer Composites in the Aerospace Industry.* Elsevier, 2020.

[2]   C. Silvestre, D. Duraccio, and S. Cimmino, "Food packaging based on polymer nanomaterials", *Progress in Polymer Science.,* Progress in Polymer Science: Oxford, 2011.
[http://dx.doi.org/10.1016/j.progpolymsci.2011.02.003]

[3]   S. Thomas, J. Kuruvilla, S. K. Malhotra, K. Goda, and M. S. Sreekala, *Polymer Composites.* Wiley-VCH Verlag GmbH & Co. KGaA: Weinheim, Germany, 2012.

[4]   X. Chen, X. Xu, W. Li, B. Sun, J. Yan, C. Chen, J. Liu, J. Qian, and D. Sun, "Effective Drug Carrier Based on Polyethylenimine-Functionalized Bacterial Cellulose with Controllable Release Properties", *ACS Appl. Bio Mater.,* vol. 1, no. 1, pp. 42-50, 2018.
[http://dx.doi.org/10.1021/acsabm.8b00004]

[5]   R. Langer, "Drug delivery and targeting", *Nature,* vol. 392, no. 6679, suppl. Suppl., pp. 5-10, 1998.
[http://dx.doi.org/10.1016/S0378-5173(02)00260-0] [PMID: 9579855]

[6]   S. Li, "Fabrication of pH-electroactive Bacterial Cellulose/Polyaniline Hydrogel for the Development of a Controlled Drug Release System", *ES Mater. Manuf,* pp. 41-49, 2018.
[http://dx.doi.org/10.30919/esmm5f120]

[7]   R. Langer, and J. P. Vacanti, "Tissue engineering", *Science, 80,* 1993.
[http://dx.doi.org/10.1126/science.8493529]

[8]   M. Ul-Islam, S. Khan, M.W. Ullah, and J.K. Park, "Bacterial cellulose composites: Synthetic strategies and multiple applications in bio-medical and electro-conductive fields", *Biotechnol. J.,* vol. 10, no. 12, pp. 1847-1861, 2015.
[http://dx.doi.org/10.1002/biot.201500106] [PMID: 26395011]

[9]   R. Portela, C.R. Leal, P.L. Almeida, and R.G. Sobral, "Bacterial cellulose: a versatile biopolymer for wound dressing applications", *Microb. Biotechnol.,* vol. 12, no. 4, pp. 586-610, 2019.
[http://dx.doi.org/10.1111/1751-7915.13392] [PMID: 30838788]

[10]   S. Dubey, R. Mishra, P. Roy, and R.P. Singh, "3-D macro/microporous-nanofibrous bacterial cellulose scaffolds seeded with BMP-2 preconditioned mesenchymal stem cells exhibit remarkable potential for bone tissue engineering", *Int. J. Biol. Macromol.,* no. Nov, 2020.
[http://dx.doi.org/10.1016/j.ijbiomac.2020.11.049] [PMID: 33189758]

[11]   X. Jiang, S. Wu, M. Kuss, Y. Kong, W. Shi, P.N. Streubel, T. Li, and B. Duan, "3D printing of multilayered scaffolds for rotator cuff tendon regeneration", *Bioact. Mater.,* vol. 5, no. 3, pp. 636-643, 2020.
[http://dx.doi.org/10.1016/j.bioactmat.2020.04.017] [PMID: 32405578]

[12]   L. Bacakova, J. Pajorova, M. Bacakova, A. Skogberg, P. Kallio, K. Kolarova, and V. Svorcik, "Versatile application of nanocellulose: From industry to skin tissue engineering and wound healing", *Nanomaterials (Basel),* vol. 9, no. 2, p. 164, 2019.
[http://dx.doi.org/10.3390/nano9020164] [PMID: 30699947]

[13]   L. Lamboni, C. Xu, J. Clasohm, J. Yang, M. Saumer, K.H. Schäfer, and G. Yang, "Silk sericin-enhanced microstructured bacterial cellulose as tissue engineering scaffold towards prospective gut

repair", *Mater. Sci. Eng. C,* vol. 102, pp. 502-510, 2019.
[http://dx.doi.org/10.1016/j.msec.2019.04.043] [PMID: 31147021]

[14]   F.C.A. Silveira, F.C.M. Pinto, S. da Silva Caldas Neto, M. de Carvalho Leal, J. Cesário, and J.L. de Andrade Aguiar, "Treatment of tympanic membrane perforation using bacterial cellulose: a randomized controlled trial", *Rev. Bras. Otorrinolaringol. (Engl. Ed.),* 2015.
[http://dx.doi.org/10.1016/j.bjorl.2015.03.015] [PMID: 26631330]

[15]   R. Langer, and N.A. Peppas, "Advances in Biomaterials", *Drug Delivery, and Bionanotechnology,* 2003.
[http://dx.doi.org/10.1002/aic.690491202]

[16]   U. Farooq, M.W. Ullah, Q. Yang, A. Aziz, J. Xu, L. Zhou, and S. Wang, "High-density phage particles immobilization in surface-modified bacterial cellulose for ultra-sensitive and selective electrochemical detection of Staphylococcus aureus", *Biosens. Bioelectron.,* vol. 157, p. 112163, 2020.
[http://dx.doi.org/10.1016/j.bios.2020.112163] [PMID: 32250935]

[17]   A. Jasim, M.W. Ullah, Z. Shi, X. Lin, and G. Yang, "Fabrication of bacterial cellulose/polyaniline/single-walled carbon nanotubes membrane for potential application as biosensor", *Carbohydr. Polym.,* vol. 163, pp. 62-69, 2017.
[http://dx.doi.org/10.1016/j.carbpol.2017.01.056] [PMID: 28267519]

[18]   K. Jung, N. Corrigan, M. Ciftci, J. Xu, S.E. Seo, C.J. Hawker, and C. Boyer, "Designing with Light: Advanced 2D, 3D, and 4D Materials", *Adv. Mater.,* vol. 32, no. 18, p. 1903850, 2020.
[http://dx.doi.org/10.1002/adma.201903850] [PMID: 31788850]

[19]   A. Bagheri, and J. Jin, "Photopolymerization in 3D Printing", *ACS Appl. Polym. Mater.,* vol. 1, no. 4, pp. 593-611, 2019.
[http://dx.doi.org/10.1021/acsapm.8b00165]

[20]   T.M. Lovestead, A.K. O'Brien, and C.N. Bowman, "Models of multivinyl free radical photopolymerization kinetics", *J. Photochem. Photobiol. Chem.,* vol. 159, no. 2, pp. 135-143, 2003.
[http://dx.doi.org/10.1016/S1010-6030(03)00178-3]

[21]   M.D. Goodner, and C.N. Bowman, "Development of a comprehensive free radical photopolymerization model incorporating heat and mass transfer effects in thick films", *Chem. Eng. Sci.,* vol. 57, no. 5, pp. 887-900, 2002.
[http://dx.doi.org/10.1016/S0009-2509(01)00287-1]

[22]   N.C. Strandwitz, Y. Nonoguchi, S.W. Boettcher, and G.D. Stucky, *In situ* photopolymerization of pyrrole in mesoporous TiO_2, *Langmuir,* vol. 26, no. 8, pp. 5319-5322, 2010.
[http://dx.doi.org/10.1021/la100913e] [PMID: 20329722]

[23]   K. D. Dorkenoo, S. Klein, J. P. Bombenger, A. Barsella, L. Mager, and A. Fort, *Functionalized photopolymers for integrated optical components.,* 2006.
[http://dx.doi.org/10.1080/15421400500383410]

[24]   J.P. Fouassier, X. Allonas, and D. Burget, "Photopolymerization reactions under visible lights: principle, mechanisms and examples of applications", *Prog. Org. Coat.,* vol. 47, no. 1, pp. 16-36, 2003.
[http://dx.doi.org/10.1016/S0300-9440(03)00011-0]

[25]   C. Decker, "UV-Curing Chemistry: Past, Present, and Future", *J. Coatings Technol.,* 1987.

[26]   K.T. Nguyen, and J.L. West, "Photopolymerizable hydrogels for tissue engineering applications", *Biomaterials,* vol. 23, no. 22, pp. 4307-4314, 2002.
[http://dx.doi.org/10.1016/S0142-9612(02)00175-8] [PMID: 12219820]

[27]   I. Ullah, "Impact of structural features of Sr/Fe co-doped HAp on the osteoblast proliferation and osteogenic differentiation for its application as a bone substitute", *Mater. Sci. Eng. C.,* vol. 110, no. December, p. 110633, 2020.
[http://dx.doi.org/10.1016/j.msec.2020.110633]

[28] I. Anton-Sales, J.C. D'Antin, J. Fernández-Engroba, V. Charoenrook, A. Laromaine, A. Roig, and R. Michael, "Bacterial nanocellulose as a corneal bandage material: a comparison with amniotic membrane", *Biomater. Sci.,* vol. 8, no. 10, pp. 2921-2930, 2020.
[http://dx.doi.org/10.1039/D0BM00083C] [PMID: 32314754]

[29] I. Ullah, W. Li, S. Lei, Y. Zhang, W. Zhang, U. Farooq, S. Ullah, M.W. Ullah, and X. Zhang, "Simultaneous co-substitution of Sr2+/Fe3+ in hydroxyapatite nanoparticles for potential biomedical applications", *Ceram. Int.,* vol. 44, no. 17, pp. 21338-21348, 2018.
[http://dx.doi.org/10.1016/j.ceramint.2018.08.187]

[30] J.A. Burdick, and K.S. Anseth, "Photoencapsulation of osteoblasts in injectable RGD-modified PEG hydrogels for bone tissue engineering", *Biomaterials,* vol. 23, no. 22, pp. 4315-4323, 2002.
[http://dx.doi.org/10.1016/S0142-9612(02)00176-X] [PMID: 12219821]

[31] Z. Shi, X. Gao, M.W. Ullah, S. Li, Q. Wang, and G. Yang, "Electroconductive natural polymer-based hydrogels", *Biomaterials,* vol. 111, pp. 40-54, 2016.
[http://dx.doi.org/10.1016/j.biomaterials.2016.09.020] [PMID: 27721086]

[32] Z. Shi, X. Shi, M.W. Ullah, S. Li, V.V. Revin, and G. Yang, "Fabrication of nanocomposites and hybrid materials using microbial biotemplates", *Adv. Compos. Hybrid Mater.,* 2017.
[http://dx.doi.org/10.1007/s42114-017-0018-x]

[33] J. Shao, Z. Ding, L. Li, Y. Chen, J. Zhu, and Q. Qian, "Improved accumulation of TGF-β by photopolymerized chitosan/silk protein bio-hydrogel matrix to improve differentiations of mesenchymal stem cells in articular cartilage tissue regeneration", *J. Photochem. Photobiol. B,* vol. 203, p. 111744, 2020.
[http://dx.doi.org/10.1016/j.jphotobiol.2019.111744] [PMID: 31887637]

[34] X. Li, B. Li, M.W. Ullah, R. Panday, J. Cao, Q. Li, Y. Zhang, L. Wang, and G. Yang, "Water-stable and finasteride-loaded polyvinyl alcohol nanofibrous particles with sustained drug release for improved prostatic artery embolization — In vitro and in vivo evaluation"., *Mater. Sci. Eng. C,* vol. 115, no. April, p. 111107, 2020.
[http://dx.doi.org/10.1016/j.msec.2020.111107] [PMID: 32600710]

[35] A. Ottenhall, J. Henschen, J. Illergård, and M. Ek, "Cellulose-based water purification using paper filters modified with polyelectrolyte multilayers to remove bacteria from water through electrostatic interactions", *Environ. Sci. Water Res. Technol.,* vol. 4, no. 12, pp. 2070-2079, 2018.
[http://dx.doi.org/10.1039/C8EW00514A]

[36] A. Shoukat, F. Wahid, T. Khan, M. Siddique, S. Nasreen, G. Yang, M.W. Ullah, and R. Khan, "Titanium oxide-bacterial cellulose bioadsorbent for the removal of lead ions from aqueous solution", *Int. J. Biol. Macromol.,* vol. 129, pp. 965-971, 2019.
[http://dx.doi.org/10.1016/j.ijbiomac.2019.02.032] [PMID: 30738165]

[37] L. Shi, Y. Hu, M.W. Ullah, I. ullah, H. Ou, W. Zhang, L. Xiong, and X. Zhang, "Cryogenic free-form extrusion bioprinting of decellularized small intestinal submucosa for potential applications in skin tissue engineering", *Biofabrication,* vol. 11, no. 3, p. 035023, 2019.
[http://dx.doi.org/10.1088/1758-5090/ab15a9] [PMID: 30943455]

[38] M. Ul-Islam, F. Subhan, S.U. Islam, S. Khan, N. Shah, S. Manan, M.W. Ullah, and G. Yang, "Development of three-dimensional bacterial cellulose/chitosan scaffolds: Analysis of cell-scaffold interaction for potential application in the diagnosis of ovarian cancer", *Int. J. Biol. Macromol.,* vol. 137, pp. 1050-1059, 2019.
[http://dx.doi.org/10.1016/j.ijbiomac.2019.07.050] [PMID: 31295500]

[39] N.A. Peppas, J.Z. Hilt, A. Khademhosseini, and R. Langer, "Hydrogels in biology and medicine: From molecular principles to bionanotechnology", *Adv. Mater.,* vol. 18, no. 11, pp. 1345-1360, 2006.
[http://dx.doi.org/10.1002/adma.200501612]

[40] W. Aljohani, M.W. Ullah, W. Li, L. Shi, X. Zhang, and G. Yang, "Three-dimensional printing of alginate-gelatin-agar scaffolds using free-form motor assisted microsyringe extrusion system", *J.*

Polym. Res., vol. 25, no. 3, p. 62, 2018.
[http://dx.doi.org/10.1007/s10965-018-1455-0]

[41]   M.W. Tibbitt, and K.S. Anseth, "Dynamic microenvironments: the fourth dimension", *Sci. Transl. Med.,* vol. 4, no. 160, p. 160ps24, 2012.
[http://dx.doi.org/10.1126/scitranslmed.3004804] [PMID: 23152326]

[42]   M.W. Ullah, L. Fu, L. Lamboni, Z. Shi, and G. Yang, "Current trends and biomedical applications of resorbable polymers", *Materials for Biomedical Engineering.,* Elsevier, pp. 41-86, 2019.
[http://dx.doi.org/10.1016/B978-0-12-818415-8.00003-6]

[43]   J. Stampfl, R. Liska, A. Ovsianikov, Ed., *Multiphoton Lithography.* Wiley-VCH Verlag GmbH & Co. KGaA: Weinheim, Germany, 2016.
[http://dx.doi.org/10.1002/9783527682676]

[44]   D. Bratton, D. Yang, J. Dai, and C.K. Ober, "Recent progress in high resolution lithography", *Polym. Adv. Technol.,* vol. 17, no. 2, pp. 94-103, 2006.
[http://dx.doi.org/10.1002/pat.662]

[45]   X. Zhang, X.N. Jiang, and C. Sun, "Micro-stereolithography of polymeric and ceramic microstructures", *Sens. Actuators A Phys.,* vol. 77, no. 2, pp. 149-156, 1999.
[http://dx.doi.org/10.1016/S0924-4247(99)00189-2]

[46]   M.W. Ullah, Z. Shi, S. Manan, and G. Yang, "Introduction to Science and Engineering Principles for the Development of Bioinspired Materials", *Bioinspired Materials Science and Engineering.,* John Wiley & Sons, Inc.: Hoboken, NJ, USA, pp. 1-16, 2018.
[http://dx.doi.org/10.1002/9781119390350.ch0]

[47]   Y. Xia, and G.M. Whitesides, "Soft Lithography", *Angew. Chem. Int. Ed.,* vol. 37, no. 5, pp. 550-575, 1998.
[http://dx.doi.org/10.1002/(SICI)1521-3773(19980316)37:5<550::AID-ANIE550>3.0.CO;2-G]
[PMID: 29711088]

[48]   J. Guan, N. Ferrell, L. James Lee, and D.J. Hansford, "Fabrication of polymeric microparticles for drug delivery by soft lithography", *Biomaterials,* vol. 27, no. 21, pp. 4034-4041, 2006.
[http://dx.doi.org/10.1016/j.biomaterials.2006.03.011] [PMID: 16574217]

[49]   A. Khademhosseini, R. Langer, J. Borenstein, and J.P. Vacanti, "Microscale technologies for tissue engineering and biology", *Proceedings of the National Academy of Sciences of the United States of America,* 2006.
[http://dx.doi.org/10.1073/pnas.0507681102]

[50]   R.S. Kane, S. Takayama, E. Ostuni, D.E. Ingber, and G.M. Whitesides, "Patterning proteins and cells using soft lithography", *The Biomaterials.,* Silver Jubilee Compendium, 1999.

[51]   S.R. Marder, J.L. Brédas, and J.W. Perry, "Materials for multiphoton 3D microfabrication", *MRS Bull.,* vol. 32, no. 7, pp. 561-565, 2007.
[http://dx.doi.org/10.1557/mrs2007.107]

[52]   M. Malinauskas, M. Farsari, A. Piskarskas, and S. Juodkazis, "Ultrafast laser nanostructuring of photopolymers: A decade of advances", *Phys. Rep.,* vol. 533, no. 1, pp. 1-31, 2013.
[http://dx.doi.org/10.1016/j.physrep.2013.07.005]

[53]   M. Pawlicki, H.A. Collins, R.G. Denning, and H.L. Anderson, "Two-photon absorption and the design of two-photon dyes", *International Edition,* Angewandte Chemie, 2009.
[http://dx.doi.org/10.1002/anie.200805257]

[54]   W.R. Zipfel, R.M. Williams, and W.W. Webb, "Nonlinear magic: multiphoton microscopy in the biosciences", *Nat. Biotechnol.,* vol. 21, no. 11, pp. 1369-1377, 2003.
[http://dx.doi.org/10.1038/nbt899] [PMID: 14595365]

[55]   P. Golvari, and S.M. Kuebler, "Fabrication of Functional Microdevices in SU-8 by Multi-Photon Lithography", *Micromachines (Basel),* vol. 12, no. 5, p. 472, 2021.

[http://dx.doi.org/10.3390/mi12050472] [PMID: 33919437]

[56]   W. Denk, J. H. Strickler, and W. W. Webb, "Two-photon laser scanning fluorescence microscopy", *Science, 80,* 1990.
[http://dx.doi.org/10.1126/science.2321027]

[57]   C. Xu, W. Zipfel, J.B. Shear, R.M. Williams, and W.W. Webb, "Multiphoton fluorescence excitation: new spectral windows for biological nonlinear microscopy", *Proc. Natl. Acad. Sci. USA,* vol. 93, no. 20, pp. 10763-10768, 1996.
[http://dx.doi.org/10.1073/pnas.93.20.10763] [PMID: 8855254]

[58]   P. Mahou, M. Zimmerley, K. Loulier, K.S. Matho, G. Labroille, X. Morin, W. Supatto, J. Livet, D. Débarre, and E. Beaurepaire, "Multicolor two-photon tissue imaging by wavelength mixing", *Nat. Methods,* vol. 9, no. 8, pp. 815-818, 2012.
[http://dx.doi.org/10.1038/nmeth.2098] [PMID: 22772730]

[59]   A. Majewska, A. Tashiro, and R. Yuste, "Regulation of spine calcium dynamics by rapid spine motility", *J. Neurosci.,* vol. 20, no. 22, pp. 8262-8268, 2000.
[http://dx.doi.org/10.1523/JNEUROSCI.20-22-08262.2000] [PMID: 11069932]

[60]   S. Kumar, I.Z. Maxwell, A. Heisterkamp, T.R. Polte, T.P. Lele, M. Salanga, E. Mazur, and D.E. Ingber, "Viscoelastic retraction of single living stress fibers and its impact on cell shape, cytoskeletal organization, and extracellular matrix mechanics", *Biophys. J.,* vol. 90, no. 10, pp. 3762-3773, 2006.
[http://dx.doi.org/10.1529/biophysj.105.071506] [PMID: 16500961]

[61]   M. Farsari, M. Vamvakaki, and B.N. Chichkov, "Multiphoton polymerization of hybrid materials", *J. Opt.,* vol. 12, no. 12, p. 124001, 2010.
[http://dx.doi.org/10.1088/2040-8978/12/12/124001]

[62]   H.B. Sun, Y. Xu, S. Juodkazis, K. Sun, M. Watanabe, S. Matsuo, H. Misawa, and J. Nishii, "Arbitrary-lattice photonic crystals created by multiphoton microfabrication", *Opt. Lett.,* vol. 26, no. 6, pp. 325-327, 2001.
[http://dx.doi.org/10.1364/OL.26.000325] [PMID: 18040312]

[63]   S. Kawata, H.B. Sun, T. Tanaka, and K. Takada, "Finer features for functional microdevices", *Nature,* vol. 412, no. 6848, pp. 697-698, 2001.
[http://dx.doi.org/10.1038/35089130] [PMID: 11507627]

[64]   Z. Li, J. Torgersen, A. Ajami, S. Mühleder, X. Qin, W. Husinsky, W. Holnthoner, A. Ovsianikov, J. Stampfl, and R. Liska, "Initiation efficiency and cytotoxicity of novel water-soluble two-photon photoinitiators for direct 3D microfabrication of hydrogels", *RSC Advances,* vol. 3, no. 36, p. 15939, 2013.
[http://dx.doi.org/10.1039/c3ra42918k]

[65]   F. Klein, B. Richter, T. Striebel, C.M. Franz, G. Freymann, M. Wegener, and M. Bastmeyer, "Two-component polymer scaffolds for controlled three-dimensional cell culture", *Adv. Mater.,* vol. 23, no. 11, pp. 1341-1345, 2011.
[http://dx.doi.org/10.1002/adma.201004060] [PMID: 21400593]

[66]   G.S. He, L.S. Tan, Q. Zheng, and P.N. Prasad, "Multiphoton absorbing materials: molecular designs, characterizations, and applications", *Chem. Rev.,* vol. 108, no. 4, pp. 1245-1330, 2008.
[http://dx.doi.org/10.1021/cr050054x] [PMID: 18361528]

[67]   J.C. Culver, J.C. Hoffmann, R.A. Poché, J.H. Slater, J.L. West, and M.E. Dickinson, "Three-dimensional biomimetic patterning in hydrogels to guide cellular organization", *Adv. Mater.,* vol. 24, no. 17, pp. 2344-2348, 2012.
[http://dx.doi.org/10.1002/adma.201200395] [PMID: 22467256]

[68]   Y.C. Chen, R.Z. Lin, H. Qi, Y. Yang, H. Bae, J.M. Melero-Martin, and A. Khademhosseini, "Functional human vascular network generated in photocrosslinkable gelatin methacrylate hydrogels", *Adv. Funct. Mater.,* vol. 22, no. 10, pp. 2027-2039, 2012.
[http://dx.doi.org/10.1002/adfm.201101662] [PMID: 22907987]

[69]   G. Odian, *Principles of Polymerization.* John Wiley & Sons, Inc.: Hoboken, NJ, USA, 2004.
       [http://dx.doi.org/10.1002/047147875X]

[70]   I. Aujard, C. Benbrahim, M. Gouget, O. Ruel, J.B. Baudin, P. Neveu, and L. Jullien, "o-nitrobenzyl
       photolabile protecting groups with red-shifted absorption: syntheses and uncaging cross-sections for
       one- and two-photon excitation", *Chemistry,* vol. 12, no. 26, pp. 6865-6879, 2006.
       [http://dx.doi.org/10.1002/chem.200501393] [PMID: 16763952]

[71]   N.I. Kiskin, R. Chillingworth, J.A. McCray, D. Piston, and D. Ogden, "The efficiency of two-photon
       photolysis of a "caged" fluorophore, o-1-(2-nitrophenyl)ethylpyranine, in relation to photodamage of
       synaptic terminals", *Eur. Biophys. J.,* vol. 30, no. 8, pp. 588-604, 2002.
       [http://dx.doi.org/10.1007/s00249-001-0187-x] [PMID: 11908850]

[72]   A.B. Scranton, C.N. Bowman, J. Klier, and N.A. Peppas, "Polymerization reaction dynamics of
       ethylene glycol methacrylates and dimethacrylates by calorimetry", *Polymer (Guildf.),* vol. 33, no. 8,
       pp. 1683-1689, 1992.
       [http://dx.doi.org/10.1016/0032-3861(92)91067-C]

[73]   J.P. Fouassier, and J. Lalevée, "Photoinitiators for Polymer Synthesis: Scope", *Reactivity and
       Efficiency,* 2012.
       [http://dx.doi.org/10.1002/9783527648245]

[74]   N.E. Fedorovich, M.H. Oudshoorn, D. van Geemen, W.E. Hennink, J. Alblas, and W.J.A. Dhert, "The
       effect of photopolymerization on stem cells embedded in hydrogels", *Biomaterials,* vol. 30, no. 3, pp.
       344-353, 2009.
       [http://dx.doi.org/10.1016/j.biomaterials.2008.09.037] [PMID: 18930540]

[75]   S.J. Bryant, C.R. Nuttelman, and K.S. Anseth, "Cytocompatibility of UV and visible light
       photoinitiating systems on cultured NIH/3T3 fibroblasts *in vitro.*", *J. Biomater. Sci. Polym. Ed.,* vol.
       11, no. 5, pp. 439-457, 2000.
       [http://dx.doi.org/10.1163/156856200743805] [PMID: 10896041]

[76]   A. Nakagawa, N. Kobayashi, T. Muramatsu, Y. Yamashina, T. Shirai, M.W. Hashimoto, M. Ikenaga,
       and T. Mori, "Three-dimensional visualization of ultraviolet-induced DNA damage and its repair in
       human cell nuclei", *J. Invest. Dermatol.,* vol. 110, no. 2, pp. 143-148, 1998.
       [http://dx.doi.org/10.1046/j.1523-1747.1998.00100.x] [PMID: 9457909]

[77]   J. Cadet, M. Berger, T. Douki, B. Morin, S. Raoul, J.L. Ravanat, and S. Spinelli, "Effects of UV and
       visible radiation on DNA-final base damage", *Biol. Chem.,* vol. 378, no. 11, pp. 1275-1286, 1997.
       [PMID: 9426187]

[78]   T. Majima, W. Schnabel, and W. Weber, "Phenyl-2,4,6-trimethylbenzoylphosphinates as water-
       soluble photoinitiators. Generation and reactivity of O-$\dot{P}$(C_6H_5)(O-) radical anions", *Makromol.
       Chem.,* vol. 192, no. 10, pp. 2307-2315, 1991.
       [http://dx.doi.org/10.1002/macp.1991.021921010]

[79]   B.D. Fairbanks, M.P. Schwartz, C.N. Bowman, and K.S. Anseth, "Photoinitiated polymerization of
       PEG-diacrylate with lithium phenyl-2,4,6-trimethylbenzoylphosphinate: polymerization rate and
       cytocompatibility", *Biomaterials,* vol. 30, no. 35, pp. 6702-6707, 2009.
       [http://dx.doi.org/10.1016/j.biomaterials.2009.08.055] [PMID: 19783300]

[80]   J.L. Ifkovits, and J.A. Burdick, "Review: photopolymerizable and degradable biomaterials for tissue
       engineering applications", *Tissue Eng.,* vol. 13, no. 10, pp. 2369-2385, 2007.
       [http://dx.doi.org/10.1089/ten.2007.0093] [PMID: 17658993]

[81]   J. Jakubiak and J. F. Rabek, *Photoinitiators for visible light polymerization,* 1999.
       [http://dx.doi.org/10.14314/polimery.1999.447]

[82]   W. Shi, and B. Rånby, "Photopolymerization of dendritic methacrylated polyesters. I. Synthesis and
       properties", *J. Appl. Polym. Sci.,* 1996.
       [http://dx.doi.org/10.1002/(SICI)1097-4628(19960321)59:12<1937::AID-APP16>3.0.CO;2-O]

[83] C.E. Hoyle, and C.N. Bowman, "Thiol-ene click chemistry", *Angewandte Chemie - International Edition,* 2010.
[http://dx.doi.org/10.1002/anie.200903924]

[84] C. Dworak, T. Koch, F. Varga, and R. Liska, "Photopolymerization of biocompatible phosphorus-containing vinyl esters and vinyl carbamates", *J. Polym. Sci. A Polym. Chem.,* vol. 48, no. 13, pp. 2916-2924, 2010.
[http://dx.doi.org/10.1002/pola.24072]

[85] C. Heller, M. Schwentenwein, G. Russmüller, T. Koch, D. Moser, C. Schopper, F. Varga, J. Stampfl, and R. Liska, "Vinylcarbonates and vinylcarbamates: Biocompatible monomers for radical photopolymerization", *J. Polym. Sci. A Polym. Chem.,* vol. 49, no. 3, pp. 650-661, 2011.
[http://dx.doi.org/10.1002/pola.24476]

[86] C. Heller, M. Schwentenwein, G. Russmueller, F. Varga, J. Stampfl, and R. Liska, "Vinyl esters: Low cytotoxicity monomers for the fabrication of biocompatible 3D scaffolds by lithography based additive manufacturing", *J. Polym. Sci. A Polym. Chem.,* vol. 47, no. 24, pp. 6941-6954, 2009.
[http://dx.doi.org/10.1002/pola.23734]

[87] B.V. Slaughter, S.S. Khurshid, O.Z. Fisher, A. Khademhosseini, and N.A. Peppas, "Hydrogels in regenerative medicine", *Adv. Mater.,* vol. 21, no. 32-33, pp. 3307-3329, 2009.
[http://dx.doi.org/10.1002/adma.200802106] [PMID: 20882499]

[88] W. Aljohani, M.W. Ullah, X. Zhang, and G. Yang, "Bioprinting and its applications in tissue engineering and regenerative medicine", *Int. J. Biol. Macromol.,* vol. 107, no. Pt A, pp. 261-275, 2018.
[http://dx.doi.org/10.1016/j.ijbiomac.2017.08.171] [PMID: 28870749]

[89] Y. Wang, P. Liu, Y. Duan, X. Yin, Q. Wang, X. Liu, X. Wang, J. Zhou, W. Wang, L. Qiu, and W. Di, "Specific cell targeting with APRPG conjugated PEG–PLGA nanoparticles for treating ovarian cancer", *Biomaterials,* vol. 35, no. 3, pp. 983-992, 2014.
[http://dx.doi.org/10.1016/j.biomaterials.2013.09.062] [PMID: 24176193]

[90] X. Li, X. Ji, K. Chen, M.W. Ullah, X. Yuan, Z. Lei, J. Cao, J. Xiao, and G. Yang, "Development of finasteride/PHBV@polyvinyl alcohol/chitosan reservoir-type microspheres as a potential embolic agent: from *in vitro* evaluation to animal study", *Biomater. Sci.,* vol. 8, no. 10, pp. 2797-2813, 2020.
[http://dx.doi.org/10.1039/C9BM01775E] [PMID: 32080688]

[91] X. Li, X. Ji, K. Chen, M.W. Ullah, B. Li, J. Cao, L. Xiao, J. Xiao, and G. Yang, "Immobilized thrombin on X-ray radiopaque polyvinyl alcohol/chitosan embolic microspheres for precise localization and topical blood coagulation", *Bioact. Mater.,* vol. 6, no. 7, pp. 2105-2119, 2021.
[http://dx.doi.org/10.1016/j.bioactmat.2020.12.013] [PMID: 33511310]

[92] G.D. Mogoşanu, and A.M. Grumezescu, "Natural and synthetic polymers for wounds and burns dressing", *Int. J. Pharm.,* vol. 463, no. 2, pp. 127-136, 2014.
[http://dx.doi.org/10.1016/j.ijpharm.2013.12.015] [PMID: 24368109]

[93] W. Zhang, I. Ullah, L. Shi, Y. Zhang, H. Ou, J. Zhou, M.W. Ullah, X. Zhang, and W. Li, "Fabrication and characterization of porous polycaprolactone scaffold *via* extrusion-based cryogenic 3D printing for tissue engineering"., *Mater. Des.,* vol. 180, p. 107946, 2019.
[http://dx.doi.org/10.1016/j.matdes.2019.107946]

[94] Z. Di, Z. Shi, M.W. Ullah, S. Li, and G. Yang, "A transparent wound dressing based on bacterial cellulose whisker and poly(2-hydroxyethyl methacrylate)", *Int. J. Biol. Macromol.,* vol. 105, no. Pt 1, pp. 638-644, 2017.
[http://dx.doi.org/10.1016/j.ijbiomac.2017.07.075] [PMID: 28716748]

[95] A.S. Hoffman, "Hydrogels for biomedical applications", *Adv. Drug Deliv. Rev.,* vol. 64, pp. 18-23, 2012.
[http://dx.doi.org/10.1016/j.addr.2012.09.010] [PMID: 11755703]

[96] W. Aljohani, wenchao, M.W. Ullah, X. Zhang, and G. Yang, "Application of Sodium Alginate

Hydrogel", *IOSR J. Biotechnol. Biochem.,* vol. 3, no. 3, pp. 19-31, 2017.
[http://dx.doi.org/10.9790/264X-03031931]

[97]   C. Wang, J. Lai, K. Li, S. Zhu, B. Lu, J. Liu, Y. Tang, and Y. Wei, "Cryogenic 3D printing of dual-delivery scaffolds for improved bone regeneration with enhanced vascularization", *Bioact. Mater.,* vol. 6, no. 1, pp. 137-145, 2021.
[http://dx.doi.org/10.1016/j.bioactmat.2020.07.007] [PMID: 32817920]

[98]   S. Khan, M. Ul-Islam, M. Ikram, M.W. Ullah, M. Israr, F. Subhan, Y. Kim, J.H. Jang, S. Yoon, and J.K. Park, "Three-dimensionally microporous and highly biocompatible bacterial cellulose–gelatin composite scaffolds for tissue engineering applications", *RSC Advances,* vol. 6, no. 112, pp. 110840-110849, 2016.
[http://dx.doi.org/10.1039/C6RA18847H]

[99]   L. Wei, S. Wu, M. Kuss, X. Jiang, R. Sun, P. Reid, X. Qin, and B. Duan, "3D printing of silk fibroin-based hybrid scaffold treated with platelet rich plasma for bone tissue engineering", *Bioact. Mater.,* vol. 4, pp. 256-260, 2019.
[http://dx.doi.org/10.1016/j.bioactmat.2019.09.001] [PMID: 31667442]

[100] X. Li, X. Ji, K. Chen, X. Yuan, Z. Lei, M.W. Ullah, J. Xiao, and G. Yang, "Preparation and evaluation of ion-exchange porous polyvinyl alcohol microspheres as a potential drug delivery embolization system", *Mater. Sci. Eng. C,* vol. 121, no. January, p. 111889, 2021.
[http://dx.doi.org/10.1016/j.msec.2021.111889] [PMID: 33579501]

[101] X. Li, K. Chen, X. Ji, X. Yuan, Z. Lei, M.W. Ullah, J. Xiao, and G. Yang, "Microencapsulation of Poorly Water-soluble Finasteride in Polyvinyl Alcohol/chitosan Microspheres as a Long-term Sustained Release System for Potential Embolization Applications", *Engineered Science,* pp. 1-15, 2020.
[http://dx.doi.org/10.30919/es8d1159]

[102] R. R. McCarthy, M. W. Ullah, E. Pei, and G. Yang, "Antimicrobial Inks: The Anti-Infective Applications of Bioprinted Bacterial Polysaccharides", *Trends in Biotechnology,* vol. 37, Elsevier Ltd, no. 11, pp. 1153-1155, 2019.
[http://dx.doi.org/10.1016/j.tibtech.2019.05.004]

[103] R.R. McCarthy, M.W. Ullah, P. Booth, E. Pei, and G. Yang, "The use of bacterial polysaccharides in bioprinting", *Biotechnol. Adv.,* vol. 37, no. 8, p. 107448, 2019.
[http://dx.doi.org/10.1016/j.biotechadv.2019.107448] [PMID: 31513840]

[104] M. W. Ullah, S. Manan, S. J. Kiprono, M. Ul-Islam, and G. Yang, *Synthesis, structure, and properties of bacterial cellulose.,* 2019.
[http://dx.doi.org/10.1002/9783527807437.ch4]

[105] M. Ul-Islam, F. Ahmad, A. Fatima, N. Shah, S. Yasir, M.W. Ahmad, S. Manan, and M.W. Ullah, "*Ex situ* Synthesis and Characterization of High Strength Multipurpose Bacterial Cellulose-*Aloe vera* Hydrogels"., *Front. Bioeng. Biotechnol.,* vol. 9, p. 601988, 2021.
[http://dx.doi.org/10.3389/fbioe.2021.601988] [PMID: 33634082]

[106] K.Y. Lee, and D.J. Mooney, "Hydrogels for tissue engineering", *Chem. Rev.,* vol. 101, no. 7, pp. 1869-1880, 2001.
[http://dx.doi.org/10.1021/cr000108x] [PMID: 11710233]

[107] L. Li, and J.T. Fourkas, "Multiphoton polymerization", *Mater. Today,* vol. 10, no. 6, pp. 30-37, 2007.
[http://dx.doi.org/10.1016/S1369-7021(07)70130-X]

[108] C.A. Coenjarts, and C.K. Ober, "Two-photon three-dimensional microfabrication of poly(dimethylsiloxane) elastomers", *Chem. Mater.,* vol. 16, no. 26, pp. 5556-5558, 2004.
[http://dx.doi.org/10.1021/cm048717z]

[109] B.J. Adzima, C.J. Kloxin, C.A. DeForest, K.S. Anseth, and C.N. Bowman, "3D photofixation lithography in Diels-Alder networks", *Macromol. Rapid Commun.,* vol. 33, no. 24, pp. 2092-2096, 2012.

[http://dx.doi.org/10.1002/marc.201200599] [PMID: 23080017]

[110] J. Serbin, A. Egbert, A. Ostendorf, B.N. Chichkov, R. Houbertz, G. Domann, J. Schulz, C. Cronauer, L. Fröhlich, and M. Popall, "Femtosecond laser-induced two-photon polymerization of inorganic–organic hybrid materials for applications in photonics", *Opt. Lett.,* vol. 28, no. 5, pp. 301-303, 2003.
[http://dx.doi.org/10.1364/OL.28.000301] [PMID: 12659425]

[111] W. Lee, S.A. Pruzinsky, and P.V. Braun, "Multi-photon polymerization of waveguide structures within three-dimensional photonic crystals", *Adv. Mater.,* vol. 14, no. 4, pp. 271-274, 2002.
[http://dx.doi.org/10.1002/1521-4095(20020219)14:4<271::AID-ADMA271>3.0.CO;2-Y]

[112] S. Klein, A. Barsella, H. Leblond, H. Bulou, A. Fort, C. Andraud, G. Lemercier, J.C. Mulatier, and K. Dorkenoo, "One-step waveguide and optical circuit writing in photopolymerizable materials processed by two-photon absorption", *Appl. Phys. Lett.,* vol. 86, no. 21, p. 211118, 2005.
[http://dx.doi.org/10.1063/1.1915525]

[113] R. Guo, S. Xiao, X. Zhai, J. Li, A. Xia, and W. Huang, "Micro lens fabrication by means of femtosecond two photon photopolymerization", *Opt. Express,* vol. 14, no. 2, pp. 810-816, 2006.
[http://dx.doi.org/10.1364/OPEX.14.000810] [PMID: 19503401]

[114] S. Yokoyama, T. Nakahama, H. Miki, and S. Mashiko, *Fabrication of three-dimensional microstructure in optical-gain medium using two-photon-induced photopolymerization technique.,* 2003.
[http://dx.doi.org/10.1016/S0040-6090(03)00804-6]

[115] W.P. Daley, S.B. Peters, and M. Larsen, "Extracellular matrix dynamics in development and regenerative medicine", *J. Cell Sci.,* vol. 121, no. 3, pp. 255-264, 2008.
[http://dx.doi.org/10.1242/jcs.006064] [PMID: 18216330]

[116] M.W. Tibbitt, and K.S. Anseth, "Hydrogels as extracellular matrix mimics for 3D cell culture", *Biotechnol. Bioeng.,* vol. 103, no. 4, pp. 655-663, 2009.
[http://dx.doi.org/10.1002/bit.22361] [PMID: 19472329]

[117] J. Torgersen, X.H. Qin, Z. Li, A. Ovsianikov, R. Liska, and J. Stampfl, "Hydrogels for two-photon polymerization: A toolbox for mimicking the extracellular matrix", *Adv. Funct. Mater.,* vol. 23, no. 36, pp. 4542-4554, 2013.
[http://dx.doi.org/10.1002/adfm.201203880]

[118] A. Doraiswamy, C. Jin, R. Narayan, P. Mageswaran, P. Mente, R. Modi, R. Auyeung, D. Chrisey, A. Ovsianikov, and B. Chichkov, "Two photon induced polymerization of organic–inorganic hybrid biomaterials for microstructured medical devices", *Acta Biomater.,* vol. 2, no. 3, pp. 267-275, 2006.
[http://dx.doi.org/10.1016/j.actbio.2006.01.004] [PMID: 16701886]

[119] F. Claeyssens, E.A. Hasan, A. Gaidukeviciute, D.S. Achilleos, A. Ranella, C. Reinhardt, A. Ovsianikov, X. Shizhou, C. Fotakis, M. Vamvakaki, B.N. Chichkov, and M. Farsari, "Three-dimensional biodegradable structures fabricated by two-photon polymerization", *Langmuir,* vol. 25, no. 5, pp. 3219-3223, 2009.
[http://dx.doi.org/10.1021/la803803m] [PMID: 19437724]

[120] A. Ovsianikov, B. Chichkov, O. Adunka, H. Pillsbury, A. Doraiswamy, and R.J. Narayan, "Rapid prototyping of ossicular replacement prostheses", *Appl. Surf. Sci.,* vol. 253, no. 15, pp. 6603-6607, 2007.
[http://dx.doi.org/10.1016/j.apsusc.2007.01.062]

[121] X.H. Qin, J. Torgersen, R. Saf, S. Mühleder, N. Pucher, S.C. Ligon, W. Holnthoner, H. Redl, A. Ovsianikov, J. Stampfl, and R. Liska, "Three-dimensional microfabrication of protein hydrogels *via* two-photon-excited thiol-vinyl ester photopolymerization"., *J. Polym. Sci. A Polym. Chem.,* vol. 51, no. 22, pp. 4799-4810, 2013.
[http://dx.doi.org/10.1002/pola.26903]

[122] J. Torgersen, A. Ovsianikov, V. Mironov, N. Pucher, X. Qin, Z. Li, K. Cicha, T. Machacek, R. Liska,

V. Jantsch, and J. Stampfl, "Photo-sensitive hydrogels for three-dimensional laser microfabrication in the presence of whole organisms", *J. Biomed. Opt.,* vol. 17, no. 10, p. 1, 2012.
[http://dx.doi.org/10.1117/1.JBO.17.10.105008] [PMID: 23070525]

[123] S.J. Jhaveri, J.D. McMullen, R. Sijbesma, L.S. Tan, W. Zipfel, and C.K. Ober, "Direct three-dimensional microfabrication of hydrogels *via* two-photon lithography in aqueous solution"., *Chem. Mater.,* vol. 21, no. 10, pp. 2003-2006, 2009.
[http://dx.doi.org/10.1021/cm803174e] [PMID: 20160917]

[124] B. Kaehr, and J.B. Shear, "Multiphoton fabrication of chemically responsive protein hydrogels for microactuation", *Proc. Natl. Acad. Sci. USA,* vol. 105, no. 26, pp. 8850-8854, 2008.
[http://dx.doi.org/10.1073/pnas.0709571105] [PMID: 18579775]

[125] J.D. Pitts, P.J. Campagnola, G.A. Epling, and S.L. Goodman, "Submicron multiphoton free-form fabrication of proteins and polymers: studies of reaction efficiencies and applications in sustained release", *Macromolecules,* vol. 33, no. 5, pp. 1514-1523, 2000.
[http://dx.doi.org/10.1021/ma9910437]

[126] J.D. Pitts, A.R. Howell, R. Taboada, I. Banerjee, J. Wang, S.L. Goodman, and P.J. Campagnola, "New photoactivators for multiphoton excited three-dimensional submicron cross-linking of proteins: bovine serum albumin and type 1 collagen", *Photochem. Photobiol.,* vol. 76, no. 2, pp. 135-144, 2002.
[http://dx.doi.org/10.1562/0031-8655(2002)076<0135:NPFMET>2.0.CO;2] [PMID: 12194208]

[127] B. Kaehr, R. Allen, D.J. Javier, J. Currie, and J.B. Shear, "Guiding neuronal development with *in situ* microfabrication"., *Proc. Natl. Acad. Sci. USA,* vol. 101, no. 46, pp. 16104-16108, 2004.
[http://dx.doi.org/10.1073/pnas.0407204101] [PMID: 15534228]

[128] J.E. Leslie-Barbick, C. Shen, C. Chen, and J.L. West, "Micron-scale spatially patterned, covalently immobilized vascular endothelial growth factor on hydrogels accelerates endothelial tubulogenesis and increases cellular angiogenic responses", *Tissue Eng. Part A,* vol. 17, no. 1-2, pp. 221-229, 2011.
[http://dx.doi.org/10.1089/ten.tea.2010.0202] [PMID: 20712418]

[129] C.A. DeForest, and K.S. Anseth, "Photoreversible patterning of biomolecules within click-based hydrogels", *Angew. Chem. Int. Ed.,* vol. 51, no. 8, pp. 1816-1819, 2012.
[http://dx.doi.org/10.1002/anie.201106463] [PMID: 22162285]

[130] J.D. McCall, J.E. Luoma, and K.S. Anseth, "Covalently tethered transforming growth factor beta in PEG hydrogels promotes chondrogenic differentiation of encapsulated human mesenchymal stem cells", *Drug Deliv. Transl. Res.,* vol. 2, no. 5, pp. 305-312, 2012.
[http://dx.doi.org/10.1007/s13346-012-0090-2] [PMID: 23019539]

[131] R.G. Wylie, S. Ahsan, Y. Aizawa, K.L. Maxwell, C.M. Morshead, and M.S. Shoichet, "Spatially controlled simultaneous patterning of multiple growth factors in three-dimensional hydrogels", *Nat. Mater.,* vol. 10, no. 10, pp. 799-806, 2011.
[http://dx.doi.org/10.1038/nmat3101] [PMID: 21874004]

[132] K.A. Mosiewicz, L. Kolb, A.J. van der Vlies, M.M. Martino, P.S. Lienemann, J.A. Hubbell, M. Ehrbar, and M.P. Lutolf, "*In situ* cell manipulation through enzymatic hydrogel photopatterning"., *Nat. Mater.,* vol. 12, no. 11, pp. 1072-1078, 2013.
[http://dx.doi.org/10.1038/nmat3766] [PMID: 24121990]

[133] C.N. Salinas, and K.S. Anseth, "The enhancement of chondrogenic differentiation of human mesenchymal stem cells by enzymatically regulated RGD functionalities", *Biomaterials,* vol. 29, no. 15, pp. 2370-2377, 2008.
[http://dx.doi.org/10.1016/j.biomaterials.2008.01.035] [PMID: 18295878]

[134] C.G. Rolli, H. Nakayama, K. Yamaguchi, J.P. Spatz, R. Kemkemer, and J. Nakanishi, "Switchable adhesive substrates: Revealing geometry dependence in collective cell behavior", *Biomaterials,* vol. 33, no. 8, pp. 2409-2418, 2012.
[http://dx.doi.org/10.1016/j.biomaterials.2011.12.012] [PMID: 22197568]

[135] H.M. Lin, W.K. Wang, P.A. Hsiung, and S.G. Shyu, "Light-sensitive intelligent drug delivery systems

of coumarin-modified mesoporous bioactive glass", *Acta Biomater.*, vol. 6, no. 8, pp. 3256-3263, 2010.
[http://dx.doi.org/10.1016/j.actbio.2010.02.014] [PMID: 20152945]

[136]   A.M.L. Hossion, M. Bio, G. Nkepang, S.G. Awuah, and Y. You, "Visible light controlled release of anticancer drug through double activation of prodrug", *ACS Med. Chem. Lett.*, vol. 4, no. 1, pp. 124-127, 2013.
[http://dx.doi.org/10.1021/ml3003617] [PMID: 24900573]

[137]   N.K. Mal, M. Fujiwara, and Y. Tanaka, "Photocontrolled reversible release of guest molecules from coumarin-modified mesoporous silica", *Nature,* vol. 421, no. 6921, pp. 350-353, 2003.
[http://dx.doi.org/10.1038/nature01362] [PMID: 12540896]

[138]   M.W. Tibbitt, A.M. Kloxin, K.U. Dyamenahalli, and K.S. Anseth, "Controlled two-photon photodegradation of PEG hydrogels to study and manipulate subcellular interactions on soft materials", *Soft Matter,* vol. 6, no. 20, pp. 5100-5108, 2010.
[http://dx.doi.org/10.1039/c0sm00174k] [PMID: 21984881]

[139]   A. Heisterkamp, I.Z. Maxwell, E. Mazur, J.M. Underwood, J.A. Nickerson, S. Kumar, and D.E. Ingber, "Pulse energy dependence of subcellular dissection by femtosecond laser pulses", *Opt. Express,* vol. 13, no. 10, pp. 3690-3696, 2005.
[http://dx.doi.org/10.1364/OPEX.13.003690] [PMID: 16035172]

[140]   N. Shen, D. Datta, C.B. Schaffer, P. LeDuc, D.E. Ingber, and E. Mazur, "Ablation of cytoskeletal filaments and mitochondria in live cells using a femtosecond laser nanoscissor", *MCB Mech. Chem. Biosyst,* 2005.
[http://dx.doi.org/10.3970/mcb.2005.002.017]

Biomaterial Fabrication Techniques, 2022, 195-217

CHAPTER 8

Particulate Leaching (Salt Leaching) Technique for Fabrication of Biomaterials

Nurhasni Hasan[1,*], **Aliyah Putranto**[1], **Sumarheni**[1] and **Andi Arjuna**[1]

[1] Faculty of Pharmacy, Hasanuddin University, Jl. Perintis Kemerdekaan Km 1, Makassar 90245, Republic of Indonesia

Abstract: The most important characteristic of a scaffold used in tissue engineering is the possession of appropriate physical and mechanical properties to support or restore the biological function of damaged or degenerated tissue. Pore size, porosity, pore interconnectivity, and mechanical strength are all physical and mechanical properties that must be considered. Various fabrication techniques have been investigated to create a scaffold suitable for tissue engineering. One example is the particulate leaching (salt leaching) technique. The type of polymers and salts used, the particle size of the salt, and the fabrication technique all affect the desired physical and mechanical properties of salt leaching scaffolds. Over the past decade, there have been numerous studies on the fabrication of scaffolds for tissue engineering. This chapter reviews the different types of materials used, the basic salt leaching process, and its new modifications. It also discusses the advantages and disadvantages of the salt leaching technique and its future prospects.

Keywords: Interconnectivity, Mechanical strength, Polymers, Porosity, Salt leaching, Scaffold, Tissue engineering.

INTRODUCTION

Tissue engineering is a discipline of biomedical engineering that aims to facilitate cell ingrowth or replace damaged or diseased tissue with a combination of bioactive molecules, biomaterials, and cells or engineered cells [1]. To achieve these goals, scaffolds are commonly used in tissue engineering. Various biomaterials, from biopolymers to bioceramics to biodegradable metals, have been shown to be useful in the fabrication process [2].

The most important characteristics of a scaffold for tissue engineering are sufficient mechanical strength to support biological function by promoting cell

* **Corresponding author Nurhasni Hasan:** Faculty of Pharmacy, Hasanuddin University, Jl. Perintis Kemerdekaan Km 10, Makassar 90245, Republic of Indonesia; E-mail: nurhasni.hasan@unhas.ac.id

Adnan Haider & Sajjad Haider (Eds.)

adhesion, differentiation, and proliferation [3, 4]. Various techniques have been investigated to fabricate such scaffolds, including particle leaching (salt leaching), freeze-drying, solvent casting, self-assembly, phase separation, electrospinning, rapid prototyping, melt molding, gas foaming, and membrane lamination [5]. This chapter deals exclusively with the fabrication of a framework by particle leaching (salt leaching).

Particulate leaching (salt leaching/porogen leaching) is one of the most common, long-established conventional techniques for preparing porous biomaterials for tissue engineering. It involves dispersion of salts/porogens in a polymeric or monomeric solution, followed by gelation or fixation in the template and removal of salts/porogens to form an interconnecting porous architecture. The method has several advantages and disadvantages, which are also discussed in this chapter.

The main goal of preparing biomaterials for tissue engineering is to create a well-designed three-dimensional (3D) scaffold. The scaffold is an important tool to facilitate tissue formation both *in vitro* and *in vivo*. To regenerate tissue, tissue engineering uses biodegradable or non-biodegradable polymers, with or without the inclusion of molecules or biological cells. Many scaffolds for tissue engineering have been fabricated using the particle leaching technique (salt leaching). However, different tissues require different scaffold properties. For example, scaffolds for bone engineering may have different desirable properties than scaffolds for skin substitutes or retinal neural progenitor cells. Therefore, selecting the right polymers, salts, and salt leaching techniques (simple or modified) is critical, especially if the scaffold is designed to allow the target cells to function in the manner required for tissue regeneration. In this chapter, particle/salt leaching is presented for the preparation of biomaterials for tissue engineering applications. The materials and methods used and their new modifications are compared. Recent studies on scaffold materials fabricated using these techniques are summarized and discussed.

PARTICULATE LEACHING (SALT LEACHING) TECHNIQUE

The technique of particle leaching (salt leaching) involves the use of polymers or a combination of polymers and salt particles of a specific size to produce a suitable scaffold for tissue engineering. The desired physical and mechanical properties of the scaffold depend largely on the choice of the type of polymer and salt, the size of the salt particles, and the fabrication techniques. The types of polymers and salt typically used in the salt leaching technique, as well as the step-by-step approach to the basic salt leaching technique and its modifications, have been discussed in this section.

Polymers

Natural or synthetic biodegradable polymeric materials are widely used for the production of biomaterials because their properties offer greater advantages compared to other materials, such as metal or ceramics. Apart from the fact that biodegradable polymers are naturally absorbed by the human body, some of them are also suitable for tissue regeneration, which is basically helpful in injuries and reconstruction of damaged or aging tissues. Another advantage of polymers as biodegradable drug carriers is their low cost and ability to adapt to target organs or tissues. In laboratory processing, the particle leaching (or salt leaching) technique is often used in the development phase to produce biodegradable or non-biodegradable polymeric scaffolds with sufficient porosity for use in tissue engineering. The fabrication technique of this polymer can be easily extended to a larger quantity through industrial production [6].

Polymers are available with different mechanical and physical properties. Therefore, the basic properties of scaffolds, such as biocompatibility with the human body, sterilizability, and a suitable degradation profile, must be considered before fabrication. The processing of polymers into scaffolds for tissue engineering with specific properties for each application is highly dependent on the type of polymer chosen. The most commonly used biodegradable polymers for salt leaching techniques are aliphatic polyesters, such as poly(lactic acid) (PLA), polyglycolic acid (PGA), polycaprolactone (PCL) and their copolymers. However, there are also some other polymers, such as silk fibroin (SF), nylon and many others that are used to produce biomaterials for tissue engineering. Table **1** summarizes the properties of the polymers used in the production of biomaterials using the salt leaching technique.

Table 1. The properties of the polymer used in the preparation of biomaterial with the salt leaching technique.

Materials		Density (g/cm³)	E (GPa)	σ (MPa)	ε (%)	References
Biodegradable Polymers	**Non-biodegradable Polymers or Other Material**					
Poly (glycolic acid)	-	1.53	>6.9	>68.9	15-20	[7]
Poly (L-lactic acid)	-	1.210–1.430	2.4-4.2	55.2-82.7	5-10	[8]
Poly (L-lactic-co-glycolic acid)	-	1.3	1.4-2.08	41.4-55.2	3-10	[9]
Polycaprolactone	-	1.14	0.21-0.34	20.7-34.5	300-700	[8, 10]
Chitosan	-	0.15–0.3	-	30	-	[11]
Starch	-	1.5	116.42–294.98	4.48–8.14	35.41–100.34	[12]
Poly(3-hydroxybutyrate-co-3-hydroxyvalerate)	-	1.17–1.2	0.7–3.5	20–60	6–8	[10]
polymethyl methacrylate	-	1.17-1.20	1.8–3.1	48-76	2-10	[10]
Cellulose nanofiber	-	0.96–1.02	138	10	-	[13]
Silk fibroin	-	1.40	9.860	513	23.4	[10, 14]

(Table 1) cont.....

Materials		Density (g/cm³)	E (GPa)	σ (MPa)	ε (%)	References
Biodegradable Polymers	**Non-biodegradable Polymers or Other Material**					
Gelatin	-	1.3–1.4	1.09–1.57	50.68–128.37	5.86–18.4	[10]
Silk sericin	-	0.00132–0.0014	1.09–1.57	50.68–128.37	5.86–18.4	[15]
PVA	-	1.2–1.3	37–45	67–110 (98-99% hydrolyzed)	225–445	[10]
-	Nylon	1.15	2.7	82.7	10.0–86.0	[16]
-	Polyurethane	0.048–0.961	0.091–0.02	35	-	[17, 18]

E, Young's modulus; σ, tensile strength; ε, elongation at break.

Salt

A variety of salts, including sodium chloride (NaCl), sodium bicarbonate (NaHCO$_3$), sodium acetate (NaOAc), and calcium chloride (CaCl$_2$), have been used in salt leaching techniques to produce biomaterials [19]. They are also known as porogens due to their ability to form porous structures, which is an important property of scaffolds for tissue engineering. The ideal salt that is mostly used in this method is NaCl. Due to its high solubility in aqueous media, it can be easily removed by the leaching process. Since plasma normally contains the ions Na and Cl-, the remaining salt in the scaffold after leaching is not harmful to the human body [20].

Methods

1. Conventional Salt Leaching

The method involves the addition of a water-soluble porogen (leaching agent to create a porous/channel) such as salt. The salt is crushed into small particles or according to the desired size and poured into a mold. The polymers are then dissolved in an organic solvent before being cast into the mold filled with salt. After evaporation of the solvent, the salt is leached in water for two days to form a porous structure [5]. Fig. (1) shows an illustration of the basic salt leaching technique. In this method, the size of the salts significantly affects the size and shape of the pores, while the porosity of the fabricated framework depends on the amount of salt added during fabrication. It has also been reported that the pores exhibit high interconnectivity when the salt content is 70% w/w or more [21].

Kwon *et al.* recently reported the preparation of poly(-caprolactone-ran lactic acid) (PCLA-F) scaffolds by the salt leaching method. The PCLA-F polymer was dissolved in methylene chloride and the sieved NaCl particles (200-250 μm diameter) were added to the polymer solution. Then the mixture was poured into a round silicone mold (diameter 10 mm and height 5 mm). The molded slurry was then dried at room temperature for up to 24 hours, followed by leaching in deionized water at a constant stirring speed of 100 rotations per minute (rpm) for 24 hours. The resulting scaffold has an irregular and well-interconnected pore structure [22]. Fig. (**2**) shows the resulting scaffold as well as scanning electron micrographs of the pores and the interconnectivity between the pores.

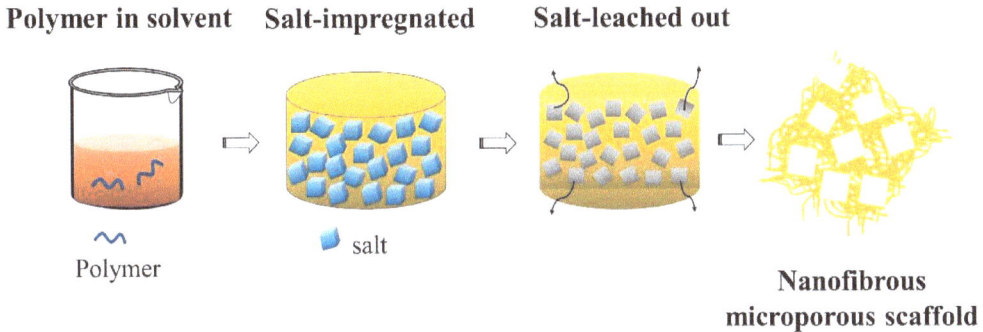

Polymer in solvent   Salt-impregnated   Salt-leached out

Nanofibrous microporous scaffold

Fig. (1). Schematic illustration of the preparation of scaffold by using conventional salt leaching technique.

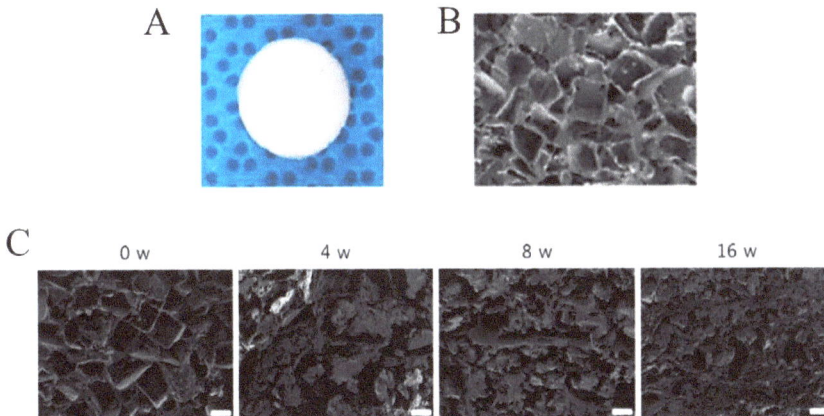

Fig. (2). (A) Macroscopic image of scaffold. **(B)** Pore structure and interconnectivity of scaffold by SEM. **(C)** The cross-sectional SEM image showed the degradation of scaffold after 16 weeks implanted in SD rats. Reprinted with permission from [22].

2. Combination of Melt Mixing and Particulate Leaching

This method is similar to the traditional salt leaching technique. The only difference is that this method uses a batch/measurement mixer. Fig. (**3**) shows the preparation of scaffold by using a combination of melt mixing and salt leaching technique In short, the polymer is mixed with the salt and placed in a batch mixer, where it is processed after a specified mixing time and temperature. The mixture is then compressed in a compression molding tool (mold diameter 10 mm and height 3 mm) at 180 bar and 190-220 °C. The salt is then leached in demineralized water under specific pH, time and temperature conditions determined by the experimental design. The porous frameworks are then dried at room temperature for 24 hours. Scaffaro *et al.* reported using this method to produce porous scaffolds with a highly interconnected pore with suitable mechanical properties for tissue engineering [23].

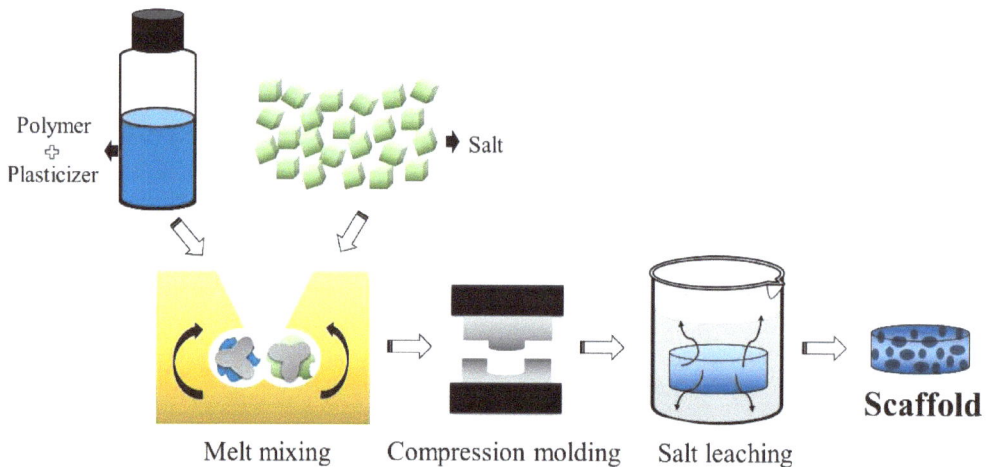

Fig. (3). Schematic illustration of the preparation of scaffold by using a combination of melt mixing and salt leaching technique.

3. High Compression Molding-Salt Leaching

This method combines salt leaching with compression molding. Fig. (**4**) shows the preparation of scaffold by using high compression molding-salt leaching. The process essentially consists of using a high-pressure compression molding machine in which temperature and pressure are controlled. This method results in a strict, regular architecture and good mechanical properties of scaffolds due to the effect of the melting process combined with the high pressure. In 2016, Zhang *et al.* developed a home-built high-pressure press molding device for the fabrication of poly(L-lactide)/poly(lactide-co-glycolide)/hydroxyapatite composite scaffolds using the combination technique with salt leaching. The high-pressure cell has a guide column, a mold core and a sample chamber within the main mold, which is heated by an electric heating jacket and tempered by a thermocouple. The scaffold is prepared by dissolving polymers in a suitable organic solvent and mixing them with sieved salt (sodium chloride) of the desired particle size. The mixture is then mechanically mixed in an internal mixer at 50 rpm and 180 °C. It is then molded under high pressure at a predetermined pressure of 640 MPa for 10-15 minutes. The interconnected pore structure is maintained after lowering the temperature to room temperature by leaching the salt with water until a constant weight is achieved. This method results in a framework with a good porous structure and good interconnectivity between pores. The porosity ranges from 81.5 to 82.7%, and the compressive modulus can reach up to 4.64 0.2 MPa, which is comparable to the modulus of human cancellous bone (2-10 MPa) [24].

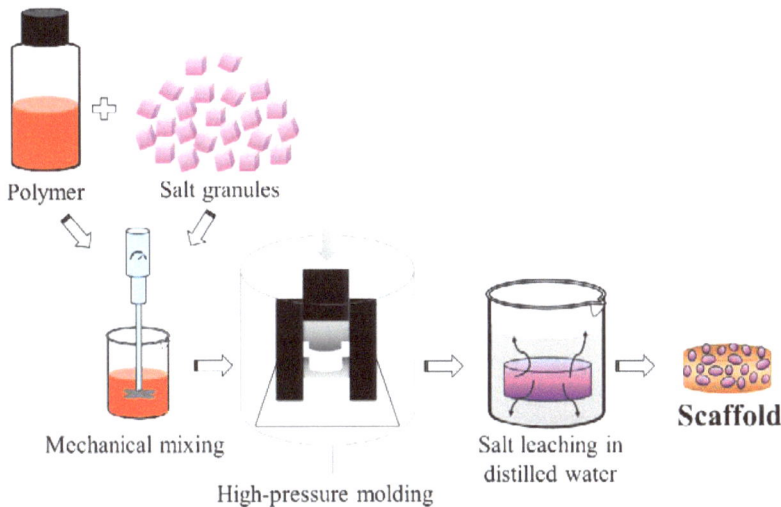

Fig. (4). Schematic illustration of the preparation of scaffold by using high compression molding-salt leaching.

4. Gas Foaming-Salt Leaching

This method offers a fast and solvent-free process. Fig. (**5**) shows the preparation of scaffold by using gas foaming-salt leaching. Gas foam salt leaching technique is divided into three steps: First, the polymer/salt mixture is prepared by melt mixing. Second, the composite is foamed with dense CO_2 gas, and finally, the salt particles are leached out of the scaffold. In 2011, Annabi *et al.* used this method to fabricate a porous PCL/elastin composite scaffold. The scaffold has an average pore size of 540 m and a porosity of 91%. They used a CO_2 pressure of 65 bar, a temperature of 70°C to melt the polymer, a depressurization rate of 15 bar/min, and a processing time of 1 h in a gas foaming reactor [25].

In another study, Bak *et al.* demonstrated the use of supercritical carbon dioxide (scCO_2) to produce a porous, solvent-free scaffold. In this method, gas foaming was combined with a salt leaching process. The polymer was melted at the glass transition temperature (Tg) and then mixed with salt at a suitable ratio. The mixture was then poured into a Teflon mold (diameter 2 cm and height 1 cm). The gas foaming process was carried out in a gas foaming reactor with controlled temperature and pressure. In their study, Bak *et al.* used a temperature of 50°C and a pressure of 8 MPa in a gas foaming reactor, and CO_2 was dissolved in the samples for 6 hours. Then, the pressure was quickly released to allow the scaffold to form a bimodal pore structure. After that, the salt was leached with distilled water for one day. The prepared scaffold has an average pore size of 427.89 m and a high degree of pore interconnectivity. The scaffold also promotes proliferation and adhesion of MC3T3-E1, indicating a potential application in bone tissue engineering [26].

It is known that the amount of gas generated and expanded in the internal structure of a scaffold affects the percentage of porosity and the pore size of the scaffold. The higher the percentage of porosity and the larger the pore size of the scaffolds, the more gas is generated and expanded. However, in the fabrication of silver nanocomposite scaffolds by El-Kady *et al.* using a combination of gas foams and salt leaching, the addition of silver nanoparticles hindered the expansion of the generated gas throughout the scaffold matrix. The percent porosity and pore size of the framework were reduced by this process [27].

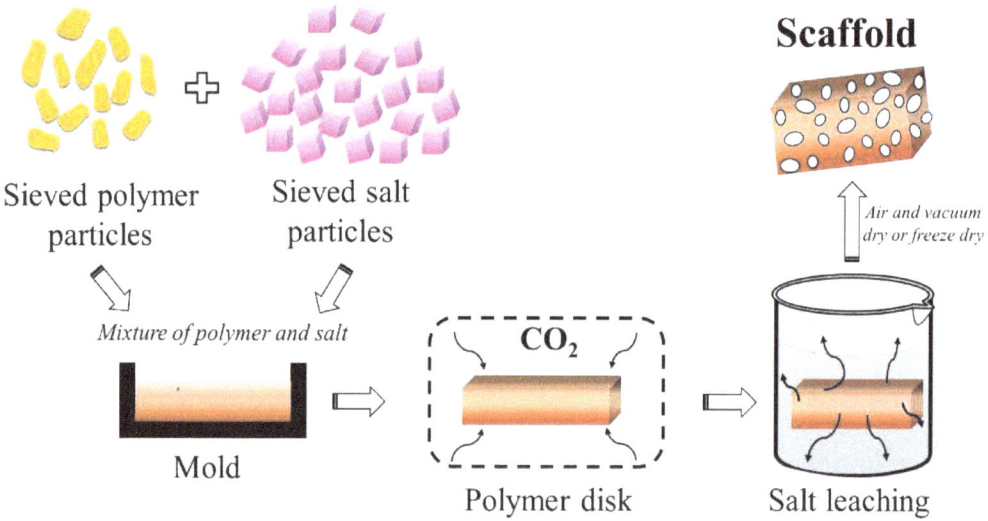

Fig. (5). Schematic illustration of the preparation of scaffold by using gas foaming-salt leaching.

5. Salt Leaching Electrospinning (SLE)

In 2016, Park *et al.* fabricated a scaffold for skin tissue engineering using a combination of salt leaching and 3D electrospinning techniques. Since the structure and physical properties of the skin scaffold fabricated by this method are similar to those of an extracellular matrix (ECM), the scaffold can mimic the ECM. The preparation of scaffold by using salt leaching electrospinning can be seen in the Fig. (**6**). To obtain stable nozzles, the polymer solution was placed in a syringe with the proper needle size, and the electrospinning device was set at an appropriate voltage and distance between the needle tip and the collector. The salt with the appropriate particle size was then released from the rotating cylinder above the drum collector at the desired rate. After the fabrication process, the salt from the nanofiber scaffold was leached in distilled water for one day and then freeze-dried. The scaffolds have advantageous properties such as high interconnectivity between pores, porosity and water uptake capacity. Therefore, the scaffold can support cell infiltration and proliferation, which could be useful for skin tissue engineering [28].

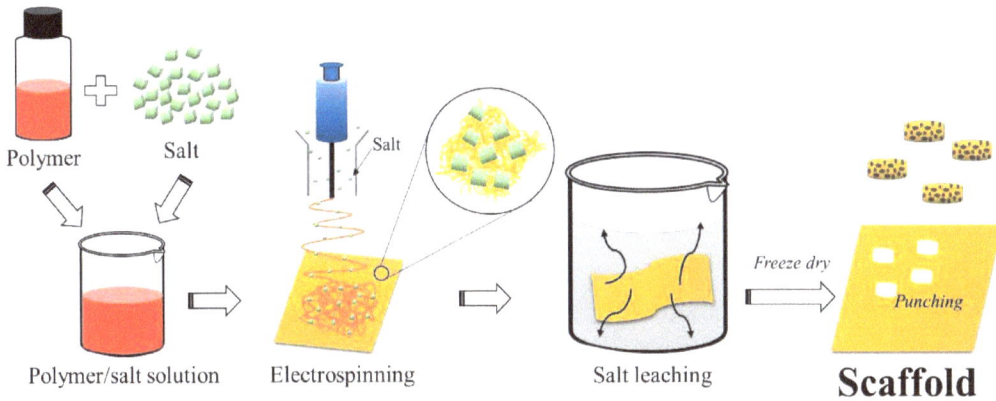

Polymer   Salt

Polymer/salt solution   Electrospinning   Salt   Salt leaching   Freeze dry   Punching   **Scaffold**

Fig. (6). Schematic illustration of the preparation of scaffold by using salt leaching electrospinning.

6. Salt Leaching Using Powder (SLUP)

Fig. (**7**) shows the preparation of scaffold by using salt leaching using powder. The heat treatment used in several previously described methods to alter salt leaching can be avoided in this method. In the SLUP method, the sieved powder (*e.g.*, Bioglass 46S6) was mixed with a polymer. The salt was then applied homogeneously and stirred continuously for 10-15 minutes. The mixture was then poured into a mold and dried. The salt was then leached from the framework with distilled water for at least two days. The finished scaffold was then allowed to dry for 24 hours. Recently, Refifi *et al*. reported the application of this method to fabricate hybrid biomaterials from bioglass (46S6) and chitosan scaffolds. The fabricated scaffold exhibited high porosity (90%) and well-interconnected pores in the internal networks of the scaffold [29].

Advantage and Disadvantage of Scaffolds Produced by Salt Leaching Technique

Scaffolds produced by the salt leaching technique have several merits and demerits that can be seen in Table **2**.

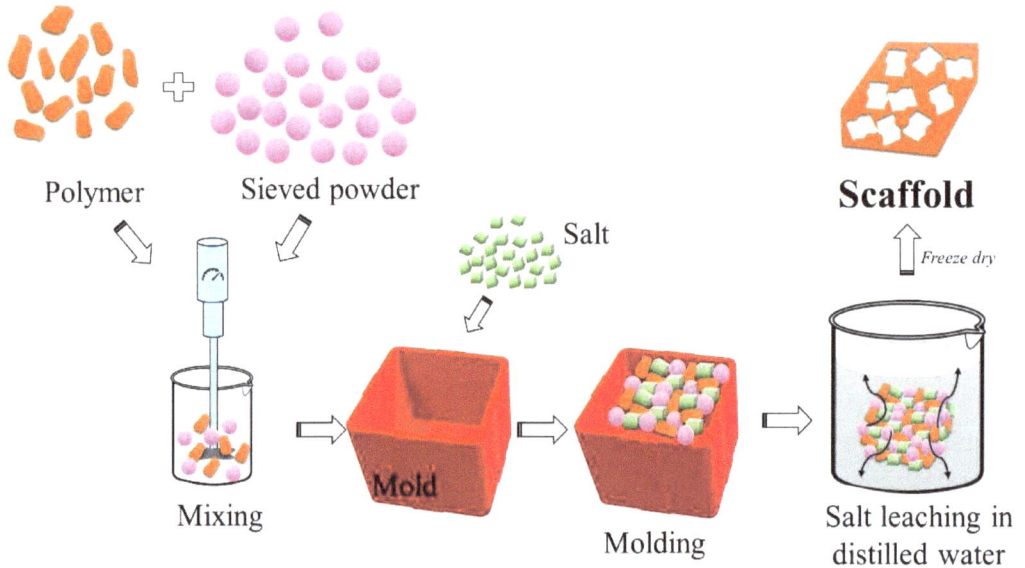

Fig. (7). Schematic illustration of the preparation of scaffold by using salt leaching using powder.

Table 2. Advantage and disadvantage of particulate leaching (salt leaching) technique [5, 30, 31].

Advantage	Disadvantage
Relatively easy technique	Result in Thin Membranes (up to 3 mm Thick)
Porous 3D structure	A brief period of use
Low cost	Limited porous size
High porosity (50%–90%)	Toxicity due to the use of solvent (solvent residue)
Capable of tailoring 3D cell growth by tuning pore size	Time consuming (evaporation of solvent takes days to weeks)
A variety of polymers can be us	Particle entrapment in the polymer matrix prevents the formation of open cell structures
Possibility of controlling porosity and crystallinity	Irregular shaped pores
-	Inter-pore openings are difficult to control

PHYSICAL AND MECHANICAL CHARACTERISTICS

1. Pore Size

A scaffold produced by particulate leaching (salt leaching) was reported to have a thickness of less than 3 mm. The average pore diameter has been reported to be up to 500 μm [32]. The larger porous matrix can be used, for example, to fabricate complex geometric tissue scaffolds such as bone models and ear models [33]. The pores in tissue engineering scaffolds serve several important functions, including facilitating the transfer of metabolites and nutrients for cells during cell development, improving cell visibility and adhesion, and accelerating scaffold degradation. The preferred pore size of scaffolds that provide a suitable structure for tissue engineering is on the microscale [34]. Scaffolds with a pore diameter of 50-300 μm are generally well suited for cell penetration and neovascularization, while smaller pore diameters of 0.5-10 μm are required to facilitate the transport of nutrients and physical stimuli that can stimulate cell function and accelerate healing [35 - 37]. Based on this fact, several studies have proposed the development of a scaffold with a hierarchical porous structure that is more advantageous, especially for tissue engineering. Lao *et al*. fabricated hierarchically porous PLLA/-TCP nanocomposites that promoted MG -63 osteoblast proliferation, penetration, and deposition of ECM, suggesting the possibility of bone tissue engineering [38]. Zhang *et al*. proposed a hierarchical 3D poly(-caprolactone) scaffold (pore size 3-250 μm), which showed potential for the neo-tissue formation and uniform cell distribution in another study [39].

2. Porosity

One of the most important properties of scaffolds for tissue engineering, especially for tissue engineering of cartilage, is their porosity. Porosity serves a number of functions, including ensuring uniform distribution of physiological cells, facilitating the organization of the new extracellular matrix (ECM) formed by cells, and effectively facilitating nutrient supply and waste removal during *in vitro* cell culture [40].

The salt leaching method has been used to produce scaffolds with porosity ranging from 50% to over 90% [41]. It has been reported that porosity greater than 70% is highly recommended for the regeneration of bone tissue, especially cartilage. This allows ECM distribution similar to that in physiological tissue [42]. In 2014, Nasri-Nasrabadi *et al*. discovered that increasing the sodium chloride content leads to increased interconnectivity of the pores. A scaffold containing 90% salt had a higher percentage of porosity than a scaffold containing 70% salt

[21]. However, it has also been reported that higher porosity can lead to poor mechanical properties and is therefore unsuitable for bone tissue engineering [43].

3. Mechanical Properties

The mechanical properties of a prepared scaffold are determined by the host tissue to be regenerated. Therefore, the more similar the mechanical properties are, the more suitable the scaffold is for tissue engineering. It is known that the mechanical strength of a scaffold has a major impact on inter/intracellular signaling, motility, cell ingrowth and response to stimulation or inhibition [44].

4. Interconnected 3D Structures

The interconnectivity of pores in the internal structures of the scaffold is another important feature of tissue engineering. A good or well-connected pore can facilitate cell ingrowth or cell proliferation. If a scaffold has pore structures but they are not interconnected, it is useless and unsuitable for tissue engineering. A scaffold with 100 percent contiguous pore volume can support cell migration and proliferation while enhancing the diffusion and exchange of nutrients within the scaffold's pore networks. ECM infiltration of the target tissue is also supported by the highly interconnected pores [45].

BIOMATERIALS PREPARED WITH PARTICULATE LEACHING TECHNIQUE

A modification of the conventional salt leaching technique was developed to improve the current scaffold design by controlling the parameters affecting pore size, porosity, pore interconnectivity, and pore properties for better and appropriate cell attachment and nutrient diffusion. Table **3** summarizes the latest scaffold-based salt leaching techniques that can be considered for tissue engineering applications. Based on the scaffolds listed in the table, readers can learn about the types of polymers and salts used, potential applications, scaffold properties, and toxicological experiments and results.

Table 3. Biomaterials-based salt leaching techniques and toxicological evaluations.

Biomaterials	Material Component		Porous 3D Scaffold Applications	Outcomes	Toxicological Experiment	Toxicology Result	References
	Polymer/Other Materials	Salt Particle					
Bioglass®45S5-Polycaprolactone (PCL) composite scaffolds	PCL	Sodium chloride (NaCl) + Sodium bicarbonate (NaHCO₃)	Bone regeneration and vascularization	-a high degree of interconnected porosity -excellent mechanical properties -the addition of bio glass had no effect on mechanical properties -simple to use -good cell proliferation -good cytoplasmic continuity	-	-	[19]
Porous HA/nylon 6,6 scaffold	Nylon	NaCl	Bone regeneration	-a well-developed interconnected porosity -pore size is around 200-500µm -ideal for bone regeneration and vascularization -the mechanical strength is comparable to cortical bone	-	-	[20]
Porous starch/cellulose nanofibers composite	Starch + cellulose nanofiber	NaCl	Tissue engineering	-the diameter of nanofibers is between 40-90 nm -interconnected porous morphology -good hydrophilicity -uniform porous structure -good mechanical properties -good biodegradability	MTT assay	No toxicity	[21]
Polylactic acid/hydroxyapatite porous nanocomposites (PLA/HAp)	PLA	Ammonium bicarbonate (NH₄HCO₃)	Bone engineering (bone implant)	-↑porosity by 39% -↑hydrophilicity of nanocomposite -↑surface area capable of improving mechanical properties -after immersing the nanocomposite in simulated body fluid, hydroxyapatite shaped on its surfaces	-	-	[46]
Poros Poly (Lactide-c--glycolic acid) (PLGA) scaffolds	PLGA	Sodium chloride (NaCl)	Pluripotent stem cells and neural retinal precursor cells	-Pore size is less than 10 µm -Scaffolds have amorphous smooth and irregularly shaped pores -cells can attach to the scaffolds and proliferate well -successful early retinal development -beneficial to neural phenotypes	-	-	[47]
Porous Poly(3-hydroxybutyrate-co-3-hydroxyvalerate) (PHBV) scaffolds	PHBV	Sieved sodium chloride (NaCl)	Tissue engineering (particularly bone engineering)	-high porosity (88.8%) -good mechanical properties for 3D cell culture -elevate the proliferation of MG-63 human osteoblast-like cells and MC3T3-E1 pre-osteoblast cells -up-regulated transcription of extracellular matrix and growth factors genes	-	-	[48]

(Table 3) cont.....

Biomaterials	Material Component		Porous 3D Scaffold Applications	Outcomes	Toxicological Experiment	Toxicology Result	References
	Polymer/Other Materials	Salt Particle					
Solvent-casting particulate leaching (SCPL) polymer scaffolds	polymethyl methacrylate (PMMA), Polyurethane (PU)	NaCl	Bone marrow	-well-interconnected and high porosity (82.1vol%–91.3vol%) -easy control of pore size -mechanical properties are determined by the polymer matrix and the architecture of the scaffold -improved 3D stromal cell support -capable of stimulating the bone marrow microenvironment	-	-	[49]
Porous PLGA scaffolds-impregnated small intestinal submucosa (SIS)	PLGA	NaCl particles	Tissue-engineered bio discs	-good mechanical properties -seeding of cells uniformly -good attachment of cells -↑cell growth and ECM synthesis -easy adjustment of pore size and scaffold area for nucleus pulposus regeneration.	-	-	[50]
3D electrospun silk-fibroin nanofiber	Silk fibroin	NaCl crystal	Skin tissue engineering (Skin substitutes)	-larger pores -high interconnectivity between pores -high porosity and water uptake -↑Proliferation of fibroblast in the deep layer -↑keratinocytes differentiation in the superficial layer	-	-	[28]
Porous polycaprolactone (PCL)/elastin composite scaffold	PCL	NaCl	Tissue engineering (bone regeneration)	-high porosity (91%) -large pores size (average 540μm) -high degree of interconnectivity and homogenous 3D scaffolds	-	-	[25]
PCL scaffolds with dual leaching method	PCL	NaCl	Bone tissue engineering	-high interconnectivity of the pores -3D structure -high porosity -high water uptake -uniform pore size -support bone cell proliferation and differentiation in culture media	MTT assay	No toxicity	[51]
Porous poly (lactic acid) (PLA) scaffolds	PLA	Stable salt stack	Tissue engineering	-PLA foams are less crystallizable than the bulk materials -the pores and caves have a diameter of about 250 μm -polymer foam can be crystallized without damaging the structure of scaffolds	-	-	[52]
Poly (ε-caprolactone-r-n-lactic acid) (PCLA-F) scaffolds	PCLA-F	Sieved NaCl	Tissue regeneration	-irregular structure -rapid bulk erosion -the resulting scaffold degrades quickly *in vivo*	-	-	[52]

(Table 3) cont.....

Biomaterials	Material Component		Porous 3D Scaffold Applications	Outcomes	Toxicological Experiment	Toxicology Result	References
	Polymer/Other Materials	Salt Particle					
Composite glass/chitosan (BG-CH) scaffold	CH	NaCl	Tissue engineering	-high porosity -no solvent residue due to SLUP method -↑pore size interconnectivity - three-layer scaffolds with varying porosity and pore size in each layer	-	-	[29]
Porous PCL scaffolds	PCL	NaCl	Bone tissue engineering	-the mean pore size is ~420 µm -high degree of pore interconnectivity -↑adhesion and proliferation of MC3T3-E1 cells	MTT assay	No toxicity	[26]
PLA nanofibrous microporous scaffolds with bioactive glass nanoparticles	PLA	NaCl	Bone tissue engineering	-high porosity with size of ~90 nm -nanofibrous structure with 90-95% porosities -↑hydrophilicity -rapid hydrolytic degradation -↑formation of apatite throughout the scaffolds	-	-	[53]
PCL/starch nanocomposite scaffold	PCL, starch	NaCl	Bone tissue engineering	-uniform pore morphology and structure -good comprehensive-loa--resisting capabilities in human cancellous bone -the porosity ranges from 50 to 90% -↑porosity results in decreased mechanical properties -appropriate pore size and pore interconnectivity -adequate mechanical properties	-	-	[54]
Silk fibroin/hyaluronic acid (SF/HA) 3D matrices	SF, HA	NaCl	Cartilage tissue engineering	-sponge-like structure -high porosity -the porosity is in the range of 61–72% -decrease pore volume due to the presence of HA -high crystallinity -scaffold porosity, micro porosity of the pore wall, and material inhomogeneity all had an effect on water absorption and mechanical strength	-	-	[40]
Biomimetic hybrid porous scaffold of chitosan/polyvinyl alcohol/carboxymethyl cellulose (CH/PVA/CMC)	CH, PVA, CMC	NaCl	Soft tissue engineering	-a pore structure that is uniformly distributed and interconnected -the pore size is 13.6 µm to 15.5 µm -the porosity is 90% -addition of CMC increased hydrophilicity -good mechanical strength -facilitate cell migration and proliferation	-	-	[55]
Polylactic acid porous scaffolds	PLA	NaCl, CaCl$_2$	Tissue engineering	-highly interconnected pores -appropriate mechanical properties	MTT assay	No toxicity	[23]

(Table 3) cont.....

Biomaterials	Material Component		Porous 3D Scaffold Applications	Outcomes	Toxicological Experiment	Toxicology Result	References
	Polymer/Other Materials	Salt Particle					
Cell-loaded gelatin/chitosan scaffolds	Gelatin/CH	NaCl	Skin tissue engineering	-high porosity -uniform pores morphology inside and on the surface of the scaffold -highly interconnected pores -support cell proliferation -appropriate mechanical properties -great potential for skin tissue engineering	MTT assay	No toxicity	[56]
Sericin/PVA/glycerin scaffolds	Silk Sericin, PVA, Glycerin	NaCl	Wound dressing	-large porous structure -good interconnectivity structure -fast biodegradation rate -less adhesive to the wound site -support fibroblast proliferation	MTT assay	No toxicity	[57]
Nanocomposite scaffold-releasing silver (Ag)	Neat polymer (Poly (L-lactide) + Ag	Sodium bicarbonate (NaHCO$_3$)	Bone treatment	-good-interconnected structures -pore size is in the range of 100-250 μm -high porosity -the scaffold induced an apatite layer on its surface	-	-	[27]
PLLA/β-TCP nanocomposite scaffolds	PLLA	NaCl	Bone tissue engineering	-hierarchical porosity -500 nm to 300 μm pore diameter -PLLA nanofibrous matrix with fiber diameters ranging from 70-300 nm -good mechanical properties -↑bioactivity of the PLLA matrix -support cells proliferation and penetration, as well as ECM deposition	-	-	[38]

POROUS 3D SCAFFOLD APPLICATIONS WITH SALT LEACHING TECHNIQUE

Bone Engineering

The new gold standard for the treatment of bone defects (*e.g.*, autologous bone grafts) is limited by the lack of donor sources, the body's response to the graft, and, most importantly, the patient's resources [58]. The advancement of tissue engineering technology can overcome this limitation. In the tissue engineering concept, the scaffold of biodegradable polymers combined with autologous cells is a promising strategy to obtain an ideal bone graft. The patient's cells were isolated and grown *in vitro* before being seeded into 3D scaffolds. After successfully culturing the cells to a specific cell number *in vitro*, the cells were transplanted to the patients. The development of native tissue is then followed by biodegradation of the scaffold [59]. In this method, newly formed tissue from patients would be used to replace the implanted scaffold and eventually treat the bone defects.

Scaffolds for bone implants must have a variety of properties, including compatibility, mechanical strength and stability, and a highly open porous structure that allows tissue regeneration. Careful consideration must also be given to the size and distribution of pores for new cell growth [60 - 62].

Neuronal Retinal Precursor Cells

Stem cells are a viable treatment option for patients suffering from neurodegenerative diseases or age-related macular degeneration, which is characterized by the death of light-sensitive photoreceptor cells in the outer neuronal retina. However, there are still limitations with this approach, such as the minimal number of surviving and integrated cells after subretinal bolus injection in patients, especially in patients with end-stage disease. The main reason is that the donor cells do not receive physical support during bolus injection [63, 64]. In this case, making a scaffold to support the donor cells during transplantation could solve the stem cell problem. Several porous PLGA scaffolds have been reported to improve the survival rates of retinal progenitor cells (RPCs) after transplantation and even promote the integration of the cells into the host retina [65, 66]. The salt leaching scaffold has also been used in the treatment of retinal degeneration as a cell replacement therapy. In 2016, Worthington *et al.* developed PLGA scaffolds using a salt leaching/solvent evaporation process. The resulting scaffold product was shown to have the ability to support and facilitate the proliferation of induced pluripotent stem cells (iPSCs) within the amorphous smooth pore network of the scaffold. Therefore, this scaffold represents a promising approach for the treatment of retinal degeneration prior to transplantation [47].

Skin Substitutes

Skin tissue engineering has emerged as a new and promising strategy for wound healing because autograft implantation is associated with limitations, such as the occurrence of infection, pain, slow healing at the donor site, scarring, and major skin loss. It must have appropriate physical, biological, and mechanical properties for nutrient and gas exchange, just like scaffolds for another tissue engineering [67]. An example of skin tissue engineering is the regeneration of the skin layer by culturing fibroblast cells on polymeric scaffolds that mimic ECM to accelerate the healing of skin wounds [68]. Park *et al.* developed a skin substitute using 3D electrospun silk fibroin nanofiber scaffolds. NIH 3T3 fibroblast cells were successfully cultured in silk fibroin scaffolds, and the metabolic activity of the cells within the pore structure of the scaffold increased significantly from day 1 to day 14. The salt leaching technique was also used in combination with electrospinning to artificially create larger pores for better cell adhesion and nutrient and gas exchange [28]. Thus, the scaffold produced by the salt leaching technique has great potential for tissue engineering of the skin.

FUTURE PROSPECT

Future research should focus on modifying the salt leaching technique to improve the physical properties of the scaffold for better application or efficiency in tissue engineering. As mentioned earlier, mechanical strength, pore size, porosity, and interconnectivity are important properties of the scaffold to provide a desirable platform for tissue engineering. According to Cheng *et al.*, the incorporation of two-dimensional nanomaterials into silk fibroin-based scaffolds enables the delivery of chondrogenic growth factors to the target site of cartilage regeneration [69].

In addition, further studies should be conducted to completely remove the organic solvent from the scaffold without affecting the pore architecture of the scaffold. Reportedly, one of the alternative methods to remove residual solvent from biodegradable polymers is carbon dioxide extraction. However, this method had an impact on the pore architecture of the scaffold matrix [70]. Future research needs to address the potential toxicity of the remaining organic solvent without affecting the physical and mechanical properties of the scaffold by broadening the choice of polymers and solvents.

CONCLUDING REMARK

The purpose of this chapter was to describe how particulate leaching (salt leaching) is used to prepare biomaterials for tissue engineering, as well as new discoveries in the methods and polymers used. The salt leaching method is undoubtedly the most widely used and relatively simple method for preparing scaffolds for tissue engineering applications such as bone regeneration, skin replacement, and neural retinal progenitor cells. However, because this method uses organic solvents, the risk of these scaffolds being toxic due to solvent residues must be carefully considered. For example, the solvents (hexafluoroacetone and hexafluoroisopropanol) used to dissolve PGA are both extremely toxic to cells. Many of the examples discussed previously, which were from recent studies, lacked an assessment of toxicity. Therefore, before further recommendations can be made regarding their potential uses and applications in tissue engineering, scaffolds prepared using salt leaching techniques need to be evaluated for toxicology, safety risks, and potential effects on damaged tissue.

CONSENT FOR PUBLICATION

Not applicable.

CONFLICT OF INTEREST

The author declares no conflict of interest, financial or otherwise.

ACKNOWLEDGEMENT

Declared none.

REFERENCES

[1]   W.M. Saltzman, and W.L.J.N.R.D.D. Olbricht, *Building drug delivery into tissue engineering design*, 2002.
        [http://dx.doi.org/10.1038/nrd744]

[2]   N. Sultana, M.I. Hassan, and M.M. Lim, "Scaffolding Biomaterials", In: *Composite Synthetic Scaffolds for Tissue Engineering and Regenerative Medicine.* Springer, 2015, pp. 1-11.

[3]   D.V. Bax, D.R. McKenzie, A.S. Weiss, and M.M.J.B. Bilek, *The linker-free covalent attachment of collagen to plasma immersion ion implantation treated polytetrafluoroethylene and subsequent cell-binding activity*, 2010.
        [http://dx.doi.org/10.1016/j.biomaterials.2009.12.009]

[4]   N. Annabi, *Controlling the porosity and microarchitecture of hydrogels for tissue engineering*, 2010.
        [http://dx.doi.org/10.1089/ten.teb.2009.0639]

[5]   B. Subia, J. Kundu, and S.J.T.e. Kundu, *Biomaterial scaffold fabrication techniques for potential tissue engineering applications.* vol. Vol. 141. , 2010.
        [http://dx.doi.org/10.5772/8581]

[6]   B. Guo, and P.X.J.B. Ma, *Conducting polymers for tissue engineering*, 2018.
        [http://dx.doi.org/10.1021/acs.biomac.8b00276]

[7]   J.B. Park, *Biomaterials science and engineering.* Springer Science & Business Media, 2012.

[8]   C. Simões, J. Viana, and A.J.J.A.P.S. Cunha, *Mechanical properties of poly (ε-caprolactone) and poly (lactic acid) blends*, 2009.
        [http://dx.doi.org/10.1002/app.29425]

[9]   G. Chen, T. Ushida, and T. Tateishi, *A biodegradable hybrid sponge nested with collagen microsponges.* vol. 51. , 2000, no. 2, pp. 273-279.

[10]   A.C.J.P.h. Kuo, *Poly (dimethylsiloxane)*, 1999.

[11]   H. Kweon, H.C. Ha, I.C. Um, and Y.H.J.J.s. Park, *Physical properties of silk fibroin/chitosan blend films*, 2001.
        [http://dx.doi.org/10.1002/app.1172]

[12]   D. Domene-López, J. C. García-Quesada, I. Martin-Gullon, and M. G. J. P. Montalbán, *Influence of starch composition and molecular weight on physicochemical properties of biodegradable films.*, vol. 11, p. 1084, 2019.
        [http://dx.doi.org/10.3390/polym11071084]

[13]   A.N. Nakagaito, H. Takagi, and J.K.J.I.J.O.S.E. Pandey, *The processing and mechanical performance of cellulose nanofiber-based composites*, 2011.
        [http://dx.doi.org/10.5574/IJOSE.2011.1.4.180]

[14]   J.J.A.C. Warwicker, *The crystal structure of silk fibroin*, 1954.
        [http://dx.doi.org/10.1107/S0365110X54001867]

[15]   I. Korbag, and S.J.I.J.E.S. Mohamed Saleh, *Studies on mechanical and biodegradability properties of PVA/lignin blend films*, 2016.

[http://dx.doi.org/10.1080/00207233.2015.1082249]

[16]   F. Feng, and L.J.J.A.P.S. Ye, *Morphologies and mechanical properties of polylactide/thermoplastic polyurethane elastomer blends*, 2011.
[http://dx.doi.org/10.1002/app.32863]

[17]   J. Park, and R.S. Lakes, *Biomaterials: an introduction.* Springer Science & Business Media, 2007.

[18]   E. Wondu, Z. Lule, and J. J. P. Kim, Thermal conductivity and mechanical properties of thermoplastic polyurethane-/silane-modified Al_2O_3 composite fabricated *via* melt compounding,
[http://dx.doi.org/10.3390/polym11071103]

[19]   V. Cannillo, F. Chiellini, P. Fabbri, and A.J.C.S. Sola, *Production of Bioglass® 45S5–Polycaprolactone composite scaffolds via salt-leaching*, 2010.
[http://dx.doi.org/10.1016/j.compstruct.2010.01.017]

[20]   M. Mehrabanian, and M. J. I. j. o. n. Nasr-Esfahani, *HA/nylon 6, 6 porous scaffolds fabricated by salt-leaching/solvent casting technique: effect of nano-sized filler content on scaffold properties.*, vol. 6, p. 1651, 2011.

[21]   B. Nasri-Nasrabadi, M. Mehrasa, M. Rafienia, S. Bonakdar, T. Behzad, and S.J.C.p. Gavanji, *Porous starch/cellulose nanofibers composite prepared by salt leaching technique for tissue engineering*, 2014.
[http://dx.doi.org/10.1016/j.carbpol.2014.02.075]

[22]   D.Y. Kwon, J.Y. Park, B.Y. Lee, and M.S.J.P. Kim, "Comparison of Scaffolds Fabricated via 3D Printing and Salt Leaching: *in vivo*", In: *Imaging, Biodegradation, and Inflammation* vol. 12. , 2020, no. 10, p. 2210.
[http://dx.doi.org/10.3390/polym12102210]

[23]   R. Scaffaro, F. Sutera, and F. J. M. Lopresti, *Using Taguchi method for the optimization of processing variables to prepare porous scaffolds by combined melt mixing/particulate leaching.*, vol. 131, pp. 334-342, 2017.

[24]   J. Zhang, S-G. Yang, J-X. Ding, and Z-M.J.R.a. Li, *Tailor-made poly (l-lactide)/poly (lactide-c--glycolide)/hydroxyapatite composite scaffolds prepared via high-pressure compression molding/salt leaching*, 2016.
[http://dx.doi.org/10.1039/C6RA06906A]

[25]   N. Annabi, A. Fathi, S.M. Mithieux, A.S. Weiss, and F.J.T.J.S.F. Dehghani, *Fabrication of porous PCL/elastin composite scaffolds for tissue engineering applications*, 2011.
[http://dx.doi.org/10.1016/j.supflu.2011.06.010]

[26]   T-Y. Bak, M-S. Kook, S-C. Jung, and B-H.J.J.n. Kim, "Biological effect of gas plasma treatment on CO_2 gas foaming/salt leaching fabricated porous polycaprolactone scaffolds in bone tissue engineering".,

[27]   A. M. El-Kady, R. A. Rizk, B. M. Abd El-Hady, M. W. Shafaa, and M. M. J. J. o. G. E. Ahmed, *Characterization, and antibacterial properties of novel silver releasing nanocomposite scaffolds fabricated by the gas foaming/salt-leaching technique.*, vol. 10, pp. 229-238, 2012.

[28]   Y.R. Park, *Three-dimensional electrospun silk-fibroin nanofiber for skin tissue engineering.* vol. Vol. 93. , 2016, pp. 1567-1574.

[29]   J. Refifi, H. Oudadesse, O. Merdrignac-Conanec, H. El Feki, and B. J. J. o. t. A. C. S. Lefeuvre, *Salt leaching using powder (SLUP) process for glass/chitosan scaffold elaboration for biomaterial applications.*, 2020.
[http://dx.doi.org/10.1007/s41779-020-00460-6]

[30]   J. Sanz-Herrera, J. García-Aznar, and M.J.A.B. Doblaré, *On scaffold designing for bone regeneration: a computational multiscale approach*, 2009.

[31]   P. Agrawal, *Bioadhesive micelles of d-α-tocopherol polyethylene glycol succinate 1000: synergism of*

chitosan and transferrin in targeted drug delivery, 2017.

[32] A.G. Mikos, G. Sarakinos, J.P. Vacanti, R.S. Langer, and L.G. Cima, *Biocompatible polymer membranes and methods of preparation of three dimensional membrane structures.*, G. Patents, Ed., , 1996.

[33] C.J. Liao, *Fabrication of porous biodegradable polymer scaffolds using a solvent merging/particulate leaching method*, 2002.
[http://dx.doi.org/10.1002/jbm.10030]

[34] L-P. Yan, J.M. Oliveira, A.L. Oliveira, S.G. Caridade, J.F. Mano, and R.L.J.A.b. Reis, *Macro/microporous silk fibroin scaffolds with potential for articular cartilage and meniscus tissue engineering applications*, 2012.
[http://dx.doi.org/10.1016/j.actbio.2011.09.037]

[35] A.R. Studart, J. Studer, L. Xu, K. Yoon, H.C. Shum, and D.A.J.L. Weitz, *Hierarchical porous materials made by drying complex suspensions*, 2011.
[http://dx.doi.org/10.1021/la103995g]

[36] D. Khang, *Role of subnano-, nano-and submicron-surface features on osteoblast differentiation of bone marrow mesenchymal stem cells*, 2012.
[http://dx.doi.org/10.1016/j.biomaterials.2012.05.005]

[37] D. J. M. S. Yoo, *New paradigms in hierarchical porous scaffold design for tissue engineering.*, vol. 33, no. 3, pp. 1759-1772, 2013.

[38] T. Lou, X. Wang, G. Song, Z. Gu, and Z.J.I.m. Yang, *Fabrication of PLLA/β-TCP nanocomposite scaffolds with hierarchical porosity for bone tissue engineering*, 2014.
[http://dx.doi.org/10.1016/j.ijbiomac.2014.06.004]

[39] Q. Zhang, *Effect of porosity on long-term degradation of poly (ε-caprolactone) scaffolds and their cellular response*, 2013.
[http://dx.doi.org/10.1016/j.polymdegradstab.2012.10.008]

[40] C. Foss, E. Merzari, C. Migliaresi, and A.J.B. Motta, *Silk fibroin/hyaluronic acid 3D matrices for cartilage tissue engineering*, 2013.
[http://dx.doi.org/10.1021/bm301174x]

[41] Z. Li, *Recent progress in tissue engineering and regenerative medicine*, 2016.
[http://dx.doi.org/10.1166/jbt.2016.1510]

[42] J.S. Temenoff, and A.G.J.B. Mikos, *Tissue engineering for regeneration of articular cartilage*, 2000.
[http://dx.doi.org/10.1016/S0142-9612(99)00213-6]

[43] F. Shokrolahi, H. Mirzadeh, H. Yeganeh, and M. Daliri, *Fabrication of poly (urethane urea)-based scaffolds for bone tissue engineering by a combined strategy of using compression moulding and particulate leaching methods.*, 2011.

[44] G. Rijal, and W.J.B. Li, *3D scaffolds in breast cancer research*, 2016.
[http://dx.doi.org/10.1016/j.biomaterials.2015.12.016]

[45] D.W. Hutmacher, T.B. Woodfield, and P.D. Dalton, "Scaffold design and fabrication", In: *Tissue engineering.* Elsevier, 2014, pp. 311-346.
[http://dx.doi.org/10.1016/B978-0-12-420145-3.00010-9]

[46] D.T.M. Thanh, *Effects of porogen on structure and properties of poly lactic acid/hydroxyapatite nanocomposites (PLA/HAp).*, 2016.
[http://dx.doi.org/10.1166/jnn.2016.12032]

[47] K. S. Worthington, L. A. Wiley, C. A. Guymon, A. K. Salem, and B. A. J. J. o. O. P. Tucker, *Differentiation of induced pluripotent stem cells to neural retinal precursor cells on porous poly-lactic-co-glycolic acid scaffolds.*, vol. 32, pp. 310-316, 2016.

[48] L. Xia, *Icariin delivery porous PHBV scaffolds for promoting osteoblast expansion in vitro*, 2013.

[http://dx.doi.org/10.1016/j.msec.2013.04.050]

[49] A. Sola, *Development of solvent-casting particulate leaching (SCPL) polymer scaffolds as improved three-dimensional supports to mimic the bone marrow niche*, 2019.
[http://dx.doi.org/10.1016/j.msec.2018.10.086]

[50] S. H. Kim, J. E. Song, D. Lee, and G. J. J. o. t. e. Khang, *Development of poly (lactide-co-glycolide) scaffold impregnated small intestinal submucosa with pores that stimulate extracellular matrix production in disc regeneration.*, vol. 8, no. 4, pp. 279-290, 2014.

[51] N. Thadavirul, P. Pavasant, and P.J.J.B.M.R.P.A. Supaphol, *Development of polycaprolactone porous scaffolds by combining solvent casting, particulate leaching, and polymer leaching techniques for bone tissue engineering*, 2014.
[http://dx.doi.org/10.1002/jbm.a.35010]

[52] R. Huang, X. Zhu, H. Tu, and A.J.M.L. Wan, *The crystallization behavior of porous poly (lactic acid) prepared by modified solvent casting/particulate leaching technique for potential use of tissue engineering scaffold*, 2014.
[http://dx.doi.org/10.1016/j.matlet.2014.08.044]

[53] J.-J. Kim, S.-H. Bang, A. El-Fiqi, and H.-W. J. M. C. Kim, and Physics, *Fabrication of nanofibrous macroporous scaffolds of poly (lactic acid) incorporating bioactive glass nanoparticles by camphene-assisted phase separation.*, vol. 143, no. 3, pp. 1092-1101, 2014.

[54] S. Taherkhani, and F.J.J.A.P.S. Moztarzadeh, *Fabrication of a poly (ε-caprolactone)/starch nanocomposite scaffold with a solvent-casting/salt-leaching technique for bone tissue engineering applications*, 2016.
[http://dx.doi.org/10.1002/app.43523]

[55] K. Kanimozhi, S. K. Basha, V. S. Kumari, and K. J. J. o. n. Kaviyarasu, Kaviyarasu, and nanotechnology, *Development of biomimetic hybrid porous scaffold of chitosan/polyvinyl alcohol/carboxymethyl cellulose by freeze-dried and salt leached technique.*, vol. 18, no. 7, pp. 4916-4922, 2018.

[56] M. Pezeshki-Modaress, *Cell-loaded gelatin/chitosan scaffolds fabricated by salt-leaching/lyophilization for skin tissue engineering: in vitro and in vivo study*, 2014.

[57] P. Aramwit, J. Ratanavaraporn, S. Ekgasit, D. Tongsakul, and N.J.J.B.M.R.P.B.A.B. Bang, *A green salt-leaching technique to produce sericin/PVA/glycerin scaffolds with distinguished characteristics for wound-dressing applications*, 2015.
[http://dx.doi.org/10.1002/jbm.b.33264]

[58] M. Cieślik, *Parylene coatings on stainless steel 316L surface for medical applications—Mechanical and protective properties*, 2012.
[http://dx.doi.org/10.1016/j.msec.2011.09.007]

[59] M.S. Chapekar, "Tissue engineering: challenges and opportunities", *J Biomed Mater Res,* vol. 53, no. 6, pp. 617-620.

[60] S.C. Rizzi, *Biodegradable polymer/hydroxyapatite composites: surface analysis and initial attachment of human osteoblasts*, 2001.

[61] M.J.B. Wang, *Developing bioactive composite materials for tissue replacement*, 2003.
[http://dx.doi.org/10.1016/S0142-9612(03)00037-1]

[62] V. Karageorgiou, and D.J.B. Kaplan, *Porosity of 3D biomaterial scaffolds and osteogenesis*, 2005.
[http://dx.doi.org/10.1016/j.biomaterials.2005.02.002]

[63] A.C. Barber, *Repair of the degenerate retina by photoreceptor transplantation*, 2013.
[http://dx.doi.org/10.1073/pnas.1212677110]

[64] H. Klassen, D. S. Sakaguchi, and M. J. J. P. i. r. Young, Young, and e. research, *Stem cells and retinal*

repair., vol. 23, no. 2, pp. 149-181, 2004.

[65]   B.A. Tucker, *The use of progenitor cell/biodegradable MMP2–PLGA polymer constructs to enhance cellular integration and retinal repopulation,* 2010.
[http://dx.doi.org/10.1016/j.biomaterials.2009.09.015]

[66]   M. Tomita, E. Lavik, H. Klassen, T. Zahir, R. Langer, and M.J.J.S.c. Young, *Biodegradable polymer composite grafts promote the survival and differentiation of retinal progenitor cells,* 2005.
[http://dx.doi.org/10.1634/stemcells.2005-0111]

[67]   M.P. Modaress, H. Mirzadeh, and M.J.I.P.J. Zandi, *Fabrication of a porous wall and higher interconnectivity scaffold comprising gelatin/chitosan via combination of salt-leaching and lyophilization methods,* 2012.

[68]   K.A. Corin, and L.J.J.B. Gibson, *Cell contraction forces in scaffolds with varying pore size and cell density,* 2010.
[http://dx.doi.org/10.1016/j.biomaterials.2010.01.149]

[69]   G. Cheng, *Advanced silk fibroin biomaterials for cartilage regeneration,* 2018.
[http://dx.doi.org/10.1021/acsbiomaterials.8b00150]

[70]   W.S. Koegler, C. Patrick, M.J. Cima, and L.G.J.J.B.M.R.A.O.J.T.S.B. Griffith, The Japanese Society for Biomaterials, T. A. S. f. Biomaterials, and t. K. S. f. Biomaterials., *Carbon dioxide extraction of residual chloroform from biodegradable polymers.,* vol. 63, no. 5, pp. 567-576, 2002.

CHAPTER 9

Principles of Supra Molecular Self Assembly and Use of Fiber mesh Scaffolds in the Fabrication of Biomaterials

Haseeb Ahsan[1,2], Salman Ul Islam[1,3], Muhammad Bilal Ahmed[1], Adeeb Shehzad[4], Mazhar Ul Islam[5], Young Sup Lee[1] and Jong Kyung Sonn[1,*]

[1] *School of Life Sciences, College of Natural Sciences, Kyungpook National University, 41566, Korea*

[2] *Department of Pharmacy, Faculty of Life and Environmental Sciences, University of Peshawar 25120, Khyber Pakhtunkhwa, Pakistan*

[3] *Department of Pharmacy, CECOS University, Peshawar, Pakistan*

[4] *Department of Biomedical Engineering & Sciences, School of Mechanical and Manufacturing Engineering, National University of Science & Technology, Islamabad, Pakistan*

[5] *Department of Chemical Engineering, College of Engineering, Dhofar University, Salalah, Oman*

Abstract: Tissue engineering techniques aim to create a natural tissue architecture using biomaterials that have all the histological and physiological properties of human cells to replace or regenerate damaged tissue or organs. Nanotechnology is on the rise and expanding to all fields of science, including engineering, medicine, diagnostics and therapeutics. Nanostructures (biomaterials) specifically designed to mimic the physiological signals of the cellular/extracellular environment may prove to be indispensable tools in regenerative medicine and tissue engineering. In this chapter, we have discussed biomaterial design from two different perspectives. Supramolecular self-assembly is the bottom-up approach to biomaterials design that takes advantage of all the forces and interactions present in biomolecules and are responsible for their functional organization. This approach has the potential for one of the greatest breakthroughs in tissue engineering technology because it mimics the natural, complex process of coiling and folding biomolecules. In contrast, a fiber mesh scaffold is a top-down approach in which cells are seeded. The scaffolds form the cellular scaffold while the cells produce and release the desired chemical messengers to support the regeneration process. Therefore, both techniques, if efficiently explored, may lead to the development of ideal biomaterials produced by self-assembly or by the fabrication of optimal scaffolds with long shelf life and minimal adverse reactions.

* **Corresponding author Jong Kyung Sonn:** School of Life Sciences, College of Natural Sciences, Kyungpook National University, 41566, Korea; Tel: 0539507368; E-mail: sonnjk@knu.ac.kr

Keywords: Carbon nanotubes, Chitosan, Hydrogen bonding, Peptide amphiphiles, Polycaprolactone, Regenerative medicine, Tissue engineering, Self-assembly.

INTRODUCTION TO SELF ASSEMBLY

In the recent past, nanotechnology has emerged as a potential area for the development of advanced, innovative techniques in various fields, including tissue engineering and regenerative medicine. Recent studies in nanomedicine have focused on its application in the production of biomaterials. To this end, nanotech-based biomaterials are being developed and intensively studied for their safety, efficacy, and long- and short-term effects on the human body. Nanofibers and nanotubes have been described in many studies as vehicles for drug delivery. Nanostructures specifically designed to mimic the physiological signals of the natural cellular and extracellular environment may prove to be indispensable tools in regenerative medicine.

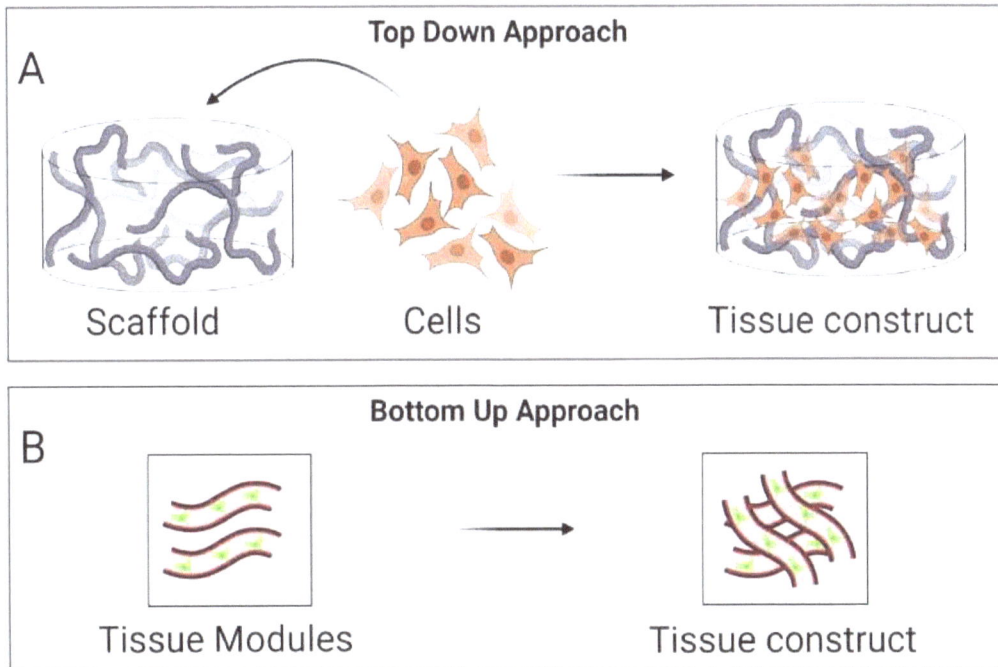

Fig. (1). Approaches for tissue engineering in regenerative medicine **A.** Traditional scaffold-based top-down approach where cells are seeded into fully formed porous scaffolds **B.** Recent bottom-up approach which involves cellular seeding in self-assembling tissue modules, capable of forming a complex three-dimensional network.

Traditionally, regenerative tissue engineering has used the top-down approach, in which the desired cells are incorporated into a scaffold in which they proliferate and differentiate into the desired tissue/organ while supported by the scaffold material (Fig. **1**). This method has some weaknesses, such as the difficulty in constructing complex vital organs with intricate architecture, such as liver and kidneys. To overcome these drawbacks, tissue engineering scientists have explored the relevance and feasibility of other approaches. One of the mechanisms used to produce such biomaterials is the bottom-up approach of self-assembly (Fig. **1B**). In this approach, cells are incorporated into modules that can spontaneously fold and form complex scaffolds. The tendency toward self-assembly is driven by the need for molecules/modules to achieve thermodynamic stability [1, 2]. The design of complex nanostructures by supramolecular self-assembly of simple biological/synthetic building blocks is one of the attractive mechanisms for the fabrication of biomaterials for various applications in biomedical sciences [3 - 5].

Self-assembly is a natural phenomenon that leads to the formation of complex macromolecules. Understanding the principles of self-assembly of natural molecules has greatly helped us in the synthesis of biomaterials using the same bottom-up approach. Molecular and supramolecular self-assembly is a spontaneous process driven by various interactions of chemical entities (charge, size, orientation, bonds) that are in close proximity to each other. The forces underlying the phenomena of self-assembly are weak (non-covalent) forces that come into play when the distance between molecules is reduced. These forces include hydrophobic interactions, weak Van der Waals forces, electrostatic interactions between dipoles, ion-dipole interactions, and hydrogen bonding (Fig. **2**). Although these forces are weak individual forces, they are collectively responsible for the formation of the unique, intricate three-dimensional biological structures with varying complexity and multiple levels of 2-organisation (Fig. **3**) [6 - 8].

Molecular Forces Responsible for Self-Assembly

Electrostatic Forces

Most macromolecules carry functional groups with charged moieties (polar groups in side chains of amino acids). The interaction between such charged groups of a macromolecule generates electrostatic attraction/repulsion, which leads to the folding of the macromolecule into supramolecular structures (ion-ion interaction, ion-dipole interaction, and dipole-dipole interaction). The self-assembly triggered by such interactions is found in polypeptides and lipids [9, 10].

① Ion-Ion interaction (250 kJ/mol)

Repulsive Forces

┌ Non-directional forces

├ Long range interaction (1/r)

└ Highly dependent on the dielectric
  constant of the medium

Attractive Forces

Na + Cl → IONIC LATTICE

② Ion-Dipole interaction 50-250 (kJ/mol) and Dipole-Dipole interaction (5.50 kJ/mol)

Repulsive Forces

┌ Non-directional forces

├ Medium range interaction ($1/r^2$)

└ Significantly weaker than ion-ion
  interactions

Attractive Forces

Repulsive Forces

┌ Somewhat directional forces

├ Short range interaction ($1/r^3$)

├ Occurs between molecules that
  have net dipoles (polar molecules)

└ Significantly weaker than ion-dipole
  interactions

Attractive Forces

③ Hydrphobic effects (Difficult to assess the interaction strength)

┌ The tendency of hydrocarbons to form
│ intermolecular aggregates in an aqueous
│ medium and analogous intramolecular
│ interactions

└ Repulsion of solute by the solvent
  (non-polar molecules tend to avoid aqueous
  surroundings)

Folding of proteins

Aggregation of
amphiphiles
into micelles

④ π-π-interaction (0-50 KJ/mol)

Weak electroststic
interactions between aromatic
rings

Sandwich   T-shaped   Parallel-dispaced

Carbon-nanotubes

⑤ Hydrogen bonding (10-50 KJ/mol)

Pyrimidines: G Guanine C Cytosine

H-bond

Complementary nucleobase pairing

→ Directional Force

→ Short range interaction (2.5-3.5A)

→ Forms when a hydrogen atom is positioned between two
  electronegative atoms, mainly O and N.

→ Special case of dipole-dipole interactions,but can be
  significantly stronger than typical dipole-dipole.

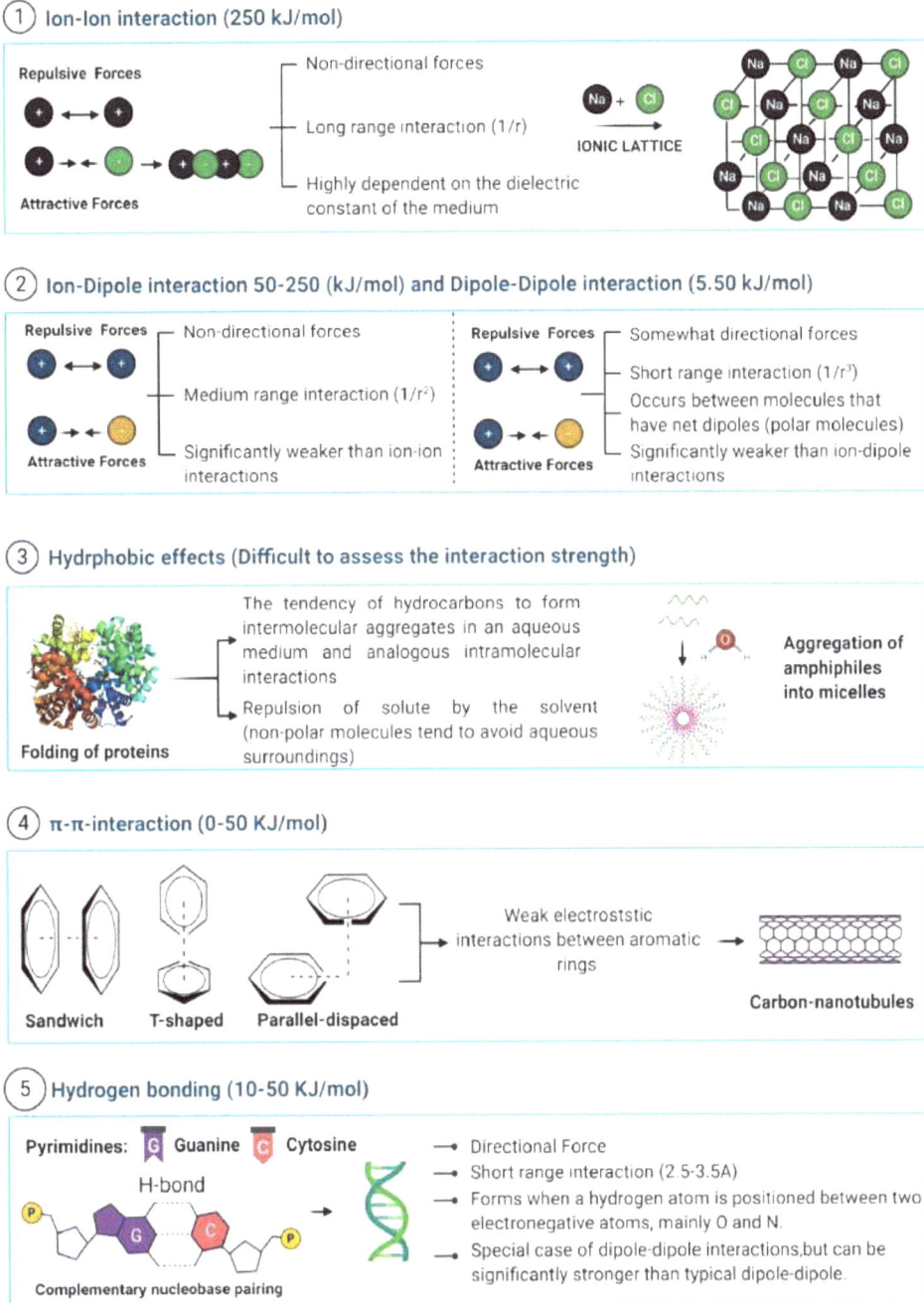

Fig. (2). Non-covalent interactions responsible for supra molecular self-assembly.

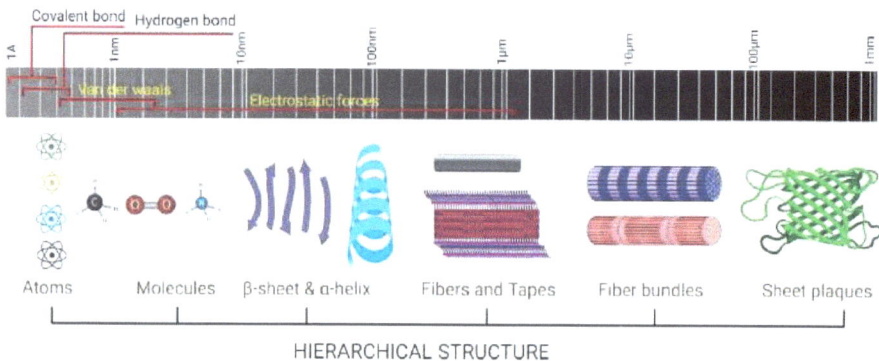

Fig. (3). Various levels of supramolecular organization in biological structures formed by self-assembly.

Hydrophobic Interactions

Macromolecules carrying certain exclusively nonpolar parts in their structure tend to move away from the aqueous media in which they are dispersed. The nonpolar parts of the molecule interact with each other to form a hydrophobic core. Thus, the self-assembly triggered by these forces creates a three-dimensional structure in which the core of the molecule is concentrated with hydrophobic groups, while the peripheral regions have polar groups oriented toward interaction with the aqueous environment of the cell or system. Such folding of the molecule reduces the surface free energy and makes it thermodynamically stable [11, 12].

Aromatic Stacking (Pi-Pi Stacking)

These are non-covalent interactions found in molecules with multiple aromatic ring systems. Pi bonds are present in the conjugated aromatic ring system. Pi stacking is inherently attractive and leads to the stacking of aromatic rings in the sandwich, T-shaped, or parallel-shifted configurations. These interactions are important for peptide self-assembly because many aromatic amino acids (phenylalanine, tyrosine) occur repeatedly in proteins. These interactions have been reported to facilitate the stability of DNA and the tertiary organizational structure of proteins [13 - 17].

Hydrogen Bonding

This is a non-covalent interaction between an electronegative atom of one compound with the hydrogen atom of another compound. In this other compound, the hydrogen atom is also covalently bonded to an electronegative atom. In water molecules, there are extensive intermolecular hydrogen bonds. It is one of the most important forces that organize and fold into supramolecular structures. These forces are directional and can lead to the folding of a single macromolecular chain by intramolecular hydrogen bonds or to the folding of multiple chains by intermolecular hydrogen bonds.

In nature, hydrogen bonding is the main force responsible for organizing and maintaining the secondary structure of proteins/polypeptides (earlier stages of self-assembly). Even the less common hydrogen bonds, known as carbon-hydrogen-oxygen bonds, have been studied because of their significant role in enzymatic catalysis and protein structure [18]. Kuhn and colleagues have shown that intramolecular hydrogen bonds form a temporary ring system that creates more polar and less polar ends in a molecule. These interactions correlate strongly with the binding affinity of the ligand (biomaterials) with its specific receptor. If this intramolecular hydrogen bonding is absent, the affinity of the ligand/biomaterial to its cellular receptor is lower. Therefore, the synthesis of biomaterials through such interactions could be a potential breakthrough in the development of new drugs and drug delivery systems [19]. Many studies have linked hydrogen bonds to the stability of β-sheets. The integration of several such bonds leads to functionally stable biological molecules as well as biomaterials for use in broad areas of applied science [20 - 24].

DESIGNING SUPRAMOLECULAR BIOMATERIALS

Designing and synthesizing biomaterials by self-assembly is a simple and inexpensive technique [25]. The structural design and shape of biomaterials can be manipulated by external factors, such as the pH of the medium, which can transform an existing molecule into a newly formed biomaterial with slightly or completely different properties from its precursor [26]. Table **1** illustrates the number of building blocks that can be used to form a variety of self-assembled biomaterials with different medical and pharmaceutical applications. The interaction between the different building blocks involves numerous non-covalent interactions to maintain the stable conformational structure. The extent of self-assembly depends on the size, solubility, components, and response of the monomeric building blocks to external stimuli. These simpler forms aggregate into a variety of organized structures such as nanotubes, dendritic molecules, and amphiphilic peptides [27].

Dendritic Molecules

Dendritic molecules (dendrimers and dendrons) are highly symmetric, spherical compounds with a low or high molecular weight that can modify the surface properties of biomaterials. They can serve as good drug carriers with better pharmacokinetic profiles than other drug encapsulation systems [28 - 31].

Table 1. Classification and applications of Self-assembling building blocks.

Building Blocks			Supramolecular Assemblies	Applications	Refs.
Synthetic	**Polymers**	Linear block-c--polymers (AB, ABA, ABC)	Micelles, Vesicles	Nanoreactors; Artificial organelles; Nano carriers drug delivery	[28, 71]
		Hyperbranched dendrimers (Dendrons)	Nanoparticles, Nanofibers	Nanocarriers for drug, gene delivery	[72, 73]
	Surfactants	Anionic, Cationic, Nonionic spans, tweens	Micelles, Vesicles	Drug/gene delivery systems; antimicrobial, antifungal activity	[74]
	Others	Graphene	Nanotubes, Carbon nanotubes	Nanomedicine; drug delivery; Hydrogels	[75 - 77]
Biological	Viruses: Bacteriophage	Aligned phage films, fibrils, particles	Biomaterials, Cell culture substrates		[26]
	Nucleic acids (DNA, RNA)	DNA origami	Carriers in drug delivery system, Bio sensing		[78, 79]
	Lipids (Fatty acid, Phospholipids, Cholesterol)	Lipid bilayer, Vesicles, Films, liposomes	Nano reactors, Artificial organelles, controlled drug delivery systems		[36, 44]
	Amylose, Cyclodextrin	Nanotubes, Spherical micelles	Drug delivery, biosensors		[80]
	Peptides	β-sheet, random coils	Hydrogel biomaterials, Drug delivery, Tissue engineering, 3D cell culture		[81 - 84]

Surfactants

Surfactants are another broad spectrum of molecules that have both an aliphatic hydrophobic part and a polar hydrophilic part. There are several types of surfactants, *e.g.*, anionic, cationic and nonionic (tweens, spans). Several studies have shown that they can aggregate into a well-defined micellar structure in a dispersion medium at a certain concentration, called the critical micellar concentration (CMC). These self-assembling micellar forms of surfactants can be used as drug-carrying biomaterials [32 - 34].

Lipids

Another class of macromolecules that can be used to produce biomaterials by self-assembly is lipids. Lipids tend to aggregate and organize their nonpolar regions away from the aqueous environment of the cell. They are a diverse group of compounds whose properties can be altered by changing their hydrophobicity (changing alkyl chain length) and hydrophilicity (polar groups such as phosphate, amino). Phospholipids are compound lipids that have both hydrophilic and hydrophobic moieties. They are able to self-assemble into micelles and vesicles. The main driving force for achieving these shapes is folding under the influence of hydrophobic interactions. Schnur and co-workers have developed biomaterials that take advantage of the self-assembling properties of lipid tubules [35 - 37]. Recently, Suzuki *et al.* reported the lipid-assisted self-assembly of DNA nanostructures [38, 39].

Amphiphilic Peptides

This group, which includes polar regions, contains O_2 types of biomolecules: surfactant-like peptides and peptide amphiphiles [40].

Surfactant-Like Peptides

These were designed by Zhang's group, with the sequence of amino acids being the most important factor in determining the degree of their amphiphilic nature. These head (polar) and tail (nonpolar) structures were able to self-assemble into vesicles, micelles, and other nanostructures for use as biomaterials [41, 42]. The tail consisted of six amino acids with nonpolar side chains such as glycine, alanine, phenylalanine, and 03 branched-chain amino acids. The head consisted of polar amino acids (aspartic acid and lysine). Zhang *et al.* demonstrated that varying the number of glycines in the tail and aspartic acid in the head resulted in the production of nanotubes and nanovesicles in aqueous media. These nanostructures exhibited different properties and orientations due to their relative hydrophilic and hydrophobic groups. In addition, the peptides are surfactant-like. The increase in glycine residues increased the polydisperse behavior of the peptides [43].

Peptide Amphiphiles (PAs)

The structural basis of PAs consists of a hydrophobic lipid chain covalently attached to a peptide. Since most cellular environments are naturally aqueous, PAs tend to self-assemble into a hydrophobic core and a hydrophilic outer periphery. PAs tend to self-assemble into a variety of structures, such as spherical micelles and nanofibers. A pioneering contribution to the understanding, design and use of PAs was made by Stupp and his coworkers. They designed a PA, in which the peptide unit near the lipid/alkyl chain tended to organize into a secondary structure of β-sheets. This secondary organization of the peptide formed the driving force for PAs to preferentially assemble into cylindrical nanofibers [44 - 46]. The similarly charged polar molecules in the peptide unit, the presence of electrolytes, and the pH of the medium also contribute to the formation of cylindrical structures. The fibers formed combine and interweave to form a gel matrix. They have also argued that a bioactive recognition sequence can be added to another side of PA so that it can identify and bind other molecules in the system and *vice versa*. This design of PAs thus allows targeted, directional targeting of biological signals to them, facilitating cell signaling and downstream responses at lower levels [47]. Van Hest's group also designed PAs using a peptide sequence derived from the CS protein of Plasmodium. They found that linking this peptide to short-chain alkyl groups did not stimulate the self-assembly of PAs into nanofibers. In contrast, when the peptide sequence was linked to long-chain hydrophobic alkyl groups, it formed stable nanofiber self-assemblies [48 - 50]. Castello's group developed a more complex form of PAs by linking a hydrophilic pentapeptide head to a hexadecyl lipid chain, which has been shown to self-assemble into nanotapes when dispersed in aqueous media [51, 52].

Drug Amphiphiles

This is a new class of PAs that has been recently reported. Drug amphiphiles are developed by combining a hydrophobic drug (camptothecin) with a peptide that tends to form β-sheets, such as valine-glutamine-isoleucine-tyrosine-lysine, which are linked together. Amphiphilic drugs have been shown to self-assemble into filamentous nanostructures, such as nanofibers and nanotubes. The main advantage of amphiphilic drugs was that no additional delivery system or carrier was required to deliver the drug molecules to their site of action in precise amounts [53 - 55].

Multi-Domain Peptides

These amphiphilic peptides were developed by Hartgerink and his collaborators. Multi-domain peptides have a distinct ABA motif in which the B part is composed of different hydrophilic and hydrophobic amino acid sequences surrounded by polar amino acids (Glu, Lys) on the flanks of block A. The B part is composed of different hydrophilic and hydrophobic amino acid sequences. Block A provides interaction with the surrounding aqueous environment. This interaction allows charged amino acids, such as serine and glutamine, to be physically separated from non-polar amino acids, such as leucine. Hydrophobic groups form a sandwich in the core of the peptides due to their lower affinity for the aqueous system. After the peptides are organized into a β-fold secondary structure, hydrogen bonds are formed between the folds to strengthen and stabilize the supramolecular structures. Many peptides with multiple domains and different epitopes have been synthesized. They have also been found to form self-supporting gels when the peptide charge is dampened and shielded by the addition of multivalent ions or radicals, such as phosphate [56 - 60].

Aromatic Peptides and Derivatives

This is another class of small peptides that carry a diphenylalanine recognition motif. These peptides tend to aggregate into nanotubes when added to a solution. The mechanism of self-assembly involves the accumulation of individual peptides through hydrogen bonding and aromatic pi-pi stacking. Ulijn *et al.* reported the formation of these peptides using aromatic groups, such as fluoroenylmethoxycarbonyl (Fmoc), conjugated to various combinations of dipeptides. The advantage of this peptide design was the synthesis of stable hydrogels. This hydrogel system contains small and simpler amino acid sequences that make it cost effective when marketed to encapsulate cells [61 - 65].

Nanogels

Nanogels are cross-linked networks of polymers that contain both hydrophilic and hydrophobic groups in the monomers. Nanogels can change their volume in response to changes in the internal and external environment. Because of their ability to restore and regenerate tissues, nanogels are recommended as important materials for tissue engineering. Their network simulates the biological tissue and therefore provides a suitable environment for cell survival, immobilization, and a potential drug carrier [66 - 68]. Nomura *et al.* have shown that self-assembled nanogels can support protein refolding. Nanogels are thus an example of an efficient network scaffold system for protein folding [69]. Purwada *et al.* recently developed a self-assembling nanogel as a carrier for antigen-mediated anticancer immunotherapy administered by the injectable route. The basic scaffold of the nanogel consisted of poly(hydroxyethyl methacrylate) linked with pyridine. The nanogels mediated antigen transfer for uptake by dendritic cells, resulting in stimulation and proliferation of CD8+ cells of the immune system. The authors also suggest that the main forces responsible for the stability of their nanogel are electrostatic interactions and hydrogen bonding [70].

SELF-ASSEMBLY OF CARBON-BASED NANOSTRUCTURED MATERIALS

Carbon Nanotubes (CNTs)

These are graphene-based sheets that are rolled or stacked into carbon nanotubes. These nanotubes can be either single-walled or multi-walled [85]. A longer length of the tube with good mechanical strength and effective conductivity make nanotubes good candidates for use in sugar sensors [86]. Zuber and coworkers have reported the surfactant-assisted self-assembly process to form CNT microspheres. The outer surface of CNT microspheres was smooth, with a large surface area and variable porosity. Because of these properties, CNTs can transport a large amount of the desired radioisotopes, with the added advantage of lower retention due to rapid renal excretion by glomerular filtration. This reduces the risk of toxicity caused by radioactive isotopes [87, 88]. In addition, the surface of single-walled CNTs may serve to bind and transport aromatic drugs, such as doxorubicin, to tumor tissue. The interactions involve aromatic stacking between CNTs and the aromatic groups of doxorubicin [89].

Graphene

Graphene is a single-layer assembly of carbon atoms in a honeycomb lattice [90]. The C-C bond in graphene is sp2 hybridized (presence of unsaturation). It has excellent mechanical and thermal properties [91]. In terms of biomaterials, graphene is a potentially advantageous material for medical imaging and as a carrier for antineoplastic drugs. Graphene oxide-based nanocarriers were recently synthesized for the delivery of the anticancer drug doxorubicin along with DNA sequences in liver carcinoma cells [92, 93]. The graphene nanocarrier was able to deliver the drug to the target tumor site in sufficient quantity without showing signs of toxicity at 100 µg/mL.

FIBERMESH SCAFFOLDS IN TISSUE ENGINEERING

Scaffold-based tissue engineering techniques have been shown to be quite effective in regenerating, healing, and actively repairing body parts such as skin and bone. This type of tissue engineering is a traditional approach to synthesize functionally active tissue with a microenvironment similar to the original endogenous tissue/organ. The basic technique is to incorporate the desired cell type into a polymeric, biodegradable scaffold that provides scaffolding and support for the cells as they transform and mature into the appropriate tissues. The cells can be isolated from the host organism and reincorporated into the scaffold for further use (Fig. **4A**). The basic requirement for this technology is the presence of appropriate scaffold materials, cells of a specific type, and growth factors that promote cell growth and differentiation of chondrocytes (Fig. **4B**) [94 - 98]. The selection of an appropriate scaffold material is crucial for successful regeneration and healing of the affected body parts by tissue engineering. The properties and specifications of a valuable scaffold material candidate are listed in Fig. (**5**).

To support tissue growth, there is a constant need in the field of tissue engineering for a scaffold with a three-dimensional (3D) architecture. An ideal scaffold must be biodegradable to better support new bone growth and avoid adverse effects. It is strongly recommended that the scaffold should be completely degraded when the injured site is fully regenerated [99]. Studies have shown that biodegradable materials temporarily provide a natural environment for tissue growth before degradation and elimination from the body [100, 101].

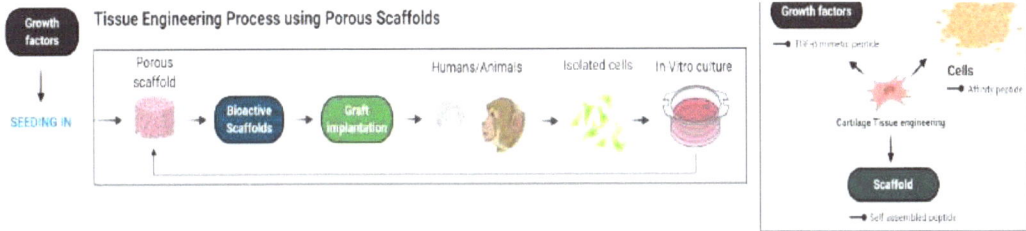

Fig. (4). A. Schematic representation of Tissue engineering cycle **B.** Basic requirements of scaffold-based tissue engineering.

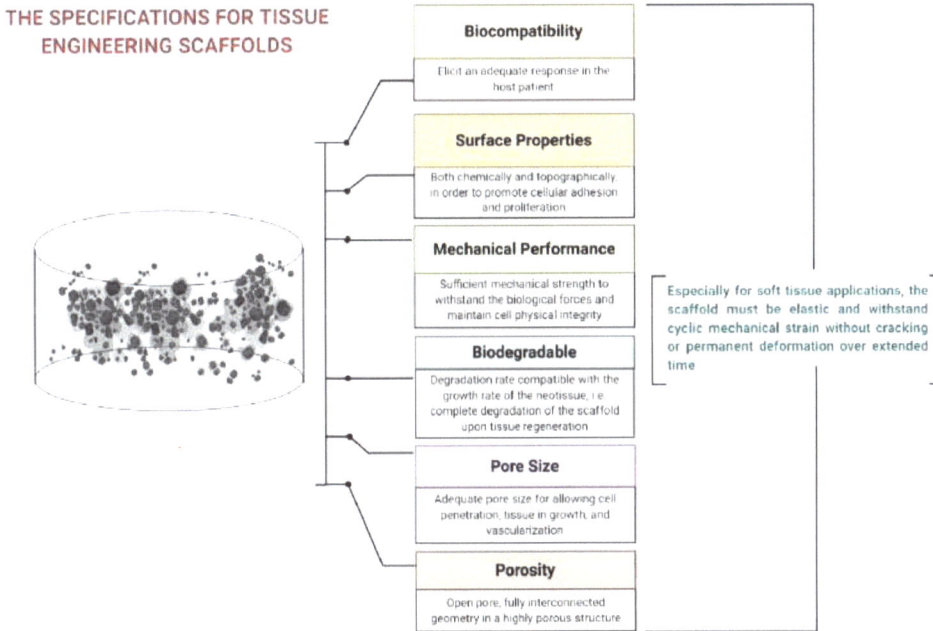

Fig. (5). Specifications of scaffolding materials used in tissue engineering and regenerative medicine.

Fiber meshes are a novel technique that either individual fibers are used to fabricate 3-D scaffolds or scaffolds are fabricated by the interweaving of fibers [99]. Polyglycolide is the first biocompatible, biodegradable material used to fabricate suture threads that are subsequently used to hold wound edges together during surgery or trauma. In this suture fabrication process, the solution of one polymer is spun onto the polymeric mesh of another nonwoven polymer and then evaporated [101]. Scaffolds fabricated by such techniques have been reported to promote cell growth and survival by providing a large surface area for adherent cells as well as an easy supply of nutrients [102]. However, fiber meshes have certain disadvantages, such as lack of stability and mechanical strength, since these two important aspects have not been sufficiently explored. Considering these two important aspects, a study reported the development of chitosan fibers and 3D fiber mesh scaffolds with a pore size of 100-500 μm by wet spinning technique. The resulting fibers showed good mechanical strength (tensile strength 204.9 MPa and maximum elongation at break (8.5%)) to be used as scaffolds. Moreover, the use of hot-dried poly(L-lactic acid) fibers solved the problem of structural stability by improving the crystallinity and keeping the structural orientation intact [103, 104].

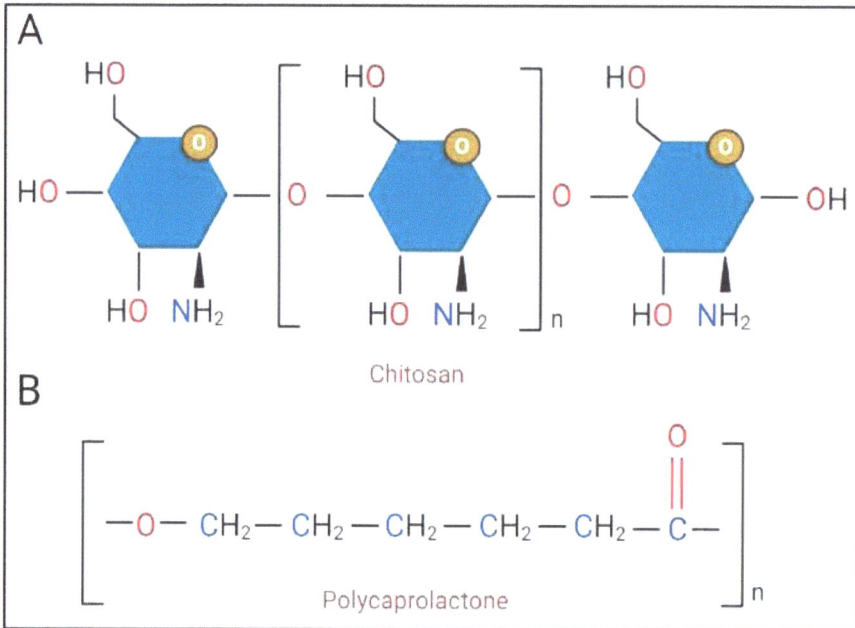

Fig. (6). Structures of representative polymers used for forming Fiber mesh scaffolds. **A.** Chitosan **B.** Polycaprolactone.

The selection of a suitable scaffold is the most important step for optimal regeneration of tissues from matrix-producing connective tissue cells or anchorage-dependent cells, such as osteoblasts. Polycaprolactone is another known polymer for weaving fibrous mesh scaffolds (Fig. **6B**). M. E. Gomes *et al.* developed and characterized SPCL (starch with epsilon-polycaprolactone, 30:70%) and SPLA [starch with poly(lactic acid), 30:70%] fibrous meshes for use in tissue engineering of bone. SPCL and SPLA scaffolds were developed by a fiber-binding process. These scaffolds were shown to effectively support cell adhesion and proliferation due to their suitable porosity and mechanical properties. In addition, the SPCL and SPLA scaffolds were found to be susceptible to enzymatic degradation, as evidenced by increased weight loss. The diameter of SPCL and SPLA fibers gradually decreased over time, which improved scaffold porosity and space for tissue and cell growth. These results suggest that SPCL and SPLA scaffolds can be used for tissue regeneration studies [105].

CHITOSAN FIBER MESH SCAFFOLDS

Chitosan (Fig. **6A**) is a well-documented, nontoxic, biocompatible, and biodegradable natural polymer [106]. Chitosan has been shown to bind excellently to a range of microbial and mammalian cells, which is why it has numerous applications for drug delivery, space-filling implants, and wound dressings [107 - 109]. It has also been reported that chitosan can be used to promote bone growth [110, 111]. For example, a study using chitosan plugs showed the formation of mineralized bone-like tissue in bone defects in sheep, dogs, and rats [110]. One study reported the production of chitosan fibers by a wet spinning process [112]. Briefly, chitosan was dissolved in aqueous 2% (v/v) acetic acid at a concentration of 5% (w/v) at room temperature. The solution was diluted with methanol to a final concentration of 3% (w/v). Then, 2.5% (w/w) glycerol was added to this solution, which served as a plasticizer. The solution was filtered and kept in an ultrasonic bath to remove the air bubbles. The solution was then placed in a coagulation bath at 40 °C containing a mixture of 10% 1M NaOH, 30% 0.5 M Na_2SO_4 and 60% distilled water. The fibers were left in this coagulation medium for one day and washed several times with distilled water. They were then suspended in 30% methanol for 4-5 hours and then in 50% methanol overnight. Finally, the filaments were wound on the cylindrical support and dried at room temperature and/or humidity for 24 h. A similar approach was adopted for the development of 3-D chitosan fiber meshes. The only difference was a different drying method and the use of a thinner needle. After the fibers formed in the coagulation medium at room temperature, they were left in this solution overnight. The fibers were then washed several times with distilled water before being suspended in 50% methanol for 1 h. In the final step, the fibers were suspended in 100% methanol for 3 h, and placed in a mold at 55 °C in an oven to dry. Various mechanical tests showed that the fibers had sufficient tensile strength to be used for the fabrication of scaffolds with efficient mechanical performance. At the same time, the developed fibers surprisingly showed bioactive behavior, which is considered very important for biomaterials to be used in tissue regeneration and bone regeneration studies [113, 114].

CHITOSAN/POLYCAPROLACTONE (CHT/PCL) BLEND FIBER MESH SCAFFOLDS

Due to its structural similarity to various glycosaminoglycans found in cartilage, chitosan remains the most important candidate [115]. Among the various potent synthetic biomaterials, poly(ε-caprolactone) (PCL) has attracted the attention of researchers due to its relatively low melting temperature (60 °C), mechanical and physicochemical properties, nontoxicity, biodegradability, and Food and Drug Administration approval for biomedical applications [116, 117]. Studies have shown that chondrocytes attach to PCL sheets, proliferate, and begin the production of a cartilaginous extracellular matrix in PCL scaffolds [118 - 120]. However, certain limitations of PCL have also been reported, including its hydrophobicity, lack of cell recognition sites, and relatively slow degradation [121]. Chitosan/poly(ε-caprolactone) blends of 3D fibers and meshes have been investigated as potential carriers for articular cartilage tissue repair (ACT). Microfibers were prepared by wet spinning three different polymer solutions: 50:50 CHT/PCL, 100:0 (100CHT), and 75:25 (75CHT), in formic acid. The surface roughness of the fibers was improved, while the swelling ratio was reduced after the incorporation of PCL into CHT. Moreover, cell spreading was improved by mixing, while cell survival and metabolic activity were not affected [122, 123]. In one study, a group of researchers developed improved gelatin-hydroxyapatite nanocomposites (HA) using a tannic acid (TA) as a cross-linking agent, which has already been reported to have antioxidant, anti-inflammatory, antiallergic, antimicrobial, antithrombotic, and cardio-protective properties [124]. Bone binding potential, compression, bio-mineralization, and degradation tests were performed to analyze the potential application of gelatin HA scaffolds in the restoration of calcified tissue. The study showed that TA (cross-linking agent) modulates the microstructure and compressive strength of the gelatin scaffold. Moreover, these properties of gelatin could also be modulated by mimicking the nanosize of HA in natural bone. Overall, these results suggest that the previously embedded HA nanoparticles have a significant effect on the epitaxial growth of the mineralized phase, mechanical behavior of the scaffold, biodegradability, and bone bonding potential.

CONCLUDING REMARKS

The fundamental need for comprehensive biomaterials research in tissue engineering is to develop replacement organs/tissues or design a scaffold of scaffolds containing cells for tissue regeneration/healing. Other indispensable properties of these materials include their ability to mimic the regenerative capacity of natural tissues. Newer, versatile, and innovative methods for synthesizing biomaterials are being introduced for use in tissue engineering to heal/repair damaged tissues, particularly in cardiovascular and musculoskeletal diseases. Biomaterials fabrication by self-assembly is known for its optimal mimicking of the natural process of macromolecule synthesis and their functional organization. Scaffolds made of fiber meshes are important for tissue reconstruction. They have emerged as one of the most promising means to overcome the challenges of 3D scaffold fabrication in tissue engineering and regenerative medicine. This field has experienced rapid growth in recent years due to its utility in a variety of fields, including electronics, engineering, materials science, imaging, diagnostics, and drug delivery systems. With the use of enormous resources and technical capabilities, the diversity of biomaterials production is growing by leaps and bounds. Although humans have successfully domesticated plants and animals, domesticating molecules that form the biochemical basis of life is proving more difficult. Moreover, much understanding of the fundamental forces of nature is still required to understand the problem of the intricate self-assembly of macromolecules such as phospholipids, polypeptides, and other similar molecules and to design analogous biomaterials that share the same cellular architecture as the original tissue.

CONSENT FOR PUBLICATION

Not applicable.

CONFLICT OF INTEREST

The author declares no conflict of interest, financial or otherwise.

ACKNOWLEDGEMENT

Declared none.

REFERENCES

[1]   A.I. Caplan, M. Elyaderani, Y. Mochizuki, S. Wakitani, and V.M. Goldberg, "Principles of cartilage repair and regeneration", *Clin. Orthop. Relat. Res.,* no. 342, pp. 254-269, 1997.
      [http://dx.doi.org/10.1097/00003086-199709000-00033] [PMID: 9308548]

[2]   J. Wang, K. Liu, R. Xing, and X. Yan, "Peptide self-assembly: thermodynamics and kinetics", *Chem. Soc. Rev.,* vol. 45, no. 20, pp. 5589-5604, 2016.
      [http://dx.doi.org/10.1039/C6CS00176A] [PMID: 27487936]

[3]   Y. Geng, P. Dalhaimer, S. Cai, R. Tsai, M. Tewari, T. Minko, and D.E. Discher, "Shape effects of filaments *versus* spherical particles in flow and drug delivery", *Nat. Nanotechnol.,* vol. 2, no. 4, pp.

249-255, 2007.
[http://dx.doi.org/10.1038/nnano.2007.70] [PMID: 18654271]

[4]   R. Langer, and D.A. Tirrell, "Designing materials for biology and medicine", *Nature,* vol. 428, no. 6982, pp. 487-492, 2004.
[http://dx.doi.org/10.1038/nature02388] [PMID: 15057821]

[5]   S.I. Stupp, "Self-Assembly and Biomaterials", *Nano Lett.,* vol. 10, no. 12, pp. 4783-4786, 2010.
[http://dx.doi.org/10.1021/nl103567y] [PMID: 21028843]

[6]   S. Zhang, "Fabrication of novel biomaterials through molecular self-assembly", *Nat. Biotechnol.,* vol. 21, no. 10, pp. 1171-1178, 2003.
[http://dx.doi.org/10.1038/nbt874] [PMID: 14520402]

[7]   M. Gross, *The nanoworld, Miniature Machinery in Nature and Technology.* vol. 254. Plenum Trade: New York, London, 1999.

[8]   J.M. Benyus, "Biomimicry: Innovation inspired by nature", *Morrow New York,* 1997.

[9]   A. Kurut, B.A. Persson, T. Åkesson, J. Forsman, and M. Lund, "Anisotropic interactions in protein mixtures: Self assembly and phase behavior in aqueous solution", *J. Phys. Chem. Lett.,* vol. 3, no. 6, pp. 731-734, 2012.
[http://dx.doi.org/10.1021/jz201680m] [PMID: 26286281]

[10]   C.N. Lam, H. Yao, and B.D. Olsen, "The effect of protein electrostatic interactions on globular protein-polymer block copolymer self-assembly", *Biomacromolecules,* vol. 17, no. 9, pp. 2820-2829, 2016.
[http://dx.doi.org/10.1021/acs.biomac.6b00522] [PMID: 27482836]

[11]   N.A. Kotov, "Layer-by-layer self-assembly: The contribution of hydrophobic interactions", *Nanostruct. Mater.,* vol. 12, no. 5-8, pp. 789-796, 1999.
[http://dx.doi.org/10.1016/S0965-9773(99)00237-8]

[12]   S. Zhang, "Emerging biological materials through molecular self-assembly", *Biotechnol. Adv.,* vol. 20, no. 5-6, pp. 321-339, 2002.
[http://dx.doi.org/10.1016/S0734-9750(02)00026-5] [PMID: 14550019]

[13]   M.A. Cejas, W.A. Kinney, C. Chen, G.C. Leo, B.A. Tounge, J.G. Vinter, P.P. Joshi, and B.E. Maryanoff, "Collagen-related peptides: self-assembly of short, single strands into a functional biomaterial of micrometer scale", *J. Am. Chem. Soc.,* vol. 129, no. 8, pp. 2202-2203, 2007.
[http://dx.doi.org/10.1021/ja066986f] [PMID: 17269769]

[14]   B. Gabryelczyk, H. Cai, X. Shi, Y. Sun, P.J.M. Swinkels, S. Salentinig, K. Pervushin, and A. Miserez, "Hydrogen bond guidance and aromatic stacking drive liquid-liquid phase separation of intrinsically disordered histidine-rich peptides", *Nat. Commun.,* vol. 10, no. 1, p. 5465, 2019.
[http://dx.doi.org/10.1038/s41467-019-13469-8] [PMID: 31784535]

[15]   M. Reches, and E. Gazit, "Formation of closed-cage nanostructures by self-assembly of aromatic dipeptides", *Nano Lett.,* vol. 4, no. 4, pp. 581-585, 2004.
[http://dx.doi.org/10.1021/nl035159z]

[16]   C.R. Martinez, and B.L. Iverson, "Rethinking the term "pi-stacking"", *Chem. Sci. (Camb.),* vol. 3, no. 7, pp. 2191-2201, 2012.
[http://dx.doi.org/10.1039/c2sc20045g]

[17]   P.K. Mishra, and A. Ekielski, "The self-assembly of lignin and its application in nanoparticle synthesis: A short review", *Nanomaterials (Basel),* vol. 9, no. 2, p. 243, 2019.
[http://dx.doi.org/10.3390/nano9020243] [PMID: 30754724]

[18]   S. Horowitz, and R.C. Trievel, "Carbon-oxygen hydrogen bonding in biological structure and function", *J. Biol. Chem.,* vol. 287, no. 50, pp. 41576-41582, 2012.
[http://dx.doi.org/10.1074/jbc.R112.418574] [PMID: 23048026]

[19]   B. Kuhn, P. Mohr, and M. Stahl, "Intramolecular hydrogen bonding in medicinal chemistry", *J. Med. Chem.,* vol. 53, no. 6, pp. 2601-2611, 2010.
[http://dx.doi.org/10.1021/jm100087s] [PMID: 20175530]

[20]   S.M. Habermann, and K.P. Murphy, "Energetics of hydrogen bonding in proteins: A model compound study", *Protein Sci.,* vol. 5, no. 7, pp. 1229-1239, 1996.
[http://dx.doi.org/10.1002/pro.5560050702] [PMID: 8819156]

[21]   D.F. Sticke, L.G. Presta, K.A. Dill, and G.D. Rose, "Hydrogen bonding in globular proteins", *J. Mol. Biol.,* vol. 226, no. 4, pp. 1143-1159, 1992.
[http://dx.doi.org/10.1016/0022-2836(92)91058-W] [PMID: 1518048]

[22]   J.K. Myers, and C.N. Pace, "Hydrogen bonding stabilizes globular proteins", *Biophys. J.,* vol. 71, no. 4, pp. 2033-2039, 1996.
[http://dx.doi.org/10.1016/S0006-3495(96)79401-8] [PMID: 8889177]

[23]   R.W. Newberry, and R.T. Raines, "A prevalent intraresidue hydrogen bond stabilizes proteins", *Nat. Chem. Biol.,* vol. 12, no. 12, pp. 1084-1088, 2016.
[http://dx.doi.org/10.1038/nchembio.2206] [PMID: 27748749]

[24]   C. Nick Pace, J.M. Scholtz, and G.R. Grimsley, "Forces stabilizing proteins", *FEBS Lett.,* vol. 588, no. 14, pp. 2177-2184, 2014.
[http://dx.doi.org/10.1016/j.febslet.2014.05.006] [PMID: 24846139]

[25]   G.M. Whitesides, J.K. Kriebel, and B.T. Mayers, "Self-assembly and nanostructured materials", In: *Nanoscale assembly* Springer, 2005, pp. 217-239.
[http://dx.doi.org/10.1007/0-387-25656-3_9]

[26]   A.C. Mendes, E.T. Baran, R.L. Reis, and H.S. Azevedo, "Self-assembly in nature: using the principles of nature to create complex nanobiomaterials", *Wiley Interdiscip. Rev. Nanomed. Nanobiotechnol.,* vol. 5, no. 6, pp. 582-612, 2013.
[http://dx.doi.org/10.1002/wnan.1238] [PMID: 23929805]

[27]   A. Blanazs, S.P. Armes, and A.J. Ryan, "Self-assembled block copolymer aggregates: from micelles to vesicles and their biological applications", *Macromol. Rapid Commun.,* vol. 30, no. 4-5, pp. 267-277, 2009.
[http://dx.doi.org/10.1002/marc.200800713] [PMID: 21706604]

[28]   D.K. Smith, "Dendritic supermolecules - towards controllable nanomaterials", *Chem. Commun. (Camb.),* no. 1, pp. 34-44, 2006.
[http://dx.doi.org/10.1039/B507416A] [PMID: 16353086]

[29]   R.W. Scott, O.M. Wilson, and R.M. Crooks, *Synthesis, characterization, and applications of dendrimer-encapsulated nanoparticles.* A.C.S. Publications, 2005.
[http://dx.doi.org/10.1021/jp0469665]

[30]   M.T. Morgan, Y. Nakanishi, D.J. Kroll, A.P. Griset, M.A. Carnahan, M. Wathier, N.H. Oberlies, G. Manikumar, M.C. Wani, and M.W. Grinstaff, "Dendrimer-encapsulated camptothecins: increased solubility, cellular uptake, and cellular retention affords enhanced anticancer activity *in vitro*", *Cancer Res.,* vol. 66, no. 24, pp. 11913-11921, 2006.
[http://dx.doi.org/10.1158/0008-5472.CAN-06-2066] [PMID: 17178889]

[31]   R.K. Tekade, T. Dutta, V. Gajbhiye, and N.K. Jain, "Exploring dendrimer towards dual drug delivery: pH responsive simultaneous drug-release kinetics", *J. Microencapsul.,* vol. 26, no. 4, pp. 287-296, 2009.
[http://dx.doi.org/10.1080/02652040802312572] [PMID: 18791906]

[32]   I.C. Reynhout, J.J.L.M. Cornelissen, and R.J.M. Nolte, "Synthesis of polymer-biohybrids: from small to giant surfactants", *Acc. Chem. Res.,* vol. 42, no. 6, pp. 681-692, 2009.
[http://dx.doi.org/10.1021/ar800143a] [PMID: 19385643]

[33]   A. Martin, P. Sinko, and Y. Singh, *Martin's physical pharmacy and pharmaceutical sciences.* vol.

Vol. 496. Lippinscott Williams & Williams: Baltimore, 2006.

[34]   J.P. Remington, *Remington: The science and practice of pharmacy.* vol. 1. Lippincott Williams & Wilkins, 2006.

[35]   J.M. Schnur, R. Price, P. Schoen, P. Yager, J.M. Calvert, J. Georger, and A. Singh, "Lipid-based tubule microstructures", *Thin Solid Films,* vol. 152, no. 1-2, pp. 181-206, 1987.
[http://dx.doi.org/10.1016/0040-6090(87)90416-0]

[36]   J.M. Schnur, "Lipid tubules: a paradigm for molecularly engineered structures", *Science,* vol. 262, no. 5140, pp. 1669-1676, 1993.
[http://dx.doi.org/10.1126/science.262.5140.1669] [PMID: 17781785]

[37]   A.S. Rudolph, J.M. Calvert, P.E. Schoen, and J.M. Schnur, "Technological development of lipid based tubule microstructures", In: *Biotechnological Applications of lipid microstructures*, 1988, pp. 305-320.
[http://dx.doi.org/10.1007/978-1-4684-7908-9_24]

[38]   Y. Suzuki, M. Endo, and H. Sugiyama, "Lipid-bilayer-assisted two-dimensional self-assembly of DNA origami nanostructures", *Nat. Commun.,* vol. 6, no. 1, p. 8052, 2015.
[http://dx.doi.org/10.1038/ncomms9052] [PMID: 26310995]

[39]   C.V. Kulkarni, "Lipid crystallization: from self-assembly to hierarchical and biological ordering", *Nanoscale,* vol. 4, no. 19, pp. 5779-5791, 2012.
[http://dx.doi.org/10.1039/c2nr31465g] [PMID: 22899223]

[40]   I.W. Hamley, "Self-assembly of amphiphilic peptides", *Soft Matter,* vol. 7, no. 9, pp. 4122-4138, 2011.
[http://dx.doi.org/10.1039/c0sm01218a]

[41]   S. Koutsopoulos, L.D. Unsworth, Y. Nagai, and S. Zhang, "Controlled release of functional proteins through designer self-assembling peptide nanofiber hydrogel scaffold", *Proc. Natl. Acad. Sci. USA,* vol. 106, no. 12, pp. 4623-4628, 2009.
[http://dx.doi.org/10.1073/pnas.0807506106] [PMID: 19273853]

[42]   S. Vauthey, S. Santoso, H. Gong, N. Watson, and S. Zhang, "Molecular self-assembly of surfactant-like peptides to form nanotubes and nanovesicles", *Proc. Natl. Acad. Sci. USA,* vol. 99, no. 8, pp. 5355-5360, 2002.
[http://dx.doi.org/10.1073/pnas.072089599] [PMID: 11929973]

[43]   S. Santoso, W. Hwang, H. Hartman, and S. Zhang, "Self-assembly of surfactant-like peptides with variable glycine tails to form nanotubes and nanovesicles", *Nano Lett.,* vol. 2, no. 7, pp. 687-691, 2002.
[http://dx.doi.org/10.1021/nl025563i]

[44]   M.J. Webber, J.A. Kessler, and S.I. Stupp, "Emerging peptide nanomedicine to regenerate tissues and organs", *J. Intern. Med.,* vol. 267, no. 1, pp. 71-88, 2010.
[http://dx.doi.org/10.1111/j.1365-2796.2009.02184.x] [PMID: 20059645]

[45]   H. Cui, M.J. Webber, and S.I. Stupp, "Self-assembly of peptide amphiphiles: From molecules to nanostructures to biomaterials", *Biopolymers,* vol. 94, no. 1, pp. 1-18, 2010.
[http://dx.doi.org/10.1002/bip.21328] [PMID: 20091874]

[46]   J.B. Matson, R.H. Zha, and S.I. Stupp, "Peptide self-assembly for crafting functional biological materials", *Curr. Opin. Solid State Mater. Sci.,* vol. 15, no. 6, pp. 225-235, 2011.
[http://dx.doi.org/10.1016/j.cossms.2011.08.001] [PMID: 22125413]

[47]   G.A. Silva, C. Czeisler, K.L. Niece, E. Beniash, D.A. Harrington, J.A. Kessler, and S.I. Stupp, "Selective differentiation of neural progenitor cells by high-epitope density nanofibers", *Science,* vol. 303, no. 5662, pp. 1352-1355, 2004.
[http://dx.doi.org/10.1126/science.1093783] [PMID: 14739465]

[48]   D.W.P.M. Löwik, J. Garcia-Hartjes, J.T. Meijer, and J.C.M. van Hest, "Tuning secondary structure and self-assembly of amphiphilic peptides", *Langmuir,* vol. 21, no. 2, pp. 524-526, 2005.
[http://dx.doi.org/10.1021/la047578x] [PMID: 15641818]

[49]   M. van den Heuvel, D.W.P.M. Löwik, and J.C.M. van Hest, "Self-assembly and polymerization of diacetylene-containing peptide amphiphiles in aqueous solution", *Biomacromolecules,* vol. 9, no. 10, pp. 2727-2734, 2008.
[http://dx.doi.org/10.1021/bm800424x] [PMID: 18785773]

[50]   M. van den Heuvel, D.W.P.M. Löwik, and J.C.M. van Hest, "Effect of the diacetylene position on the chromatic properties of polydiacetylenes from self-assembled peptide amphiphiles", *Biomacromolecules,* vol. 11, no. 6, pp. 1676-1683, 2010.
[http://dx.doi.org/10.1021/bm100376q] [PMID: 20499861]

[51]   V. Castelletto, I.W. Hamley, J. Adamcik, R. Mezzenga, and J. Gummel, "Modulating self-assembly of a nanotape-forming peptideamphiphile with an oppositely charged surfactant", *Soft Matter,* vol. 8, no. 1, pp. 217-226, 2012.
[http://dx.doi.org/10.1039/C1SM06677C]

[52]   A. Dehsorkhi, V. Castelletto, and I.W. Hamley, "Self-assembling amphiphilic peptides", *J. Pept. Sci.,* vol. 20, no. 7, pp. 453-467, 2014.
[http://dx.doi.org/10.1002/psc.2633] [PMID: 24729276]

[53]   A.G. Cheetham, Y.C. Ou, P. Zhang, and H. Cui, "Linker-determined drug release mechanism of free camptothecin from self-assembling drug amphiphiles", *Chem. Commun. (Camb.),* vol. 50, no. 45, pp. 6039-6042, 2014.
[http://dx.doi.org/10.1039/C3CC49453E] [PMID: 24769796]

[54]   A.G. Cheetham, P. Zhang, Y. Lin, L.L. Lock, and H. Cui, "Supramolecular nanostructures formed by anticancer drug assembly", *J. Am. Chem. Soc.,* vol. 135, no. 8, pp. 2907-2910, 2013.
[http://dx.doi.org/10.1021/ja3115983] [PMID: 23379791]

[55]   R. Lin, A.G. Cheetham, P. Zhang, Y. Lin, and H. Cui, "Supramolecular filaments containing a fixed 41% paclitaxel loading", *Chem. Commun. (Camb.),* vol. 49, no. 43, pp. 4968-4970, 2013.
[http://dx.doi.org/10.1039/c3cc41896k] [PMID: 23612448]

[56]   L. Aulisa, H. Dong, and J.D. Hartgerink, "Self-assembly of multidomain peptides: sequence variation allows control over cross-linking and viscoelasticity", *Biomacromolecules,* vol. 10, no. 9, pp. 2694-2698, 2009.
[http://dx.doi.org/10.1021/bm900634x] [PMID: 19705838]

[57]   K.M. Galler, A. Cavender, V. Yuwono, H. Dong, S. Shi, G. Schmalz, J.D. Hartgerink, and R.N. D'Souza, "Self-assembling peptide amphiphile nanofibers as a scaffold for dental stem cells", *Tissue Eng. Part A,* vol. 14, no. 12, pp. 2051-2058, 2008.
[http://dx.doi.org/10.1089/ten.tea.2007.0413] [PMID: 18636949]

[58]   K.M. Galler, L. Aulisa, K.R. Regan, R.N. D'Souza, and J.D. Hartgerink, "Self-assembling multidomain peptide hydrogels: designed susceptibility to enzymatic cleavage allows enhanced cell migration and spreading", *J. Am. Chem. Soc.,* vol. 132, no. 9, pp. 3217-3223, 2010.
[http://dx.doi.org/10.1021/ja910481t] [PMID: 20158218]

[59]   L. Jiang, D. Xu, T.J. Sellati, and H. Dong, "Self-assembly of cationic multidomain peptide hydrogels: supramolecular nanostructure and rheological properties dictate antimicrobial activity", *Nanoscale,* vol. 7, no. 45, pp. 19160-19169, 2015.
[http://dx.doi.org/10.1039/C5NR05233E] [PMID: 26524425]

[60]   V.A. Kumar, N.L. Taylor, S. Shi, N.C. Wickremasinghe, R.N. D'Souza, and J.D. Hartgerink, "Self-assembling multidomain peptides tailor biological responses through biphasic release", *Biomaterials,* vol. 52, pp. 71-78, 2015.
[http://dx.doi.org/10.1016/j.biomaterials.2015.01.079] [PMID: 25818414]

[61]   S. Scanlon, and A. Aggeli, "Self-assembling peptide nanotubes", *Nano Today,* vol. 3, no. 3-4, pp. 22-30, 2008.
[http://dx.doi.org/10.1016/S1748-0132(08)70041-0]

[62] R. Orbach, I. Mironi-Harpaz, L. Adler-Abramovich, E. Mossou, E.P. Mitchell, V.T. Forsyth, E. Gazit, and D. Seliktar, "The rheological and structural properties of Fmoc-peptide-based hydrogels: the effect of aromatic molecular architecture on self-assembly and physical characteristics", *Langmuir,* vol. 28, no. 4, pp. 2015-2022, 2012.
[http://dx.doi.org/10.1021/la204426q] [PMID: 22220968]

[63] A.M. Smith, R.J. Williams, C. Tang, P. Coppo, R.F. Collins, M.L. Turner, A. Saiani, and R.V. Ulijn, "Fmoc-diphenylalanine self assembles to a hydrogel *via* a novel architecture based on π-π interlocked β-sheets", *Adv. Mater.,* vol. 20, no. 1, pp. 37-41, 2008.
[http://dx.doi.org/10.1002/adma.200701221]

[64] M. Zhou, A.M. Smith, A.K. Das, N.W. Hodson, R.F. Collins, R.V. Ulijn, and J.E. Gough, "Self-assembled peptide-based hydrogels as scaffolds for anchorage-dependent cells", *Biomaterials,* vol. 30, no. 13, pp. 2523-2530, 2009.
[http://dx.doi.org/10.1016/j.biomaterials.2009.01.010] [PMID: 19201459]

[65] H. Yokoi, T. Kinoshita, and S. Zhang, "Dynamic reassembly of peptide RADA16 nanofiber scaffold", *Proc. Natl. Acad. Sci. USA,* vol. 102, no. 24, pp. 8414-8419, 2005.
[http://dx.doi.org/10.1073/pnas.0407843102] [PMID: 15939888]

[66] K. Akiyoshi, S. Deguchi, N. Moriguchi, S. Yamaguchi, and J. Sunamoto, "Self-aggregates of hydrophobized polysaccharides in water. Formation and characteristics of nanoparticles", *Macromolecules,* vol. 26, no. 12, pp. 3062-3068, 1993.
[http://dx.doi.org/10.1021/ma00064a011]

[67] K. Kuroda, K. Fujimoto, J. Sunamoto, and K. Akiyoshi, "Hierarchical self-assembly of hydrophobically modified pullulan in water: gelation by networks of nanoparticles", *Langmuir,* vol. 18, no. 10, pp. 3780-3786, 2002.
[http://dx.doi.org/10.1021/la011454s]

[68] Y. Nomura, M. Ikeda, N. Yamaguchi, Y. Aoyama, and K. Akiyoshi, "Protein refolding assisted by self-assembled nanogels as novel artificial molecular chaperone", *FEBS Lett.,* vol. 553, no. 3, pp. 271-276, 2003.
[http://dx.doi.org/10.1016/S0014-5793(03)01028-7] [PMID: 14572636]

[69] A. Purwada, Y.F. Tian, W. Huang, K.M. Rohrbach, S. Deol, A. August, and A. Singh, "Self-assembly protein nanogels for safer cancer immunotherapy", *Adv. Healthc. Mater.,* vol. 5, no. 12, pp. 1413-1419, 2016.
[http://dx.doi.org/10.1002/adhm.201501062] [PMID: 27100566]

[70] C. Giacomelli, V. Schmidt, K. Aissou, and R. Borsali, "Block copolymer systems: from single chain to self-assembled nanostructures", *Langmuir,* vol. 26, no. 20, pp. 15734-15744, 2010.
[http://dx.doi.org/10.1021/la100641j] [PMID: 20364859]

[71] F. Zeng, and S.C. Zimmerman, "Dendrimers in supramolecular chemistry: from molecular recognition to self-assembly", *Chem. Rev.,* vol. 97, no. 5, pp. 1681-1712, 1997.
[http://dx.doi.org/10.1021/cr9603892] [PMID: 11851463]

[72] D. Kitamoto, T. Morita, T. Fukuoka, M. Konishi, and T. Imura, "Self-assembling properties of glycolipid biosurfactants and their potential applications", *Curr. Opin. Colloid Interface Sci.,* vol. 14, no. 5, pp. 315-328, 2009.
[http://dx.doi.org/10.1016/j.cocis.2009.05.009]

[73] F.J.M. Hoeben, P. Jonkheijm, E.W. Meijer, and A.P.H.J. Schenning, "About supramolecular assemblies of π-conjugated systems", *Chem. Rev.,* vol. 105, no. 4, pp. 1491-1546, 2005.
[http://dx.doi.org/10.1021/cr030070z] [PMID: 15826018]

[74] T. Yamamoto, T. Fukushima, and T. Aida, "Self-assembled nanotubes and nanocoils from ss-conjugated building blocks", In: *Self-Assembled Nanomaterials II* Springer, 2008, pp. 1-27.
[http://dx.doi.org/10.1007/12_2008_171]

[75]   M. Terrones, A.R. Botello-Méndez, J. Campos-Delgado, F. López-Urías, Y.I. Vega-Cantú, F.J. Rodríguez-Macías, A.L. Elías, E. Muñoz-Sandoval, A.G. Cano-Márquez, and J-C. Charlier, "Graphene and graphite nanoribbons: Morphology, properties, synthesis, defects and applications", *Nano Today,* vol. 5, no. 4, pp. 351-372, 2010.
[http://dx.doi.org/10.1016/j.nantod.2010.06.010]

[76]   Y. Krishnan, and F.C. Simmel, "Nucleic acid based molecular devices", *Angew. Chem. Int. Ed.,* vol. 50, no. 14, pp. 3124-3156, 2011.
[http://dx.doi.org/10.1002/anie.200907223] [PMID: 21432950]

[77]   A.V. Pinheiro, D. Han, W.M. Shih, and H. Yan, "Challenges and opportunities for structural DNA nanotechnology", *Nat. Nanotechnol.,* vol. 6, no. 12, pp. 763-772, 2011.
[http://dx.doi.org/10.1038/nnano.2011.187] [PMID: 22056726]

[78]   X. Yu, Z. Liu, J. Janzen, I. Chafeeva, S. Horte, W. Chen, R.K. Kainthan, J.N. Kizhakkedathu, and D.E. Brooks, "Polyvalent choline phosphate as a universal biomembrane adhesive", *Nat. Mater.,* vol. 11, no. 5, pp. 468-476, 2012.
[http://dx.doi.org/10.1038/nmat3272] [PMID: 22426460]

[79]   G. Gattuso, S. Menzer, S.A. Nepogodiev, J.F. Stoddart, and D.J. Williams, "Carbothdrate Nanotubes", *Angew. Chem. Int. Ed. Engl.,* vol. 36, no. 1314, pp. 1451-1454, 1997.
[http://dx.doi.org/10.1002/anie.199714511]

[80]   Z. Luo, and S. Zhang, "Designer nanomaterials using chiral self-assembling peptide systems and their emerging benefit for society", *Chem. Soc. Rev.,* vol. 41, no. 13, pp. 4736-4754, 2012.
[http://dx.doi.org/10.1039/c2cs15360b] [PMID: 22627925]

[81]   A. Lakshmanan, S. Zhang, and C.A.E. Hauser, "Short self-assembling peptides as building blocks for modern nanodevices", *Trends Biotechnol.,* vol. 30, no. 3, pp. 155-165, 2012.
[http://dx.doi.org/10.1016/j.tibtech.2011.11.001] [PMID: 22197260]

[82]   J.B. Matson, and S.I. Stupp, "Self-assembling peptide scaffolds for regenerative medicine", *Chem. Commun. (Camb.),* vol. 48, no. 1, pp. 26-33, 2012.
[http://dx.doi.org/10.1039/C1CC15551B] [PMID: 22080255]

[83]   C.E. Castro, F. Kilchherr, D.N. Kim, E.L. Shiao, T. Wauer, P. Wortmann, M. Bathe, and H. Dietz, "A primer to scaffolded DNA origami", *Nat. Methods,* vol. 8, no. 3, pp. 221-229, 2011.
[http://dx.doi.org/10.1038/nmeth.1570] [PMID: 21358626]

[84]   D. Jasinski, F. Haque, D.W. Binzel, and P. Guo, "Advancement of the emerging field of RNA nanotechnology", *ACS Nano,* vol. 11, no. 2, pp. 1142-1164, 2017.
[http://dx.doi.org/10.1021/acsnano.6b05737] [PMID: 28045501]

[85]   Y. Wang, D. Maspoch, S. Zou, G.C. Schatz, R.E. Smalley, and C.A. Mirkin, "Controlling the shape, orientation, and linkage of carbon nanotube features with nano affinity templates", *Proc. Natl. Acad. Sci. USA,* vol. 103, no. 7, pp. 2026-2031, 2006.
[http://dx.doi.org/10.1073/pnas.0511022103] [PMID: 16461892]

[86]   M. Zuberi, D.M. Sherman, and Y. Cho, "Carbon Nanotube Microspheres Produced by Surfactant-Mediated Aggregation", *J. Phys. Chem. C,* vol. 115, no. 10, pp. 3881-3887, 2011.
[http://dx.doi.org/10.1021/jp110019e]

[87]   M. Engel, J.P. Small, M. Steiner, M. Freitag, A.A. Green, M.C. Hersam, and P. Avouris, "Thin film nanotube transistors based on self-assembled, aligned, semiconducting carbon nanotube arrays", *ACS Nano,* vol. 2, no. 12, pp. 2445-2452, 2008.
[http://dx.doi.org/10.1021/nn800708w] [PMID: 19206278]

[88]   M. Karimi, A. Ghasemi, S. Mirkiani, S.M.M. Basri, and M.R. Hamblin, *Carbon Nanotubes in Drug and Gene Delivery.* Morgan & Claypool Publishers San Rafael: CA, 2017.
[http://dx.doi.org/10.1088/978-1-6817-4261-8]

[89]   K.S. Novoselov, A.K. Geim, S.V. Morozov, D. Jiang, Y. Zhang, S.V. Dubonos, I.V. Grigorieva, and

A.A. Firsov, "Electric field effect in atomically thin carbon films", *Science,* vol. 306, no. 5696, pp. 666-669, 2004.
[http://dx.doi.org/10.1126/science.1102896] [PMID: 15499015]

[90] E. Campbell, M.T. Hasan, C. Pho, K. Callaghan, G.R. Akkaraju, and A.V. Naumov, "Graphene oxide as a multifunctional platform for intracellular delivery, imaging, and cancer sensing", *Sci. Rep.,* vol. 9, no. 1, p. 416, 2019.
[http://dx.doi.org/10.1038/s41598-018-36617-4] [PMID: 30674914]

[91] X. Cao, S. Zheng, S. Zhang, Y. Wang, X. Yang, H. Duan, Y. Huang, and Y. Chen, "Functionalized graphene oxide with hepatocyte targeting as anti-tumor drug and gene intracellular transporters", *J. Nanosci. Nanotechnol.,* vol. 15, no. 3, pp. 2052-2059, 2015.
[http://dx.doi.org/10.1166/jnn.2015.9145] [PMID: 26413620]

[92] A.M. Martins, Q.P. Pham, P.B. Malafaya, R.A. Sousa, M.E. Gomes, R.M. Raphael, F.K. Kasper, R.L. Reis, and A.G. Mikos, "The role of lipase and α-amylase in the degradation of starch/poly(epsilon-caprolactone) fiber meshes and the osteogenic differentiation of cultured marrow stromal cells", *Tissue Eng. Part A,* vol. 15, no. 2, pp. 295-305, 2009.
[http://dx.doi.org/10.1089/ten.tea.2008.0025] [PMID: 18721077]

[93] R. Langer, and J.P. Vacanti, "Tissue Engineering", *Science,* vol. 260, no. 5110, pp. 920-926, 1993.
[http://dx.doi.org/10.1126/science.8493529] [PMID: 8493529]

[94] J.W. Nichol, and A. Khademhosseini, "Modular tissue engineering: engineering biological tissues from the bottom up", *Soft Matter,* vol. 5, no. 7, pp. 1312-1319, 2009.
[http://dx.doi.org/10.1039/b814285h] [PMID: 20179781]

[95] K.J. Gooch, T. Blunk, D.L. Courter, A.L. Sieminski, G. Vunjak-Novakovic, and L.E. Freed, "Bone morphogenetic proteins-2, -12, and -13 modulate *in vitro* development of engineered cartilage", *Tissue Eng.,* vol. 8, no. 4, pp. 591-601, 2002.
[http://dx.doi.org/10.1089/107632702760240517] [PMID: 12201999]

[96] M.S. Hahn, J.S. Miller, and J.L. West, "Three-dimensional biochemical and biomechanical patterning of hydrogels for guiding cell behavior", *Adv. Mater.,* vol. 18, no. 20, pp. 2679-2684, 2006.
[http://dx.doi.org/10.1002/adma.200600647]

[97] P. Martin, "Wound healing-aiming for perfect skin regeneration", *Science,* vol. 276, no. 5309, pp. 75-81, 1997.
[http://dx.doi.org/10.1126/science.276.5309.75] [PMID: 9082989]

[98] Y.M. Bello, A.F. Falabella, and W.H. Eaglstein, "Tissue-Engineered Skin", *Am. J. Clin. Dermatol.,* vol. 2, no. 5, pp. 305-313, 2001.
[http://dx.doi.org/10.2165/00128071-200102050-00005] [PMID: 11721649]

[99] D.W. Hutmacher, J.T. Schantz, C.X.F. Lam, K.C. Tan, and T.C. Lim, "State of the art and future directions of scaffold-based bone engineering from a biomaterials perspective", *J. Tissue Eng. Regen. Med.,* vol. 1, no. 4, pp. 245-260, 2007.
[http://dx.doi.org/10.1002/term.24] [PMID: 18038415]

[100] G. Matsumura, N. Nitta, S. Matsuda, Y. Sakamoto, N. Isayama, K. Yamazaki, and Y. Ikada, "Long-term results of cell-free biodegradable scaffolds for *in situ* tissue-engineering vasculature: in a canine inferior vena cava model", *PLoS One,* vol. 7, no. 4, p. e35760, 2012.
[http://dx.doi.org/10.1371/journal.pone.0035760] [PMID: 22532873]

[101] B. Subia, J. Kundu, and S. Kundu, *Biomaterial scaffold fabrication techniques for potential tissue engineering applications.* vol. Vol. 141. Tissue Eng, 2010.
[http://dx.doi.org/10.5772/8581]

[102] A. Haider, S. Haider, M. Rao Kummara, T. Kamal, A.A.A. Alghyamah, F. Jan Iftikhar, B. Bano, N. Khan, M. Amjid Afridi, S. Soo Han, A. Alrahlah, and R. Khan, "Advances in the scaffolds fabrication techniques using biocompatible polymers and their biomedical application: A technical and statistical review", *J. Saudi Chem. Soc.,* vol. 24, no. 2, pp. 186-215, 2020.

[http://dx.doi.org/10.1016/j.jscs.2020.01.002]

[103] M.E. Gomes, H.S. Azevedo, A.R. Moreira, V. Ellä, M. Kellomäki, and R.L. Reis, "Starch-poly(ε-caprolactone) and starch-poly(lactic acid) fibre-mesh scaffolds for bone tissue engineering applications: structure, mechanical properties and degradation behaviour", *J. Tissue Eng. Regen. Med.,* vol. 2, no. 5, pp. 243-252, 2008.
[http://dx.doi.org/10.1002/term.89] [PMID: 18537196]

[104] P.J. VandeVord, H.W.T. Matthew, S.P. DeSilva, L. Mayton, B. Wu, and P.H. Wooley, "Evaluation of the biocompatibility of a chitosan scaffold in mice", *J. Biomed. Mater. Res.,* vol. 59, no. 3, pp. 585-590, 2002.
[http://dx.doi.org/10.1002/jbm.1270] [PMID: 11774317]

[105] A. Lahiji, A. Sohrabi, D.S. Hungerford, and C.G. Frondoza, "Chitosan supports the expression of extracellular matrix proteins in human osteoblasts and chondrocytes", *J. Biomed. Mater. Res.,* vol. 51, no. 4, pp. 586-595, 2000.
[http://dx.doi.org/10.1002/1097-4636(20000915)51:4<586::AID-JBM6>3.0.CO;2-S]   [PMID: 10880106]

[106] J-K. Francis Suh, and H.W.T. Matthew, "Application of chitosan-based polysaccharide biomaterials in cartilage tissue engineering: a review", *Biomaterials,* vol. 21, no. 24, pp. 2589-2598, 2000.
[http://dx.doi.org/10.1016/S0142-9612(00)00126-5] [PMID: 11071608]

[107] R. Muzzarelli, G. Biagini, A. Pugnaloni, O. Filippini, V. Baldassarre, C. Castaldini, and C. Rizzoli, "Reconstruction of parodontal tissue with chitosan", *Biomaterials,* vol. 10, no. 9, pp. 598-603, 1989.
[http://dx.doi.org/10.1016/0142-9612(89)90113-0] [PMID: 2611308]

[108] S. Hsu, Y.B. Chang, C.L. Tsai, K.Y. Fu, S.H. Wang, and H.J. Tseng, "Characterization and biocompatibility of chitosan nanocomposites", *Colloids Surf. B Biointerfaces,* vol. 85, no. 2, pp. 198-206, 2011.
[http://dx.doi.org/10.1016/j.colsurfb.2011.02.029] [PMID: 21435843]

[109] R.A.A. Muzzarelli, M. Mattioli-Belmonte, C. Tietz, R. Biagini, G. Ferioli, M.A. Brunelli, M. Fini, R. Giardino, P. Ilari, and G. Biagini, "Stimulatory effect on bone formation exerted by a modified chitosan", *Biomaterials,* vol. 15, no. 13, pp. 1075-1081, 1994.
[http://dx.doi.org/10.1016/0142-9612(94)90093-0] [PMID: 7888578]

[110] T. Kawakami, M. Antoh, H. Hasegawa, T. Yamagishi, M. Ito, and S. Eda, "Experimental study on osteoconductive properties of a chitosan-bonded hydroxyapatite self-hardening paste", *Biomaterials,* vol. 13, no. 11, pp. 759-763, 1992.
[http://dx.doi.org/10.1016/0142-9612(92)90014-F] [PMID: 1391397]

[111] S. Hirano, K. Nagamura, M. Zhang, S.K. Kim, B.G. Chung, M. Yoshikawa, and T. Midorikawa, "Chitosan staple fibers and their chemical modification with some aldehydes", *Carbohydr. Polym.,* vol. 38, no. 4, pp. 293-298, 1999.
[http://dx.doi.org/10.1016/S0144-8617(98)00126-X]

[112] K. Tuzlakoglu, C.M. Alves, J.F. Mano, and R.L. Reis, "Production and characterization of chitosan fibers and 3-D fiber mesh scaffolds for tissue engineering applications", *Macromol. Biosci.,* vol. 4, no. 8, pp. 811-819, 2004.
[http://dx.doi.org/10.1002/mabi.200300100] [PMID: 15468275]

[113] A. Di Martino, M. Sittinger, and M.V. Risbud, "Chitosan: A versatile biopolymer for orthopaedic tissue-engineering", *Biomaterials,* vol. 26, no. 30, pp. 5983-5990, 2005.
[http://dx.doi.org/10.1016/j.biomaterials.2005.03.016] [PMID: 15894370]

[114] A. Sarasam, and S. Madihally, "Characterization of chitosan-polycaprolactone blends for tissue engineering applications", *Biomaterials,* vol. 26, no. 27, pp. 5500-5508, 2005.
[http://dx.doi.org/10.1016/j.biomaterials.2005.01.071] [PMID: 15860206]

[115] M.A. Woodruff, and D.W. Hutmacher, "The return of a forgotten polymer—Polycaprolactone in the 21st century", *Prog. Polym. Sci.,* vol. 35, no. 10, pp. 1217-1256, 2010.

[http://dx.doi.org/10.1016/j.progpolymsci.2010.04.002]

[116] S. Ishaug-Riley, L.E. Okun, G. Prado, M.A. Applegate, and A. Ratcliffe, "Human articular chondrocyte adhesion and proliferation on synthetic biodegradable polymer films", *Biomaterials,* vol. 20, no. 23-24, pp. 2245-2256, 1999.
[http://dx.doi.org/10.1016/S0142-9612(99)00155-6] [PMID: 10614931]

[117] M.E. Hoque, W.Y. San, F. Wei, S. Li, M.H. Huang, M. Vert, and D.W. Hutmacher, "Processing of polycaprolactone and polycaprolactone-based copolymers into 3D scaffolds, and their cellular responses", *Tissue Eng. Part A,* vol. 15, no. 10, pp. 3013-3024, 2009.
[http://dx.doi.org/10.1089/ten.tea.2008.0355] [PMID: 19331580]

[118] N. Garcia-Giralt, R. Izquierdo, X. Nogués, M. Perez-Olmedilla, P. Benito, J.L. Gómez-Ribelles, M.A. Checa, J. Suay, E. Caceres, and J.C. Monllau, "A porous PCL scaffold promotes the human chondrocytes redifferentiation and hyaline-specific extracellular matrix protein synthesis", *J. Biomed. Mater. Res. A,* vol. 85A, no. 4, pp. 1082-1089, 2008.
[http://dx.doi.org/10.1002/jbm.a.31670] [PMID: 17937412]

[119] R. van Dijkhuizen-Radersma, L. Moroni, A. van Apeldoorn, Z. Zhang, and D. Grijpma, "Degradable polymers for tissue engineering", In: *Tissue engineering* Elsevier, 2008.
[http://dx.doi.org/10.1016/B978-0-12-370869-4.00007-0]

[120] B.L. Guo, and P.X. Ma, "Synthetic biodegradable functional polymers for tissue engineering: a brief review", *Sci. China Chem.,* vol. 57, no. 4, pp. 490-500, 2014.
[http://dx.doi.org/10.1007/s11426-014-5086-y] [PMID: 25729390]

[121] S.C. Neves, L.S. Moreira Teixeira, L. Moroni, R.L. Reis, C.A. Van Blitterswijk, N.M. Alves, M. Karperien, and J.F. Mano, "Chitosan/Poly(ε-caprolactone) blend scaffolds for cartilage repair", *Biomaterials,* vol. 32, no. 4, pp. 1068-1079, 2011.
[http://dx.doi.org/10.1016/j.biomaterials.2010.09.073] [PMID: 20980050]

[122] S.M. Lien, L.Y. Ko, and T.J. Huang, "Effect of crosslinking temperature on compression strength of gelatin scaffold for articular cartilage tissue engineering", *Mater. Sci. Eng. C,* vol. 30, no. 4, pp. 631-635, 2010.
[http://dx.doi.org/10.1016/j.msec.2010.02.019]

[123] C. Peña, K. de la Caba, A. Eceiza, R. Ruseckaite, and I. Mondragon, "Enhancing water repellence and mechanical properties of gelatin films by tannin addition", *Bioresour. Technol.,* vol. 101, no. 17, pp. 6836-6842, 2010.
[http://dx.doi.org/10.1016/j.biortech.2010.03.112] [PMID: 20400296]

[124] J. Sartuqui, A.N. Gravina, R. Rial, L.A. Benedini, L.H. Yahia, J.M. Ruso, and P.V. Messina, "Biomimetic fiber mesh scaffolds based on gelatin and hydroxyapatite nano-rods: Designing intrinsic skills to attain bone reparation abilities", *Colloids Surf. B Biointerfaces,* vol. 145, pp. 382-391, 2016.
[http://dx.doi.org/10.1016/j.colsurfb.2016.05.019] [PMID: 27220014]

CHAPTER 10

Solvent Casting and Melt Molding Techniques for Fabrication of Biomaterials

Atiya Fatima[1], Md. Wasi Ahmed[1], Muhammad Wajid Ullah[2,3], Sehrish Manan[2,3], Shaukat Khan[1], Aref Ahmad Wazwaz[1] and **Mazhar Ul-Islam[1,*]**

[1] *Department of Chemical Engineering, College of Engineering, Dhofar University, Salalah 211, Sultanate of Oman*

[2] *Biofuels Institute, School of the Environment and Safety Engineering, Jiangsu University, Zhenjiang 212013, PR China*

[3] *Department of Biomedical Engineering, Huazhong University of Science and Technology, Wuhan 430074, PR China*

Abstract: Biomaterials are receiving tremendous attention, especially in the biomedical field, due to their impressive structural, physiological, and biological properties, such as nontoxicity, biocompatibility, and biodegradability. Numerous biomaterials have been used to fabricate scaffolds for applications in tissue engineering and regenerative medicine, where they are used as wound dressings, grafts, organs, and substitutes. To date, a number of techniques have been developed for the fabrication of scaffolds from biomaterials. This chapter focuses on the fabrication of scaffolds by solvent casting and melt-casting techniques. It examines the solvent casting and melt-casting techniques in terms of their application in the fabrication of biological scaffolds with tailored micro- and nanostructures for their use in tissue engineering. The merits and limitations of these techniques in fabricating biological scaffolds for desired biomedical applications are also discussed. Finally, various challenges faced by solvent and melt casting techniques are described, and solutions are proposed for future research to develop biomaterials for advanced biomedical applications.

Keywords: Biocompatibility, Biomaterials, Fabrication techniques, Scaffolds, Structural features.

INTRODUCTION

Tissue engineering provides an innovative platform focused on developing scaffolds with biological and mechanical properties to overcome serious medical problems, such as tissue loss or damage and organ failure. It is highly dependent

* **Corresponding author Mazhar Ul-Islam:** Department of Chemical Engineering, College of Engineering, Dhofar University, Salalah 211, Sultanate of Oman; E-mail: mulislam@du.edu.om

Adnan Haider & Sajjad Haider (Eds.)

on the biocompatibility, biodegradability, and bioresorbability of scaffolds, which limits access to available materials and viable techniques [1, 2]. Microstructural properties, such as porosity and pore connectivity, as well as the required mechanical strength of the scaffold, pose a major challenge for biomaterials to meet the desired properties of the target tissue or organ [3, 4]. In addition, the cost, reproducibility, and simplicity of the techniques without compromising the biocompatibility of the material are other obstacles to the fabrication method [5]. The conventional fabrication technologies, such as solvent casting and melt casting techniques, have numerous limitations; nevertheless, their simple protocols have promoted their use. These techniques are inexpensive, straightforward, and easily scalable compared to their counterparts [6 - 8]. In recent years, solvent casting technology has evolved into a high-precision technique used in the fabrication of optical and medical films, opening up potential applications in bioelectronics [9]. Melt casting techniques such as additive extrusion or injection molding techniques are widely used in the development of solid implants such as plates, rods, and screws, and are also used in dentistry. These techniques are often combined with other technologies to obtain a framework with the desired properties.

Biomaterials

Biomaterials are non-toxic substances composed of either natural or synthetic components, that do not induce immunogenic and inflammatory reactions, and are frequently used in medical applications [10, 11]. A biomaterial interacts with the biological system and supplements or replaces a natural function. Generally, biomaterials are classified into two broad categories, namely natural and synthetic biomaterials. Natural biomaterials are mostly composed of natural polymers, including proteins such as collagen [12], fibrin [13], and silk [14], and polysaccharides such as cellulose [15 - 17], chitosan [18], alginate [19, 20], and hyaluronan [21, 22]. Synthetic biomaterials include three categories with polymers such as peptides and ceramic-based biomaterials [23, 24]. Examples of commonly used synthetic polymers in the development of biomaterials include poly (lactic-co-glycolic acid) (PLGA) [25], poly (ε-caprolactone) (PCL) [26], poly (ethylene glycol) (PEG) [27], poly (vinyl alcohol) (PVA) [28, 29], and others. Peptide-based materials include amino acids and peptides [29], ceramic-based biomaterials include hydroxyapatite (HAp) [30], and ceramic-based biomaterials include hydroxyapatite (HAp) [31] and bioactive glass [32]. Biomaterials are used together with cells and bioactive substances to synthesize new tissues using tissue engineering techniques [33]. Recently, biological scaffolds have become an important medical substitute for synthetic implants and tissue grafts [34]. A well-designed three-dimensional (3D) scaffold should have certain important properties to allow the cells to regenerate the tissues and organs in the desired shape and size. They should be biocompatible with the host tissue and have the required porosity and mechanical strength. Their surface should have adhesive properties that allow cell attachment, growth, migration and differentiation. Controlled biodegradability and safe implantation, as well as suitable mechanical properties, are other structural and chemical requirements for the successful development of biological scaffolds [18, 22, 35, 36, 37].

FABRICATION TECHNIQUES

Living tissue comprises different cell types and extracellular matrix organized into a complex architecture performing cellular and mechanical functions. Designing of scaffolds requires a strategical analysis of the microcellular structure of native tissue and its functioning at the cellular level enabling the proliferation and migration of cells. The engineering of scaffolds requires techniques to deliver scaffolds with the best regenerative performance with respect to the native tissue requirements. There are several methods for the fabrication of scaffolds, such as solvent casting, melt molding, phase separation, freeze-drying, gas foaming, phase separation, and membrane lamination, to name a few. In this section, the main techniques for solvent casting and melt molding are discussed in detail.

Solvent Casting

Solvent casting is a simple and inexpensive technique that requires a mold and a polymer dissolved in an organic solution to produce scaffolds. The polymer is dissolved in an organic solvent, and the scaffold is obtained by simply evaporating the solvent. The desired scaffold is obtained either by immersing the mold in the polymeric solvent or by adding the polymeric solution to the mold. In the first method, the mold is immersed in the solution and then dried to form a mold from the polymer membrane. In the second method, the solution is added to the mold, and the solution is allowed sufficient time to dry so that a layer of the polymer membrane forms on the mold [38]. Fig. (**1**) shows a generalized diagram depicting the technique of solvent casting with particle leaching to develop scaffolds.

Solvent casting is a widely used technique because it allows uniform distribution of polymer throughout the framework and provides changeable reaction conditions [39]. The role of the solvent is a critical factor in the preparation of the polymer surface. The heterogeneity of the surface, the swelling behavior, and the deformation rates of the scaffold affect its application [40]. The main advantages of the solvent casting technique are its simplicity, convenience, and easy fabrication of scaffolds. The degeneration of the scaffold does not affect the regeneration rate of the native tissue. However, the long drying time of the molds, the toxicity of the organic solvents used in the fabrication of the scaffolds, and the limitation to tubular and plate molds are the major drawbacks of this technique. The organic solvents can reduce the activity of the biopolymers by denaturing the proteins and affecting other solvents, and they can be toxic to the cells [41]. Vacuum processing can remove the traces of organic solvents to some extent. However, some progress has been made by researchers combining different techniques to overcome the drawbacks of this technique in the fabrication of 3D scaffolds. Particulate leaching techniques are combined with the solvent casting process to produce scaffolds with improved properties [42].

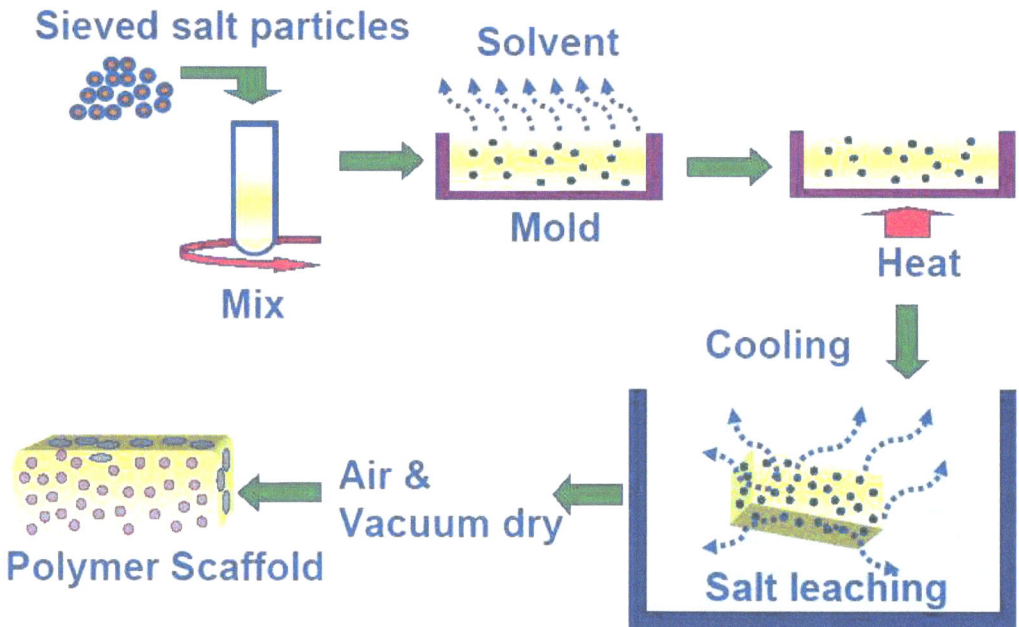

Fig. (1). Solvent-casting particle leaching technique. Figure reproduced from [43].

The technique of solvent casting with particle leaching provides a scaffold of high porosity and a wide range of pore sizes, depending on the pore size of the porogen used [44]. It is used to fabricate 3D polymer scaffolds with pores or channels using porogens such as salts or polysaccharides like crystals, gelatin beads or kerosene [45]. Various 3D scaffolds have been developed for tissue engineering applications using poly- and chitosan-modified montmorillonite (CS-MMT) with NaCl as porogen, resulting in scaffolds with varying degrees of biodegradability and bioresorbability [46]. The particle size of the porogen directly affects the porosity as well as the pore size in the final matrix. In this process, the salt is ground into small particles, and the desired size of the salt is mixed with the polymer solution. Evaporation of the solvent leaves a matrix that is immersed in water, which leaches out the salt particles and leaves a polymer matrix with high porosity [47]. This process results in scaffolds with a porosity of 50%-90%, an average pore diameter of 500 μm, a high surface area, and the desired crystallinity [48]. Different solvents such as hexafluoroisopropanol (HFIP), dichloromethane (DCM) and chloroform (CF) are used to prepare poly(lactic acid) scaffolds with different densities. The PLA scaffolds prepared using CF have high porosity (93.3%), while the PLA scaffolds prepared using DCM as solvent have the best thermal stability [49]. This technique finds application in the preparation of scaffolds containing bioactive compounds and natural polymers [50]. It has the advantage that little polymer is needed to prepare the matrix.

Poly (L-lactic acid) (PLLA), a synthetic biodegradable polymer, PLA and PLA-dextran blend scaffolds were synthesized using a combination of SCPL techniques. The resulting scaffolds exhibited micropores of size 5-10 μm with higher porosity and an open-pore structure. IPLA-dextran blends were better than PLA-only scaffolds in terms of hydrophilicity. Biocompatibility and increased cell affinity facilitated cell adhesion efficiency and penetration into the scaffolds [51]. In another study, salt/gelatin (particle size of 100-180 μm) was dispersed in a PLGA/CF solution and poured into a Teflon container. Evaporation and vacuum drying were used to remove the solvent. The particles were then dissolved in warm water (40 °C) to prepare the scaffolds. These scaffolds were then seeded with cultured chondrocytes (cells from knee cartilage) and smooth muscle cells (cells from urinary bladder) and cultured for 3 weeks. It was observed that gelatin scaffolds exhibited better adhesion of cells in the initial phase, which could be attributed to better pore connectivity at the same porosity [52]. The nanocomposite films with antibacterial activity and negligible cytotoxicity towards fibroblast cell line (NIH-3T3) were synthesized using cellulose nanocrystals (CNCs) and reduced graphene oxide (rGO) with PLA matrix using CF as solvent. These nanofilms have strong applications in the field of biomedicine [53, 54]. Another study provided a viable nano-hydroxyapatite (n-HAp)/chitosan (CS) cross-linking composite membrane that can be used as support for bone tissue engineering. In this study, researchers used genipin as a cross-linking agent for the synthesis of composite membranes with acetic acid as a solvent [55]. Two different polymers, *i.e.*, a rigid poly(methyl methacrylate) (PMMA) and flexible polyurethane (PU) with NaCl as porogen, were used to prepare a polymer-based scaffold mimicking the bone marrow niche *in vitro*. The appropriate ratio of polymer to salt resulted in a rich and cross-linked porous scaffold, and the addition of fine-grained or coarse-grained porogen allowed controlled pore size. *In vitro* studies using human stromal cells, HS -5 cultured with leukemic cells, yielded results favoring the use of scaffolds in novel therapies to mimic the bone marrow microenvironment [56].

Bionanocomposite materials composed of high aspect ratio polymer-coated copper-cystine structures (CuHARS) and cellulose fibers were used by Darder *et al*. to synthesize films by solvent casting and freeze-drying. The resulting films showed good mechanical behavior and water resistance, while the foams exhibited a network of interconnected macropores with diameters of about 130 μm. These composites are promising candidates for potential biomedical applications and can be used as implants or wound dressings. These biomaterials exhibit antimicrobial activity with the lowest cytotoxicity [57 - 60]. In a comparative study, the solvent casting process with particle leaching and the robocasting process was used to develop PCL scaffolds for tissue engineering applications. In the solvent casting process with salt leaching, NaCl was used as the porogen, and the graphene nanopowders (nanoflakes) were incorporated into the polymer matrix at concentrations of 1, 3, 5, and 10 wt%. DCM was used as a solvent for leaching the NaCl crystals, which formed an interconnected porous network within the composite scaffolds. In the robocasting method, acetone was used as a solvent, and PCL solutions were prepared at 20 wt% to produce scaffolds with a lattice-like structure. The robocast composite constructs showed a better biological response than solvent-cast scaffolds under similar conditions [61].

The main disadvantage of solvent casting with the particle leaching technique is that it can only be used to fabricate thin membranes to dissolve soluble particles from the matrix. Other disadvantages include the time required to produce the layers and the uncontrolled cross-linking of the pores and their shapes [62, 63]. The use of organic solvents, which may have toxic effects, and the uncontrolled agglomeration of porogen particles, leading to random distribution of porogen, are some of the other disadvantages of this technique [64].

Melt Casting/Molding

In the melt molding technique, a scaffold is fabricated by melting polymer powder and porogen and filling it into a mold. Polymer powder/ceramic is mixed with porogen (sodium chloride/sugar crystals/gelatin microspheres), heated above the glass transition temperature, and then poured into a mold under pressure [65]. This causes the polymer particles to bind and form scaffolds with the desired specific shapes. Once the mixture has cooled, the mold is removed, the porogen is leached by immersion in water, and the scaffolds are dried. Disadvantages limiting the use of melt casting are the high processing temperature required for non-amorphous polymers and the possibility of residual porogen remaining in the scaffolds [66, 67]. The melt casting process is used to produce implants from biodegradable polymers. The compression molding process, the injection molding process, and the extrusion process constitute the three main aspects of melt molding technology. All these techniques differ in the way the polymer is heated and in the shapes of the elements used (dye or mold). These techniques can be used to produce frameworks with complex external geometries and porosity. They enable cost-effective and reproducible scaffolds in the field of orthopedics. Scaffolds for regenerative medicine are fabricated with high porosity and good interconnectivity.

Compression Molding Technologies

In compression molding, pressure is applied to the polymer at the time of heating to release the trapped air and compress the particles. The mold containing the polymer is heated above the glass transition temperature or the melting temperature for amorphous and semi-crystalline polymers, resulting in scaffolds with high density and low material shrinkage. In this method, it is easier to introduce bioactive factors into the polymers because it operates at relatively low temperatures [68]. Complex and robust 3D shapes can be obtained, which are less susceptible to material and processing parameters [69]. This technique is also used in several pore-generating processes to produce scaffolds with the desired porosity.

Compression Molding/Particulate Leaching

Particle leaching molding is one of the most commonly used methods. It involves dissolving the porogen from compressed polymer scaffolds into solvent-releasing porous scaffolds. The advantage of this method is the production of 3D scaffolds, absence of solvents, autonomous control of porosity, pore size, macroshape, interconnectivity of pores, and geometry, which is an important factor for the exchange of nutrients or waste materials from pore to pore. Various porogens, including inorganic materials and polymers, can be used in this method. In this technique, materials with one or two porogens, including inorganic and polymeric porogens, are used to produce scaffolds with different pore morphologies and mechanical properties. Scaffolds made from PLGA/PVA blends resulted in scaffolds with high porosity and open cell pores that facilitated cell growth and proliferation of bone cells in both *in vitro* and *in vivo* studies [25]. Similar studies using poly(dioxanone-b-caprolactone) copolymer (PDOCL) scaffolds prepared by the melt casting method provided a suitable environment for cell-polymer interactions and extracellular matrix regeneration [70]. NaCl is used as a porogen to generate scaffolds with porosities from 50% to 90% and pore sizes from 10 to 1000 μm in the synthesis of biodegradable polymers based on starch [71] and poly (D, L-lactic acid) (PDLLA) and 1000PEOT70PBT30 poly (ether ester) block copolymer [72]. Gelatin microspheres are used as porogens to generate porosity from 36% to 70% and pore sizes from 106 to 710 μm in porous PLGA scaffolds

[73]. The use of double porogens can also be used in this process to obtain scaffolds with high porosity, good interconnectivity, and reasonable mechanical properties. Shokrolahi and co-workers used NaCl in combination with PEG at a ratio of 60/25 to produce poly(urethaneurea) scaffolds that are morphologically suitable for osteoblast cell growth and attachment [74]. In another experiment on modified scaffold synthesis for tissue engineering applications, a mixture of water-soluble porogens containing poly(ethylene oxide) (PEO) particles and PCL in different ratios was heated for 45 min and then compressed to obtain the desired scaffold with open and interconnected pore networks of different lengths [69]. Heatless compression is another method used in compression molding of PLLA and salt particles. These compressed molds were then heated, and the salt particles were leached [75]. Composite scaffolds can also be synthesized using CM / PL techniques. PLGA and PVA (0-20 wt% PVA) were cryogenically mixed and sandwiched between two NaCl layers and compressed into a mold. After thermal compression and leaching of salt particles, the molds were freeze-dried, resulting in porous PLGA/PVA scaffolds [25]. Gradual mixing of powders followed by gradual heating and compression/leaching of particles was also used for scaffold synthesis. In this method, researchers placed a homogeneous mixture of PLLA/β-tricalcium phosphate and salt in a mold, heated the mixture, and continued compression for 5 minutes. In the next step, the mold was heated and pressed again to obtain composite scaffolds having improved interfacial compatibility and high mechanical properties [76]. Fig. (**2**) shows a schematic diagram of melt casting with particle leaching to produce polymer-based scaffold.

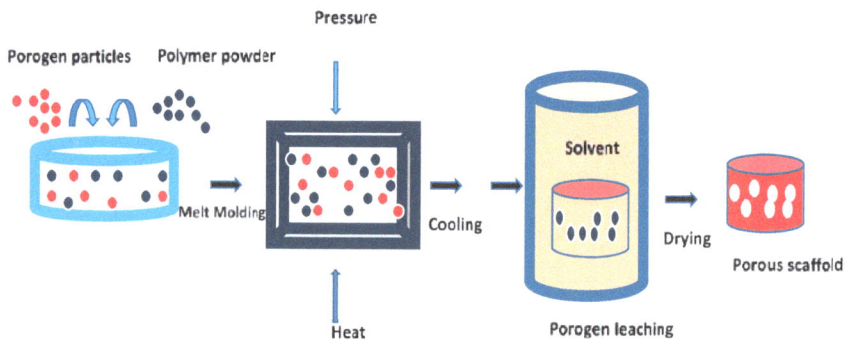

Fig. (2). A schematic representation of the melt molding-particulate-leaching scaffold fabrication technique. Figure modified from [77].

Compression Molding/Phase Separation

Phase separation by compression molding (Fig. **3**) involves phase separation of immiscible polymers with selective leaching of the porogen, resulting in a porous scaffold. In this technique, two immiscible polymers are homogeneously blended through cryomilling and then heated, resulting in segregation and coarsening to produce a co-continuous mixed morphology. This selective leaching of a polymeric phase results in interconnected pores and stronger scaffolds. Numerous parameters, such as the duration of cryogenic grinding, the composition of the mixture, and the molding temperature, can be adjusted to achieve the desired porosity and pore size in the resulting scaffolds. The resulting cylindrical pores facilitate cell spreading compared to angular pores. With this technique, multi-walled carbon nanotubes (MWCNTs) were used in combination with PCL to fabricate novel porous nanocomposite scaffolds. A homogeneous mixture of MWCNTs and PCL/PCL/poly(glycolic acid) (PGA) blends prepared by cryomilling followed by compression molding/polymer leaching resulted in scaffolds with an average pore size of about 3-5 μm. The scaffolds exhibited sustained pore interconnectivity, good mechanical strength, and good water absorption capacity [78].

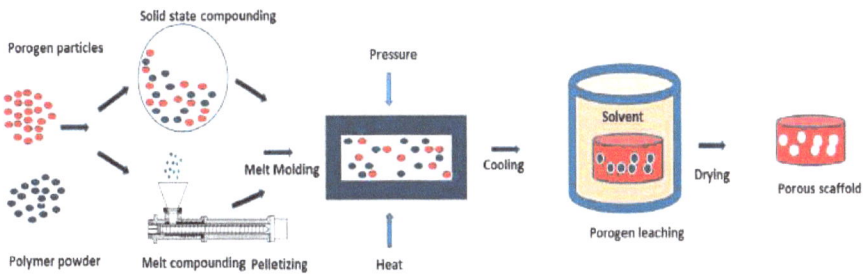

Fig. (3). A schematic representation of the compression molding/phase-separation (CM/PS) scaffold fabrication technique. Figure reproduced from [77].

Similarly, porous PCL and HAp scaffolds were prepared using PEO as porogen. The PEO and PCL mixture was prepared using a vortex mixer and then heated and compacted. PEO was washed out with deionized water. Scaffolds with PCL/PEO/ HA were prepared in two steps: In the first step, PCL/HA disks were prepared by mixing HA and PCL in a vortex mixer and then heating in a mold to form disks. These disks were cooled, ground and sieved, and mixed with PEO. This mixture was heated in a mold, and the porogen was leached. This process was able to produce scaffolds with 90% pores larger than 106 μm [79]. A schematic diagram of the melt forming/phase separation for the preparation of polymer-based scaffolds is shown in Fig. (**3**).

Compression Molding/Solvent Casting/Particulate Leaching

In this method, CM is used in combination with SC /PL techniques to synthesize porous scaffolds with a more homogeneous morphology than other common fabrication techniques such as CM /PL and SC /PL. This method has the advantages of thermal processing techniques together with particulate leaching methods. It ensures the homogeneous distribution of porogens in the polymer matrix. Wu *et al.* fabricated PDLLA- and PLGA-based auricle-like and hinge-like porous scaffolds exhibiting porosity <of 90%, highly interconnected and uniformly distributed pores [80]. Similarly, another group designed porous scaffolds from polymers such as poly(D, L-lactic acid) (PDLLA), PCL and 1000PEOT70PBT30 (segmented poly(ether ester) based on PEO and polybutylene terephthalate. The polymer solution is added to an organic solvent with dispersed water-soluble particles and precipitated into a non-solvent. It is then thermally processed, resulting in the desired porous scaffolds with 70 to 95% porosity and homogeneous interconnected pore networks [72].

Wire-Network Molding

Wire Network Molding is a simple technique that uses a network of wires inside the mold to achieve significant distribution of pores throughout the scaffold and control their size. It is advantageous in many ways, including being solvent-free, economical, and providing scaffolds with well-connected pores. A mold is fitted with a network of wires, and the molten polymer is poured into it. In the next step, the mold is cooled, and the wired network is removed, resulting in a superbly cross-linked porous structure. This technique is not limited to the use of thermoplastic polymers. In one study, alginate gel in which human mesenchymal stem cells (hMSC) were embedded was injected into a mold containing a wire mesh of ETPCS-S wires, and the mold was then soaked in $CaCl_2$ solution. After the wires and alginate gel were removed from the mold, an alginate scaffold with interconnected pores corresponding to those of the wire network [81]. Combined with the salt leaching technique, it was used in a study to synthesize PCL structures with dual porosity (hollow channels of 450-550 µm size and local pores of 50-250 µm size). A wired network mold was filled with a powdered mixture of PCL and NaCl (ratio 30/70) and then pressurized and heated. After the PCL powder was melted, the needles were removed, and the NaCl was leached out [82].

Injection Molding Technologies

Injection molding is a process that can produce complex porous 3D porous scaffolds with extreme dimensional accuracy. It is an effective technique for fabricating highly repeatable porous scaffolds with complex external geometries

[83]. The molten polymer is injected into a mold cavity where it solidifies into the shape of the mold. This technique is used for the fabrication of orthopedic parts and various other components.

Injection Molding/Particulate Leaching

In this process, the mixture of polymer powder and porogen particles is granulated in a mixer or twin-screw extruder and injected into the mold, followed by the leaching of the porogens (Fig. **4**). Tubular and ear-shaped porous PLGA scaffolds with porosity up to 94% were produced using this technology. A wet composite of polymer-porogen solvent was injected into the mold at low pressure and room temperature to avoid thermal degradation of the polymers. This approach proved to be suitable for the synthesis of highly porous foams that can be formed into complicated geometric shapes [84].

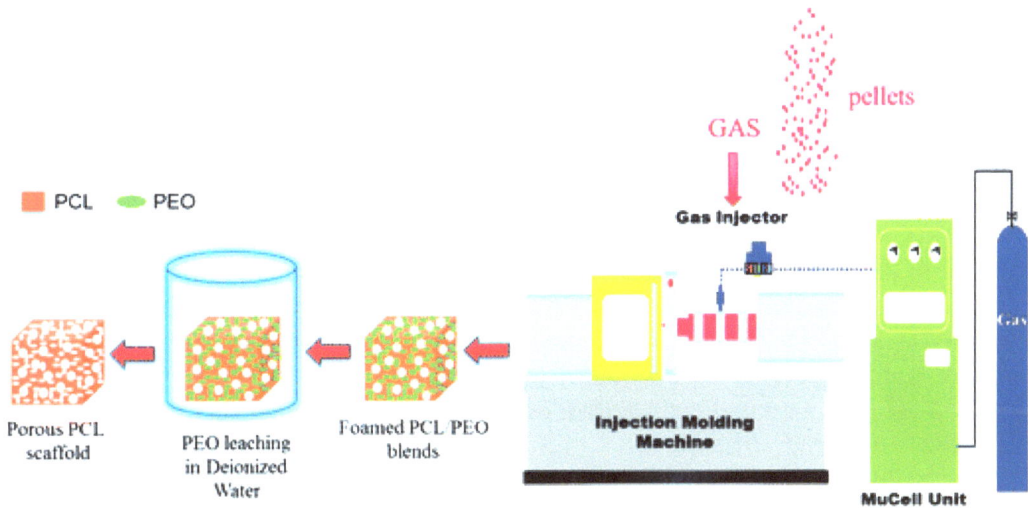

Fig. (4). Schematic presentation of the PCL tissue scaffold fabrication using microcellular injection molding and polymer leaching. Figure reproduced from [85].

Similarly, the NaCl-PLGA mixture was granulated and then injection molded into L-shaped nasal scaffolds. Then, the NaCl was leached with deionized water to obtain the porous nasal scaffold [86]. In a recent study, nHAp was used to enhance the mechanical strength and thermomechanical resistance of poly(3-hydroxybutyrate-co-3-hydroxyhexanoate) [P(3HB-co-3HHx)] microbial copolyester parts by injection molding. These novel nanocomposites could be used as low-load implants that promote bone formation and thus could potentially be used for bone reconstruction. 2.5-20 wt% nHA was incorporated into the P(3HB-co-3HHx) by melt compounding and formed into the desired scaffolds by injection molding. A high concentration of nHA resulted in agglomeration of the nanoparticles, which affected the toughness and thermal stability of the biocomposites. 20 wt% nHA increased the tensile strength (Et) and flexural modulus (Ef) of the scaffolds by up to 64% and 61%, respectively, while 10 wt% nHA composites were more similar to natural bone in terms of mechanical properties [87].

Injection Molding/Phase Separation

The injection molding/phase separation technique is used to fabricate scaffolds with continuous porous scaffolds by mixing two immiscible polymers. Biomaterials mixed with water-soluble materials can produce porous scaffolds with interconnected pores. The composite mixture is granulated and injection molded, followed by leaching of the porogenic phase. Based on these facts, porous lamellar PLLA scaffolds were fabricated by conventional injection molding of the PLLA-PEO composite mixture, followed by leaching of the PEO porogen phase. These scaffolds exhibited porosities of 57%-74% and pore sizes of about 50-100 μm [83].

Injection Molding/Gas Foaming

Gas-assisted injection molding creates porous scaffolds by flowing gas through the matrix and forming voids. The first studies on gas-assisted injection molding came into the limelight when a study with carboxylic acids led to the formation of macropores formed by the release of water and CO_2 after the decomposition of the carboxylic acids [88]. Similarly, water was used as a benign blowing agent to prepare porous poly(ether urethane) (PEU) scaffolds. A PEU-NaCl blend was compounded in a twin-screw extruder and allowed to absorb moisture before injection molding. Lowering the temperature and pressure leads to thermodynamic instability, which forces the water vapor to form nucleation sites and consequently form networks of round pores as the polymer melt cools. Leaching of salt particles resulted in the formation of interconnected pores [89]. Chemical blowing agents can leave cytotoxic residues in the matrix. To circumvent this problem, physical blowing agents such as propylene, methyl chloride, gaseous fluorocarbons, butylenes, nitrogen gas, carbon dioxide, or simply air are used that do not cause decomposition or side reactions in the matrix. However, the resulting matrix has unconnected pores and uncontrolled pore distribution. In another study, nanocomposites of PLA and halloysite nanotubes (PLA/HNT) were developed by foam injection molding using azodicarbonamide as a chemical blowing agent. The PLA/HNT and blowing agent were integrated into a twin-screw extruder by the melt-mixing method and formed by foam injection molding. The HNT loadings varied between 1, 3, and 5 wt%, and the results showed that the scaffold with 3 wt% HNT exhibited the highest mechanical strength with 124.2% and 79.2% increase in tensile strength and elongation, respectively. Cell viability was highest for scaffolds with 3 wt% HNT due to the improved morphology of the scaffold [90].

Microcellular Foam Injection Molding

In injection molding of microcellular foam, a supercritical fluid is used as a physical blowing agent that not only forms pores in the matrix but also fills the interstices between polymer molecules, resulting in a reduction in viscosity at low temperature and low pressure. SCF is dissolved in the polymer, forming a single-phase polymer solution SCF. After injection into the mold, the pressure drops from the microcellular process pressure to atmospheric pressure, separating SCF from the single-phase solution and forming a large number of cores. These nuclei are further filled with SCF due to the concentration difference of SCF within the matrix, increasing the gas bubbles and forming numerous microscale cells until the concentration reaches equilibrium or the melt freezes [91]. The microcellular foaming technique was used in a study to fabricate mixed thermoplastic polyurethane (TPU)/PLA scaffolds for tissue engineering, which exhibited different mechanical properties and phase morphologies that could be tailored to human tissue requirements. The fabricated scaffolds exhibited porosities of 49% - 79%, pore diameters of 115 - 252 μm, and pore densities of 1.43×10^5 to 3.93×10^5 cm^{-3} [92]. Experiments were conducted in a combined process using microcellular injection molding (MIM) and gas-assisted microcellular injection molding (GAMIM) to evaluate the weight reduction, surface appearance, cellular structure, and mechanical properties of the microcellular plastics. The results of the study show that the proposed method is a simple approach to produce foamed parts with greater weight reduction, better surface appearance, and better mechanical properties [93].

Extrusion Techniques

Extrusion techniques are used in the melt blending of a single polymer or polymer composites as a pre-mixing stage for other processing techniques [94]. It causes polymer chain orientation and is used to develop tubes, films *etc* ., with a defined cross-sectional profile.

Extrusion/Particulate Leaching

Extrusion is used in combination with the leaching of porogens to produce porous PLA scaffolds (Fig. **5**). The PLA/PEG/NaCl blends were melt compounded and crushed into small granules, which were then annealed and compression molded. It was then extruded through convergent dies, followed by leaching of NaCl and PEG. This resulted in PLA scaffolds with a connectivity of over 97% and a porosity of over 60%. The scaffolds also showed good mechanical strength [95, 96]. It is useful in the development of scaffolds using PLLA and PLGA to form tubular porous scaffolds for the regeneration of peripheral nerves, intestines, long bones, and blood vessels [97]. A combined solvent casting/extrusion/particle leaching process was used to synthesize biodegradable tubular PLGA and PLLA scaffolds for tissue regeneration applications. Polymer/salt composite films were prepared by the solvent casting technique, crushed into pieces and then extruded into tubular constructs. Subsequent leaching resulted in a channel with an open pore structure [97]. Highly elastic and porous tubular poly(lactide-co-ε-caprolactone) scaffolds were synthesized by the extrusion-particle leaching technique for mechano-active vascular tissue engineering.

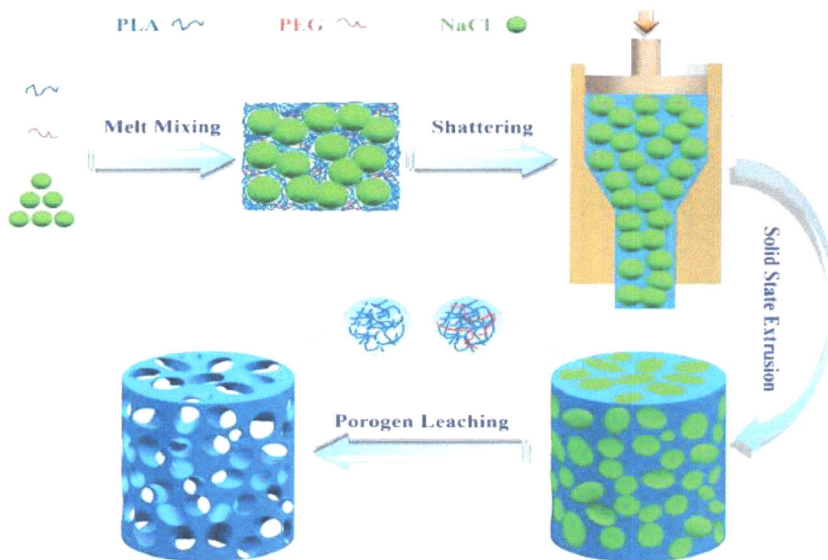

Fig. (5). Schematic of fabricating porous PLA scaffolds Solid State Extrusion/Porogen Leaching Approach. Figure reproduced from [96].

Highly porous PLCL scaffolds provided better support for adhesion and proliferation of vascular smooth muscle cells (VSMCs) than the less porous scaffolds. These highly elastic and biocompatible PLCL scaffolds could prove useful in engineering tissues such as blood vessels [98]. Although this approach has been very successful for the synthesis of tubular structures and has high feasibility, it comes with some drawbacks, such as low mechanical strength, severe thermal degradation due to high processing temperatures, and reduced rheological properties.

CONCLUSION AND FUTURE REMARKS

Scaffold development has become an important component of tissue engineering. Scaffolds must support the biological functions of the tissue and provide mechanical strength. They have a complex, mimetic, hierarchical structure and provide space for cell attachment, differentiation, proliferation, and migration. The structural morphology of scaffolds plays a critical role in promoting the diffusion of cell nutrients and metabolic wastes, cell growth, and other biological functions in tissues. High porosity, surface area to volume ratio, and pore size are critical factors that promote not only efficient diffusion of nutrients and metabolic wastes through cells, but also cell attachment, growth, migration, and neovascularization. Scaffolds with a porous structure allow rapid tissue ingrowth and minimize diffusion limitations, while structures with smaller pores are able to carry mechanical loads in the complicated biochemical environment. From the above studies, both techniques offer the advantage of controlling pore size and porosity. However, a balance should be maintained between the mechanical strength and biological properties of the scaffolds to evaluate their performance. In addition, these techniques provide a direct correlation between the bulk and surface properties of the polymer and the desired function of the scaffold. Scaffolds fabricated using these methods exhibit high biocompatibility and can be an enviable option for tissue engineering. Despite the many advantages, these methods also have some disadvantages, such as the use of toxic solvents, high processing temperatures, and residual porogens in the resulting scaffolds.

Studies have shown that the topographic features of scaffold surfaces control the regulation and stimulation of cellular activities by mimicking the micro/nanoscale features of the extracellular matrix [99]. Additive manufacturing techniques, which include solvent casting technology, melt casting, and particle leaching techniques, can be used to fabricate complex architectures in various sizes and shapes. These technologies offer an attractive approach for low-cost and large-scale fabrication of scaffolds that meet the requirements of a particular tissue.

CONSENT FOR PUBLICATION

Not applicable.

CONFLICT OF INTEREST

The author declares no conflict of interest, financial or otherwise.

ACKNOWLEDGEMENTS

This work has been supported by the "The Research Council (TRC)" Oman through Block Research Funding Program (BFP/RGP/EBR/20/261).

REFERENCES

[1]   S. Yang, K.F. Leong, Z. Du, and C.K. Chua, "The design of scaffolds for use in tissue engineering. Part I. Traditional factors", *Tissue Eng.,* vol. 7, no. 6, pp. 679-689, 2001.
[http://dx.doi.org/10.1089/107632701753337645] [PMID: 11749726]

[2]   M.W. Ullah, L. Fu, L. Lamboni, Z. Shi, and G. Yang, "Current trends and biomedical applications of resorbable polymers", In: *Materials for Biomedical Engineering.* Elsevier, 2019, pp. 41-86.
[http://dx.doi.org/10.1016/B978-0-12-818415-8.00003-6]

[3]   L. Lu, and A.G. Mikos, "The importance of new processing techniques in tissue engineering", *MRS Bull.,* vol. 21, no. 11, pp. 28-32, 1996.
[http://dx.doi.org/10.1557/S088376940003181X] [PMID: 11541498]

[4]   L. Mao, L. Wang, M. Zhang, M.W. Ullah, L. Liu, W. Zhao, Y. Li, A.A.Q. Ahmed, H. Cheng, Z. Shi, and G. Yang, *In Situ* Synthesized Selenium Nanoparticles-Decorated Bacterial Cellulose/Gelatin Hydrogel with Enhanced Antibacterial, Antioxidant, and Anti-Inflammatory Capabilities for Facilitating Skin Wound Healing., *Adv. Healthc. Mater.,* vol. 10, no. 14, p. 2100402, 2021.
[http://dx.doi.org/10.1002/adhm.202100402] [PMID: 34050616]

[5]   S. Stratton, N.B. Shelke, K. Hoshino, S. Rudraiah, and S.G. Kumbar, "Bioactive polymeric scaffolds for tissue engineering", *Bioact. Mater.,* vol. 1, no. 2, pp. 93-108, 2016.
[http://dx.doi.org/10.1016/j.bioactmat.2016.11.001] [PMID: 28653043]

[6]   L. Wang, S. Hu, M.W. Ullah, X. Li, Z. Shi, and G. Yang, "Enhanced cell proliferation by electrical stimulation based on electroactive regenerated bacterial cellulose hydrogels", *Carbohydr. Polym.,* vol. 249, p. 116829, 2020.
[http://dx.doi.org/10.1016/j.carbpol.2020.116829] [PMID: 32933675]

[7]   Z. Shi, X. Gao, M.W. Ullah, S. Li, Q. Wang, and G. Yang, "Electroconductive natural polymer-based hydrogels", *Biomaterials,* vol. 111, pp. 40-54, 2016.
[http://dx.doi.org/10.1016/j.biomaterials.2016.09.020] [PMID: 27721086]

[8]   A. Jasim, M.W. Ullah, Z. Shi, X. Lin, and G. Yang, "Fabrication of bacterial cellulose/polyaniline/single-walled carbon nanotubes membrane for potential application as biosensor", *Carbohydr. Polym.,* vol. 163, pp. 62-69, 2017.
[http://dx.doi.org/10.1016/j.carbpol.2017.01.056] [PMID: 28267519]

[9]   S. Khan, M. Ul-Islam, M.W. Ullah, Y. Kim, and J.K. Park, "Synthesis and characterization of a novel bacterial cellulose–poly(3,4-ethylenedioxythiophene)–poly(styrene sulfonate) composite for use in biomedical applications", *Cellulose,* vol. 22, no. 4, pp. 2141-2148, 2015.
[http://dx.doi.org/10.1007/s10570-015-0683-2]

[10]   W. Aljohani, M. W. Ullah, X. Zhang, and G. Yang, "Bioprinting and its applications in tissue engineering and regenerative medicine", *International Journal of Biological Macromolecules,* vol. 107, pp. 261-275, 2018.
[http://dx.doi.org/10.1016/j.ijbiomac.2017.08.171]

[11]   A. Fatima, S. Yasir, M.S. Khan, S. Manan, M.W. Ullah, and M. Ul-Islam, "Plant extract-loaded bacterial cellulose composite membrane for potential biomedical applications", *Journal of Bioresources and Bioproducts,* vol. 6, no. 1, pp. 26-32, 2021.
[http://dx.doi.org/10.1016/j.jobab.2020.11.002]

[12]   O. Akturk, A. Tezcaner, H. Bilgili, M.S. Deveci, M.R. Gecit, and D. Keskin, "Evaluation of sericin/collagen membranes as prospective wound dressing biomaterial", *J. Biosci. Bioeng.,* vol. 112,

no. 3, pp. 279-288, 2011.
[http://dx.doi.org/10.1016/j.jbiosc.2011.05.014] [PMID: 21697006]

[13] T.A.E. Ahmed, E.V. Dare, and M. Hincke, "Fibrin: a versatile scaffold for tissue engineering applications", *Tissue Eng. Part B Rev.,* vol. 14, no. 2, pp. 199-215, 2008.
[http://dx.doi.org/10.1089/ten.teb.2007.0435] [PMID: 18544016]

[14] K.S. Lee, B.Y. Kim, D.H. Kim, and B.R. Jin, "Recombinant spider silk fibroin protein produces a non-cytotoxic and non-inflammatory response", *J. Asia Pac. Entomol.,* vol. 19, no. 4, pp. 1015-1018, 2016.
[http://dx.doi.org/10.1016/j.aspen.2016.09.004]

[15] M.W. Ullah, S. Manan, S.J. Kiprono, M. Ul-Islam, and G. Yang, "Synthesis, Structure, and Properties of Bacterial Cellulose", In: *Nanocellulose,* 2019.
[http://dx.doi.org/10.1002/9783527807437.ch4]

[16] S. Khan, M. Ul-Islam, M. Ikram, S.U. Islam, M.W. Ullah, M. Israr, J.H. Jang, S. Yoon, and J.K. Park, "Preparation and structural characterization of surface modified microporous bacterial cellulose scaffolds: A potential material for skin regeneration applications *in vitro* and *in vivo*"., *Int. J. Biol. Macromol.,* vol. 117, pp. 1200-1210, 2018.
[http://dx.doi.org/10.1016/j.ijbiomac.2018.06.044] [PMID: 29894790]

[17] S. Khan, M. Ul-Islam, M. Ikram, M.W. Ullah, M. Israr, F. Subhan, Y. Kim, J.H. Jang, S. Yoon, and J.K. Park, "Three-dimensionally microporous and highly biocompatible bacterial cellulose–gelatin composite scaffolds for tissue engineering applications", *RSC Advances,* vol. 6, no. 112, pp. 110840-110849, 2016.
[http://dx.doi.org/10.1039/C6RA18847H]

[18] M. Ul-Islam, F. Subhan, S.U. Islam, S. Khan, N. Shah, S. Manan, M.W. Ullah, and G. Yang, "Development of three-dimensional bacterial cellulose/chitosan scaffolds: Analysis of cell-scaffold interaction for potential application in the diagnosis of ovarian cancer", *Int. J. Biol. Macromol.,* vol. 137, pp. 1050-1059, 2019.
[http://dx.doi.org/10.1016/j.ijbiomac.2019.07.050] [PMID: 31295500]

[19] K. Markstedt, A. Mantas, I. Tournier, H. Martínez Ávila, D. Hägg, and P. Gatenholm, "3D bioprinting human chondrocytes with nanocellulose-alginate bioink for cartilage tissue engineering applications", *Biomacromolecules,* vol. 16, no. 5, pp. 1489-1496, 2015.
[http://dx.doi.org/10.1021/acs.biomac.5b00188] [PMID: 25806996]

[20] R. Takahama, H. Kato, K. Tajima, S. Tagawa, and T. Kondo, "Biofabrication of a Hyaluronan/Bacterial Cellulose Composite Nanofibril by Secretion from Engineered Gluconacetobacter", *Biomacromolecules,* vol. 22, no. 11, pp. 4709-4719, 2021.
[http://dx.doi.org/10.1021/acs.biomac.1c00987] [PMID: 34705422]

[21] R. R. McCarthy, M. W. Ullah, E. Pei, and G. Yang, "Antimicrobial Inks: The Anti-Infective Applications of Bioprinted Bacterial Polysaccharides", *Trends in Biotechnology,* vol. 37, pp. 1153-1155, 2019.
[http://dx.doi.org/10.1016/j.tibtech.2019.05.004]

[22] R.R. McCarthy, M.W. Ullah, P. Booth, E. Pei, and G. Yang, "The use of bacterial polysaccharides in bioprinting", *Biotechnol. Adv.,* vol. 37, no. 8, p. 107448, 2019.
[http://dx.doi.org/10.1016/j.biotechadv.2019.107448] [PMID: 31513840]

[23] I. Ullah, "Impact of structural features of Sr/Fe co-doped HAp on the osteoblast proliferation and osteogenic differentiation for its application as a bone substitute", In: *Mater. Sci. Eng. C* vol. 110. , 2020, p. 110633.
[http://dx.doi.org/10.1016/j.msec.2020.110633]

[24] I. Ullah, "Synthesis and Characterization of Sintered Sr/Fe-Modified Hydroxyapatite Bioceramics for Bone Tissue Engineering Applications", *ACS Biomater. Sci. Eng,* 2019.
[http://dx.doi.org/10.1021/acsbiomaterials.9b01666]

[25] S. Oh, S.G. Kang, E.S. Kim, S.H. Cho, and J.H. Lee, "Fabrication and characterization of hydrophilic

poly(lactic-co-glycolic acid)/poly(vinyl alcohol) blend cell scaffolds by melt-molding particulate-leaching method", *Biomaterials,* vol. 24, no. 22, pp. 4011-4021, 2003.
[http://dx.doi.org/10.1016/S0142-9612(03)00284-9] [PMID: 12834596]

[26]   W. Zhang, I. Ullah, L. Shi, Y. Zhang, H. Ou, J. Zhou, M.W. Ullah, X. Zhang, and W. Li, "Fabrication and characterization of porous polycaprolactone scaffold *via* extrusion-based cryogenic 3D printing for tissue engineering"., *Mater. Des.,* vol. 180, p. 107946, 2019.
[http://dx.doi.org/10.1016/j.matdes.2019.107946]

[27]   C. Shao, M. Wang, H. Chang, F. Xu, and J. Yang, "A Self-Healing Cellulose Nanocrystal-Poly(ethylene glycol) Nanocomposite Hydrogel *via* Diels–Alder Click Reaction"., *ACS Sustain. Chem.& Eng.,* vol. 5, no. 7, pp. 6167-6174, 2017.
[http://dx.doi.org/10.1021/acssuschemeng.7b01060]

[28]   X. Li, B. Li, M.W. Ullah, R. Panday, J. Cao, Q. Li, Y. Zhang, L. Wang, and G. Yang, "Water-stable and finasteride-loaded polyvinyl alcohol nanofibrous particles with sustained drug release for improved prostatic artery embolization — *in vitro* and *in vivo* evaluation"., *Mater. Sci. Eng. C,* vol. 115, p. 111107, 2020.
[http://dx.doi.org/10.1016/j.msec.2020.111107] [PMID: 32600710]

[29]   X. Li, X. Ji, K. Chen, M.W. Ullah, X. Yuan, Z. Lei, J. Cao, J. Xiao, and G. Yang, "Development of finasteride/PHBV@polyvinyl alcohol/chitosan reservoir-type microspheres as a potential embolic agent: from *in vitro* evaluation to animal study"., *Biomater. Sci.,* vol. 8, no. 10, pp. 2797-2813, 2020.
[http://dx.doi.org/10.1039/C9BM01775E] [PMID: 32080688]

[30]   J. Kundu, F. Pati, Y. Hun Jeong, and D.W. Cho, "Biomaterials for Biofabrication of 3D Tissue Scaffolds", *Patterning Assem,* no. March, pp. 23-46, 2013.
[http://dx.doi.org/10.1016/B978-1-4557-2852-7.00002-0]

[31]   I. Ullah, W. Li, S. Lei, Y. Zhang, W. Zhang, U. Farooq, S. Ullah, M.W. Ullah, and X. Zhang, "Simultaneous co-substitution of Sr2+/Fe3+ in hydroxyapatite nanoparticles for potential biomedical applications", *Ceram. Int.,* vol. 44, no. 17, pp. 21338-21348, 2018.
[http://dx.doi.org/10.1016/j.ceramint.2018.08.187]

[32]   S. Anil, E.P. Chalisserry, S.Y. Nam, and J. Venkatesan, "Biomaterials for craniofacial tissue engineering and regenerative dentistry", In: *Advanced Dental Biomaterials*, 2019.
[http://dx.doi.org/10.1016/B978-0-08-102476-8.00025-6]

[33]   T.C. Flanagan, B. Wilkins, A. Black, S. Jockenhoevel, T.J. Smith, and A.S. Pandit, "A collagen-glycosaminoglycan co-culture model for heart valve tissue engineering applications", *Biomaterials,* vol. 27, no. 10, pp. 2233-2246, 2006.
[http://dx.doi.org/10.1016/j.biomaterials.2005.10.031] [PMID: 16313955]

[34]   S.J. Hollister, "Porous scaffold design for tissue engineering", *Nat. Mater.,* vol. 4, no. 7, pp. 518-524, 2005.
[http://dx.doi.org/10.1038/nmat1421] [PMID: 16003400]

[35]   Z. Di, Z. Shi, M.W. Ullah, S. Li, and G. Yang, "A transparent wound dressing based on bacterial cellulose whisker and poly(2-hydroxyethyl methacrylate)", *Int. J. Biol. Macromol.,* vol. 105, no. Pt 1, pp. 638-644, 2017.
[http://dx.doi.org/10.1016/j.ijbiomac.2017.07.075] [PMID: 28716748]

[36]   M. Ul-Islam, S. Khan, M.W. Ullah, and J.K. Park, "Bacterial cellulose composites: Synthetic strategies and multiple applications in bio-medical and electro-conductive fields", *Biotechnol. J.,* vol. 10, no. 12, pp. 1847-1861, 2015.
[http://dx.doi.org/10.1002/biot.201500106] [PMID: 26395011]

[37]   M. Elsawy, and A. de Mel, "Biofabrication and biomaterials for urinary tract reconstruction", *Res. Rep. Urol.,* vol. 9, pp. 79-92, 2017.
[http://dx.doi.org/10.2147/RRU.S127209] [PMID: 28546955]

[38]   A. Pandey, R. K. Sharma, and K. Balani, "Introduction to Biomaterials", *Biosurfaces: A Materials*

Science and Engineering Perspective, 2015.
[http://dx.doi.org/10.1002/9781118950623.ch1]

[39]   A. Jayakumar, H. K v, S. T S, M. Joseph, S. Mathew, P. G, I.C. Nair, and R. e K, "Starch-PVA composite films with zinc-oxide nanoparticles and phytochemicals as intelligent pH sensing wraps for food packaging application", *Int. J. Biol. Macromol.,* vol. 136, pp. 395-403, 2019.
[http://dx.doi.org/10.1016/j.ijbiomac.2019.06.018] [PMID: 31173829]

[40]   S.C. Tjong, "Structural and mechanical properties of polymer nanocomposites", *Mater. Sci. Eng. Rep.,* vol. 53, no. 3-4, pp. 73-197, 2006.
[http://dx.doi.org/10.1016/j.mser.2006.06.001]

[41]   M.S. Watson, M.J. Whitaker, S.M. Howdle, and K.M. Shakesheff, "Incorporation of proteins into polymer materials by a novel supercritical fluid processing method", *Adv. Mater.,* vol. 14, no. 24, pp. 1802-1804, 2002.
[http://dx.doi.org/10.1002/adma.200290003]

[42]   A.G. Mikos, G. Sarakinos, S.M. Leite, J.P. Vacant, and R. Langer, "Laminated three-dimensional biodegradable foams for use in tissue engineering", *Biomaterials,* vol. 14, no. 5, pp. 323-330, 1993.
[http://dx.doi.org/10.1016/0142-9612(93)90049-8] [PMID: 8507774]

[43]   H. Janik, and M. Marzec, "A review: Fabrication of porous polyurethane scaffolds", *Mater. Sci. Eng. C,* vol. 48, pp. 586-591, 2015.
[http://dx.doi.org/10.1016/j.msec.2014.12.037] [PMID: 25579961]

[44]   P. Bajaj, R.M. Schweller, A. Khademhosseini, J.L. West, and R. Bashir, "3D biofabrication strategies for tissue engineering and regenerative medicine", *Annu. Rev. Biomed. Eng.,* vol. 16, no. 1, pp. 247-276, 2014.
[http://dx.doi.org/10.1146/annurev-bioeng-071813-105155] [PMID: 24905875]

[45]   G. Turnbull, J. Clarke, F. Picard, P. Riches, L. Jia, F. Han, B. Li, and W. Shu, "3D bioactive composite scaffolds for bone tissue engineering", *Bioact. Mater.,* vol. 3, no. 3, pp. 278-314, 2018.
[http://dx.doi.org/10.1016/j.bioactmat.2017.10.001] [PMID: 29744467]

[46]   V. Mkhabela, and S.S. Ray, "Biodegradation and bioresorption of poly(ε-caprolactone) nanocomposite scaffolds", *Int. J. Biol. Macromol.,* vol. 79, pp. 186-192, 2015.
[http://dx.doi.org/10.1016/j.ijbiomac.2015.04.056] [PMID: 25952165]

[47]   A.G. Mikos, A.J. Thorsen, L.A. Czerwonka, Y. Bao, R. Langer, D.N. Winslow, and J.P. Vacanti, "Preparation and characterization of poly(l-lactic acid) foams", *Polymer (Guildf.),* vol. 35, no. 5, pp. 1068-1077, 1994.
[http://dx.doi.org/10.1016/0032-3861(94)90953-9]

[48]   Z. Li, M-B. Xie, Y. Li, Y. Ma, J.S. Li, and F.Y. Dai, "Recent progress in tissue engineering and regenerative medicine", *J. Biomater. Tissue Eng.,* vol. 6, no. 10, pp. 755-766, 2016.
[http://dx.doi.org/10.1166/jbt.2016.1510]

[49]   M.C. Smita Mohanty, "Effect of Different Solvents in Solvent Casting of Porous caffolds – In Biomedical and Tissue Engineering Applications", *J. Tissue Sci. Eng.,* vol. 6, no. 1, 2015.
[http://dx.doi.org/10.4172/2157-7552.1000142]

[50]   E. Jahed, M.A. Khaledabad, H. Almasi, and R. Hasanzadeh, "Physicochemical properties of Carum copticum essential oil loaded chitosan films containing organic nanoreinforcements", *Carbohydr. Polym.,* vol. 164, pp. 325-338, 2017.
[http://dx.doi.org/10.1016/j.carbpol.2017.02.022] [PMID: 28325333]

[51]   Q. Cai, J. Yang, J. Bei, and S. Wang, "A novel porous cells scaffold made of polylactide–dextran blend by combining phase-separation and particle-leaching techniques", *Biomaterials,* vol. 23, no. 23, pp. 4483-4492, 2002.
[http://dx.doi.org/10.1016/S0142-9612(02)00168-0] [PMID: 12322968]

[52]   S.W. Suh, J.Y. Shin, J. Kim, J. Kim, C.H. Beak, D.I. Kim, H. Kim, S.S. Jeon, and I.W. Choo, "Effect

of different particles on cell proliferation in polymer scaffolds using a solvent-casting and particulate leaching technique", *ASAIO J.,* vol. 48, no. 5, pp. 460-464, 2002.
[http://dx.doi.org/10.1097/00002480-200209000-00003] [PMID: 12296562]

[53]   N. Pal, P. Dubey, P. Gopinath, and K. Pal, "Combined effect of cellulose nanocrystal and reduced graphene oxide into poly-lactic acid matrix nanocomposite as a scaffold and its anti-bacterial activity", *Int. J. Biol. Macromol.,* vol. 95, pp. 94-105, 2017.
[http://dx.doi.org/10.1016/j.ijbiomac.2016.11.041] [PMID: 27856322]

[54]   M.W. Ahmad, B. Dey, G. Sarkhel, D.S. Bag, and A. Choudhury, "Exfoliated graphene reinforced polybenzimidazole nanocomposite with improved electrical, mechanical and thermal properties", *Mater. Chem. Phys.,* vol. 223, pp. 426-433, 2019.
[http://dx.doi.org/10.1016/j.matchemphys.2018.11.026]

[55]   X. Li, K. Nan, S. Shi, and H. Chen, "Preparation and characterization of nano-hydroxyapatite/chitosan cross-linking composite membrane intended for tissue engineering", *Int. J. Biol. Macromol.,* vol. 50, no. 1, pp. 43-49, 2012.
[http://dx.doi.org/10.1016/j.ijbiomac.2011.09.021] [PMID: 21983025]

[56]   A. Sola, J. Bertacchini, D. D'Avella, L. Anselmi, T. Maraldi, S. Marmiroli, and M. Messori, "Development of solvent-casting particulate leaching (SCPL) polymer scaffolds as improved three-dimensional supports to mimic the bone marrow niche", *Mater. Sci. Eng. C,* vol. 96, pp. 153-165, 2019.
[http://dx.doi.org/10.1016/j.msec.2018.10.086] [PMID: 30606521]

[57]   M. Darder, A. Karan, G. Real, and M.A. DeCoster, "Cellulose-based biomaterials integrated with copper-cystine hybrid structures as catalysts for nitric oxide generation", *Mater. Sci. Eng. C,* vol. 108, p. 110369, 2020.
[http://dx.doi.org/10.1016/j.msec.2019.110369] [PMID: 31923961]

[58]   K. Kattel, J.Y. Park, W. Xu, B.A. Bony, W.C. Heo, T. Tegafaw, C.R. Kim, M.W. Ahmad, S. Jin, J.S. Baeck, Y. Chang, T.J. Kim, J.E. Bae, K.S. Chae, J.Y. Jeong, and G.H. Lee, "Surface coated Eu(OH)3 nanorods: a facile synthesis, characterization, MR relaxivities and *in vitro* cytotoxicity"., *J. Nanosci. Nanotechnol.,* vol. 13, no. 11, pp. 7214-7219, 2013.
[http://dx.doi.org/10.1166/jnn.2013.8081] [PMID: 24245232]

[59]   M.Y. Ahmad, M.W. Ahmad, H. Cha, I-T. Oh, T. Tegafaw, X. Miao, S.L. Ho, S. Marasini, A. Ghazanfari, H. Yue, H-K. Ryeom, J. Lee, K.S. Chae, Y. Chang, and G.H. Lee, "Cyclic RGD-Coated Ultrasmall Gd_2O_3 Nanoparticles as Tumor-Targeting Positive Magnetic Resonance Imaging Contrast Agents"., *Eur. J. Inorg. Chem.,* vol. 2018, no. 26, pp. 3070-3079, 2018.
[http://dx.doi.org/10.1002/ejic.201800023]

[60]   M. Ul-Islam, A. Shehzad, S. Khan, W.A. Khattak, M.W. Ullah, and J.K. Park, "Antimicrobial and biocompatible properties of nanomaterials", *J. Nanosci. Nanotechnol.,* vol. 14, no. 1, pp. 780-791, 2014.
[http://dx.doi.org/10.1166/jnn.2014.8761] [PMID: 24730297]

[61]   A.M. Deliormanlı, and H. Atmaca, "Effect of pore architecture on the mesenchymal stem cell responses to graphene/polycaprolactone scaffolds prepared by solvent casting and robocasting", *J. Porous Mater.,* vol. 27, no. 1, pp. 49-61, 2020.
[http://dx.doi.org/10.1007/s10934-019-00791-1]

[62]   T. Johnson, R. Bahrampourian, A. Patel, and K. Mequanint, "Fabrication of highly porous tissue-engineering scaffolds using selective spherical porogens", *Biomed. Mater. Eng.,* vol. 20, no. 2, pp. 107-118, 2010.
[http://dx.doi.org/10.3233/BME-2010-0621] [PMID: 20592448]

[63]   C.J. Liao, C.F. Chen, J.H. Chen, S.F. Chiang, Y.J. Lin, and K.Y. Chang, "Fabrication of porous biodegradable polymer scaffolds using a solvent merging/particulate leaching method", *J. Biomed. Mater. Res.,* vol. 59, no. 4, pp. 676-681, 2002.
[http://dx.doi.org/10.1002/jbm.10030] [PMID: 11774329]

[64]   D. Puppi, F. Chiellini, A.M. Piras, and E. Chiellini, *Polymeric materials for bone and cartilage repair.* Progress in Polymer Science: Oxford, 2010.
       [http://dx.doi.org/10.1016/j.progpolymsci.2010.01.006]

[65]   R.C. Thomson, M.C. Wake, M.J. Yaszemski, and A.G. Mikos, "Biodegradable polymer scaffolds to regenerate organs", *Adv. Polym. Sci.,* vol. 122, pp. 245-274, 1995.
       [http://dx.doi.org/10.1007/3540587888_18]

[66]   K.F. Leong, C.M. Cheah, and C.K. Chua, "Solid freeform fabrication of three-dimensional scaffolds for engineering replacement tissues and organs", *Biomaterials,* vol. 24, no. 13, pp. 2363-2378, 2003.
       [http://dx.doi.org/10.1016/S0142-9612(03)00030-9] [PMID: 12699674]

[67]   C.T. Lee, P.H. Kung, and Y-D. Lee, "Preparation of poly(vinyl alcohol)-chondroitin sulfate hydrogel as matrices in tissue engineering", *Carbohydr. Polym.,* vol. 61, no. 3, pp. 348-354, 2005.
       [http://dx.doi.org/10.1016/j.carbpol.2005.06.018]

[68]   D. Jing, L. Wu, and J. Ding, "Solvent-assisted room-temperature compression molding approach to fabricate porous scaffolds for tissue engineering", *Macromol. Biosci.,* vol. 6, no. 9, pp. 747-757, 2006.
       [http://dx.doi.org/10.1002/mabi.200600079] [PMID: 16967479]

[69]   D. Yao, A. Smith, P. Nagarajan, A. Vasquez, L. Dang, and G.R. Chaudhry, "Fabrication of polycaprolactone scaffolds using a sacrificial compression-molding process", *J. Biomed. Mater. Res. B Appl. Biomater.,* vol. 77B, no. 2, pp. 287-295, 2006.
       [http://dx.doi.org/10.1002/jbm.b.30419] [PMID: 16292759]

[70]   S.H. Oh, S.C. Park, H.K. Kim, Y.J. Koh, J.H. Lee, M.C. Lee, and J.H. Lee, "Degradation behavior of 3d porous polydioxanone-b-polycaprolactone scaffolds fabricated using the melt-molding particulate-leaching method", *J. Biomater. Sci. Polym. Ed.,* vol. 22, no. 1-3, pp. 225-237, 2011.
       [http://dx.doi.org/10.1163/092050609X12597621891620] [PMID: 20557697]

[71]   M.E. Gomes, J.S. Godinho, D. Tchalamov, A.M. Cunha, and R.L. Reis, "Alternative tissue engineering scaffolds based on starch: processing methodologies, morphology, degradation and mechanical properties", *Mater. Sci. Eng. C,* vol. 20, no. 1-2, pp. 19-26, 2002.
       [http://dx.doi.org/10.1016/S0928-4931(02)00008-5]

[72]   Q. Hou, D.W. Grijpma, and J. Feijen, "Porous polymeric structures for tissue engineering prepared by a coagulation, compression moulding and salt leaching technique", *Biomaterials,* vol. 24, no. 11, pp. 1937-1947, 2003.
       [http://dx.doi.org/10.1016/S0142-9612(02)00562-8] [PMID: 12615484]

[73]   R.C. Thomson, M.J. Yaszemski, J.M. Powers, and A.G. Mikos, "Fabrication of biodegradable polymer scaffolds to engineer trabecular bone", *J. Biomater. Sci. Polym. Ed.,* vol. 7, no. 1, pp. 23-38, 1996.
       [http://dx.doi.org/10.1163/156856295X00805] [PMID: 7662615]

[74]   F. Shokrolahi, H. Mirzadeh, H. Yeganeh, and M. Daliri, "Fabrication of poly(urethane urea)-based scaffolds for bone tissue engineering by a combined strategy of using compression moulding and particulate leaching methods", In: *Iran. Polym. J,* 2011.

[75]   S.H. Lee, B.S. Kim, S.H. Kim, S.W. Kang, and Y.H. Kim, "Thermally produced biodegradable scaffolds for cartilage tissue engineering", *Macromol. Biosci.,* vol. 4, no. 8, pp. 802-810, 2004.
       [http://dx.doi.org/10.1002/mabi.200400021] [PMID: 15468274]

[76]   Y. Kang, G. Yin, Q. Yuan, Y. Yao, Z. Huang, X. Liao, B. Yang, L. Liao, and H. Wang, "Preparation of poly(l-lactic acid)/β-tricalcium phosphate scaffold for bone tissue engineering without organic solvent", *Mater. Lett.,* vol. 62, no. 12-13, pp. 2029-2032, 2008.
       [http://dx.doi.org/10.1016/j.matlet.2007.11.014]

[77]   R.M. Allaf, "Melt-molding technologies for 3D scaffold engineering", In: *Functional 3D Tissue Engineering Scaffolds.* Materials, Technologies, and Applications, 2018.
       [http://dx.doi.org/10.1016/B978-0-08-100979-6.00004-5]

[78]   R.M. Allaf, I.V. Rivero, and I.N. Ivanov, "Fabrication and characterization of multiwalled carbon

nanotube–loaded interconnected porous nanocomposite scaffolds", *Int. J. Polym. Mater.,* vol. 66, no. 4, pp. 183-192, 2017.
[http://dx.doi.org/10.1080/00914037.2016.1201761]

[79]   J. Minton, C. Janney, R. Akbarzadeh, C. Focke, A. Subramanian, T. Smith, J. McKinney, J. Liu, J. Schmitz, P.F. James, and A.M. Yousefi, "Solvent-free polymer/bioceramic scaffolds for bone tissue engineering: fabrication, analysis, and cell growth", *J. Biomater. Sci. Polym. Ed.,* vol. 25, no. 16, pp. 1856-1874, 2014.
[http://dx.doi.org/10.1080/09205063.2014.953016] [PMID: 25178801]

[80]   L. Wu, H. Zhang, J. Zhang, and J. Ding, *Fabrication of three-dimensional porous scaffolds of complicated shape for tissue engineering. I. Compression molding based on flexible-rigid combined mold.,* 2005.
[http://dx.doi.org/10.1089/ten.2005.11.1105]

[81]   S.H. Lee, A.R. Jo, G.P. Choi, C.H. Woo, S.J. Lee, B-S. Kim, H-K. You, and Y-S. Cho, "Fabrication of 3D alginate scaffold with interconnected pores using wire-network molding technique", *Tissue Eng. Regen. Med.,* vol. 10, no. 2, pp. 53-59, 2013.
[http://dx.doi.org/10.1007/s13770-013-0366-8]

[82]   Y.S. Cho, M.W. Hong, S.Y. Kim, S.J. Lee, J.H. Lee, Y.Y. Kim, and Y.S. Cho, "Fabrication of dual-pore scaffolds using SLUP (salt leaching using powder) and WNM (wire-network molding) techniques", *Mater. Sci. Eng. C,* vol. 45, pp. 546-555, 2014.
[http://dx.doi.org/10.1016/j.msec.2014.10.009] [PMID: 25491863]

[83]   S. Ghosh, J.C. Viana, R.L. Reis, and J.F. Mano, "Development of porous lamellar poly(l-lactic acid) scaffolds by conventional injection molding process", *Acta Biomater.,* vol. 4, no. 4, pp. 887-896, 2008.
[http://dx.doi.org/10.1016/j.actbio.2008.03.001] [PMID: 18396473]

[84]   L. Wu, D. Jing, and J. Ding, "A "room-temperature" injection molding/particulate leaching approach for fabrication of biodegradable three-dimensional porous scaffolds", *Biomaterials,* vol. 27, no. 2, pp. 185-191, 2006.
[http://dx.doi.org/10.1016/j.biomaterials.2005.05.105] [PMID: 16098580]

[85]   A. Huang, Y. Jiang, B. Napiwocki, H. Mi, X. Peng, and L.S. Turng, "Fabrication of poly(ε-caprolactone) tissue engineering scaffolds with fibrillated and interconnected pores utilizing microcellular injection molding and polymer leaching", *RSC Advances,* vol. 7, no. 69, pp. 43432-43444, 2017.
[http://dx.doi.org/10.1039/C7RA06987A]

[86]   J. Lee, "Study on Novel Porous Nasal Scaffold Using Injection Molding and Particle Leaching", *Biosensors, Singapore,* vol. 2, pp. 62-66, 2011.

[87]   J. Ivorra-Martinez, L. Quiles-Carrillo, T. Boronat, S. Torres-Giner, and J. A Covas, "Assessment of the mechanical and thermal properties of injection-molded poly(3-hydroxybutyrate-c--3-hydroxyhexanoate)/hydroxyapatite nanoparticles parts for use in bone tissue engineering", *Polymers (Basel),* vol. 12, no. 6, p. E1389, 2020.
[http://dx.doi.org/10.3390/polym12061389] [PMID: 32575881]

[88]   M.E. Gomes, A.S. Ribeiro, P.B. Malafaya, R.L. Reis, and A.M. Cunha, "A new approach based on injection moulding to produce biodegradable starch-based polymeric scaffolds: morphology, mechanical and degradation behaviour", *Biomaterials,* vol. 22, no. 9, pp. 883-889, 2001.
[http://dx.doi.org/10.1016/S0142-9612(00)00211-8] [PMID: 11311006]

[89]   H. Haugen, J. Will, W. Fuchs, and E. Wintermantel, "A novel processing method for injection-molded polyether–urethane scaffolds. Part 1: Processing", *J. Biomed. Mater. Res. B Appl. Biomater.,* vol. 77B, no. 1, pp. 65-72, 2006.
[http://dx.doi.org/10.1002/jbm.b.30396] [PMID: 16240432]

[90]   M. Eryildiz, and M. Altan, "Fabrication of polylactic acid/halloysite nanotube scaffolds by foam injection molding for tissue engineering", *Polym. Compos.,* vol. 41, no. 2, pp. 757-767, 2020.

[http://dx.doi.org/10.1002/pc.25406]

[91]   H. Guanghong, and W. Yue, "Microcellular Foam Injection Molding Process", In: *Some Critical Issues for Injection Molding*, 2012.
[http://dx.doi.org/10.5772/34513]

[92]   H.Y. Mi, M.R. Salick, X. Jing, B.R. Jacques, W.C. Crone, X.F. Peng, and L.S. Turng, "Characterization of thermoplastic polyurethane/polylactic acid (TPU/PLA) tissue engineering scaffolds fabricated by microcellular injection molding", *Mater. Sci. Eng. C,* vol. 33, no. 8, pp. 4767-4776, 2013.
[http://dx.doi.org/10.1016/j.msec.2013.07.037] [PMID: 24094186]

[93]   J. Hou, G. Zhao, G. Wang, G. Dong, and J. Xu, "A novel gas-assisted microcellular injection molding method for preparing lightweight foams with superior surface appearance and enhanced mechanical performance", *Mater. Des.,* vol. 127, pp. 115-125, 2017.
[http://dx.doi.org/10.1016/j.matdes.2017.04.073]

[94]   L. Shi, Y. Hu, M.W. Ullah, I. ullah, H. Ou, W. Zhang, L. Xiong, and X. Zhang, "Cryogenic free-form extrusion bioprinting of decellularized small intestinal submucosa for potential applications in skin tissue engineering", *Biofabrication,* vol. 11, no. 3, p. 035023, 2019.
[http://dx.doi.org/10.1088/1758-5090/ab15a9]

[95]   H. Liu, and T.J. Webster, "Bioinspired nanocomposites for orthopedic applications", In: *Nanotechnology for the Regeneration of Hard and Soft Tissues*, 2007.
[http://dx.doi.org/10.1142/9789812779656_0001]

[96]   H.M. Yin, J. Qian, J. Zhang, Z.F. Lin, J.S. Li, J.Z. Xu, and Z.M. Li, "Engineering porous poly(lactic acid) scaffolds with high mechanical performance *via* a solid state extrusion/porogen leaching approach"., *Polymers (Basel),* vol. 8, no. 6, p. 213, 2016.
[http://dx.doi.org/10.3390/polym8060213] [PMID: 30979308]

[97]   M.S. Widmer, P.K. Gupta, L. Lu, R.K. Meszlenyi, G.R.D. Evans, K. Brandt, T. Savel, A. Gurlek, C.W. Patrick Jr, and A.G. Mikos, "Manufacture of porous biodegradable polymer conduits by an extrusion process for guided tissue regeneration", *Biomaterials,* vol. 19, no. 21, pp. 1945-1955, 1998.
[http://dx.doi.org/10.1016/S0142-9612(98)00099-4] [PMID: 9863528]

[98]   S.I. Jeong, S.H. Kim, Y.H. Kim, Y. Jung, J.H. Kwon, B.S. Kim, and Y.M. Lee, "Manufacture of elastic biodegradable PLCL scaffolds for mechano-active vascular tissue engineering", *J. Biomater. Sci. Polym. Ed.,* vol. 15, no. 5, pp. 645-660, 2004.
[http://dx.doi.org/10.1163/156856204323046906] [PMID: 15264665]

[99]   J. Barthes, H. Özçelik, M. Hindié, A. Ndreu-Halili, A. Hasan, and N.E. Vrana, "Cell microenvironment engineering and monitoring for tissue engineering and regenerative medicine: the recent advances", *BioMed Res. Int.,* vol. 2014, pp. 1-18, 2014.
[http://dx.doi.org/10.1155/2014/921905] [PMID: 25143954]

SUBJECT INDEX

R

Radioactive isotopes 228
Reaction kinetics 178
Reactions 9, 26, 143, 167, 173, 177, 178, 179,
    182, 183, 218, 244
  chemical conjugation 179
  immune 9
  inflammatory 26, 244
  light-induced 182
  light-promoted thiol-ene addition 182
  nucleation 178
  photocage-assisted photoconjugation 178
Regenerative medicine 1, 2, 26, 181, 183, 218,
    219, 229, 233, 243, 249
Release 5, 7, 51, 59, 73, 111, 112, 117, 119,
    150, 218, 249, 254
  explosive 117
  sustained 112
Release profile 110, 112
  sustained drug 110
Remodelling process 9
Renal insufficiency 121
Resin 12, 136, 139, 142
  photopolymer 136
Resistance 4, 253
  thermomechanical 253
Resorbable ceramics 29
Resorcinol-formaldehyde hydrogels 51
Retinal 211, 212
  degeneration 211
  progenitor cells (RPCs) 211
  progenitor cells, neural 212
Revocable photoconjugation 180
Rheological properties 80, 256

S

Safety 72, 74, 83, 119, 212, 219
  fabrics 72
  occupational 83
  risks 212
Salt 199, 201, 206
  gas foam 201

  scaffold-based 206
  traditional 199
Salt leaching 202, 203, 204, 210, 211
  electrospinning (SLE) 202
  technique 204, 210
  using powder (SLUP) 203, 204
Scaffold(s) 5, 13, 31, 37, 50, 56, 73, 117, 166,
    167, 168, 181, 182, 195, 202, 203, 211,
    228, 247, 249, 250, 251, 253
  biocompatible 167
  biodegradable 181, 228
  chitosan 5, 50, 203
  chitosan-alginate 50
  chitosan-based 31
  for regenerative medicine 249
  leaching 195, 211
  nanofibers 73
  nasal 253
  polymer-based 37, 247, 250, 251
  polymer-based tissue 166
  polymeric hydrogel 182
  protein-loaded 56
  silk fibroin 211
  silver nanocomposite 202
  skin 203
  thermoplastic 13
  tissue-engineered 117, 168
Scaffold fabrication 9, 10, 11, 13, 251
  materials 11
  methods 10, 13
  technique 251
  technologies 9
Scanning electron microscope 113
SCPL techniques 247
Selective laser 11, 13, 136, 144
  melting (SLM) 144
  sintering (SLS) 11, 13, 136
SFD method 59
SFL 50, 59
  methods 50, 59
  process 59
Silica 51, 52, 54, 62, 142
  hierarchical porous 51
  honeycomb 54
  microparticles 52

U

Ultrasonic additive manufacturing (UAM) 136
Up-regulated transcription 207
UV 135, 139, 141, 142
    cross-linking 141, 142
    polymerization 141

V

Vacuum 38, 46, 48, 246, 247
    drying 247
    processing 246
Vascular 5, 117, 123, 141, 182, 256
    endothelial growth factor (VEGF) 5, 117,
        123, 141, 182
    smooth muscle cells (VSMCs) 256

W

Waals 80, 220
    forces 80, 220
Waste 168, 205
    products 168
    removal 205
Water 10, 11, 48, 50, 51, 54, 56, 58, 59, 60,
        61, 104, 175, 177, 198, 209, 246, 248,
        254
    absorption 209
    based systems 50
    filtration 104
    in-oil 56
Wire network molding 252
Wound dehydration 118
Wound dressings 11, 104, 118, 210, 231, 243,
        248
    efficient 118

X

X-ray 114, 115
    diffraction (XRD) 115
    photoelectron spectroscopy 114